O9-BUD-470

This book presents a new view of Robert Boyle (1627–91), the leading British scientist in the generation before Newton. It comprises a series of essays by scholars from Europe and North America which scrutinise Boyle's writings on science, philosophy and theology in detail, bringing out the subtlety of his ideas and the complexity of his relationship with his context. Particular attention is given to Boyle's interest in alchemy and to other facets of his ideas which might initially seem surprising in a leading advocate of the mechanical philosophy. Many of the essays use material from among Boyle's extensive manuscripts, which have recently been catalogued for the first time. The introduction surveys the state of Boyle studies and deploys the findings of the essays to offer a revaluation of Boyle. The book also includes a complete bibliography of writings on Boyle since 1940.

Robert Boyle reconsidered

Robert Boyle reconsidered

Edited by

Michael Hunter
Birkbeck College, University of London

CAMBRIDGE
UNIVERSITY PRESS

Published by the Press Syndicate of the University of Cambridge
The Pitt Building, Trumpington Street, Cambridge CB2 1RP
40 West 20th Street, New York, NY 10011–4211, USA
10 Stamford Road, Oakleigh, Melbourne 3166, Australia

© Cambridge University Press 1994

First published 1994

Printed in Great Britain at the University Press, Cambridge

A catalogue record for this book is available from the British Library

Library of Congress cataloguing in publication data

Robert Boyle reconsidered / edited by Michael Hunter.
p.   cm.
Includes bibliographical references and index.
ISBN 0 521 44205 2
1. Boyle, Robert, 1627–91. 2. Science – Great Britain – History.
3. Scientists – Great Britain – Biography.
I. Hunter, Michael Cyril William.
Q143.B77R6   1994
509.2–dc20   93–23081 CIP
[B]

ISBN 0 521 44205 2 hardback

Contents

Contributors

ANTONIO CLERICUZIO, Ricercatore of History of Science at the University of Cassino, Italy

EDWARD B. DAVIS, Associate Professor of Science and History at Messiah College, Pennsylvania

JOHN T. HARWOOD, Associate Professor of English and Associate Director of the Center for Academic Computing at Pennsylvania State University

JOHN HENRY, Lecturer in History of Science at the Science Studies Unit, University of Edinburgh

MICHAEL HUNTER, Professor of History at Birkbeck College, University of London

J. J. MACINTOSH, Professor of Philosophy at the University of Calgary, Canada

WILLIAM R. NEWMAN, Associate Professor of History of Science at Harvard University

MALCOLM OSTER, part-time Tutor in History and History of Science at the University of Oxford

LAWRENCE M. PRINCIPE, Lecturer in the Departments of Chemistry and History of Science at Johns Hopkins University, Baltimore

ROSE-MARY SARGENT, Professor of Philosophy at Merrimack College, Massachusetts, and Research Associate at the Center for Philosophy and History of Science at Boston University

TIMOTHY SHANAHAN, Assistant Professor of Philosophy at Loyola Marymount University, Los Angeles

JAN W. WOJCIK, Assistant Professor of Philosophy at Auburn University, Alabama

Abbreviations

General

Ann. Sci.	*Annals of Science*
BJHS	*British Journal for the History of Science*
BL	Royal Society Boyle Letters
BP	Royal Society Boyle Papers
Harwood, *Essays*	*The Early Essays and Ethics of Robert Boyle*, ed. John T. Harwood (Carbondale and Edwardsville: Southern Illinois University Press, 1991)
Hist. Sci.	*History of Science*
JHI	*Journal of the History of Ideas*
Maddison, *Life*	R. E. W. Maddison, *The Life of the Honourable Robert Boyle, F.R.S.* (London: Taylor and Francis, 1969)
NRRS	*Notes and Records of the Royal Society*
Oldenburg	*The Correspondence of Henry Oldenburg*, eds A. R. and M. B. Hall, 13 vols (Madison, Milwaukee and London: University of Wisconsin Press; Mansell; Taylor & Francis, 1965–86)
Phil. Trans.	*Philosophical Transactions*
R.S.	Royal Society
Shapin and Schaffer, *Leviathan and the Air-Pump*	Steven Shapin and Simon Schaffer, *Leviathan and the Air-Pump: Hobbes, Boyle and the Experimental Life* (Princeton University Press, 1985)
Works	*The Works of the Honourable Robert Boyle*, ed. Thomas Birch, 2nd edn, 6 vols (London, 1772)

Titles of works by Robert Boyle cited within *Works* (above)

Titles are given with date of first publication and subsequent edition if Birch's text was based on this and it differs significantly from the first.

Advices	*Some Advices About Judging of Things Said to Transcend Reason*, appended to *Things above Reason*, below

Aerial Noctiluca	*The Aerial Noctiluca: Or Some New Phenomena, and a Process of A Factitious Self-shining Substance* (1680)
Christian Virtuoso, I	*The Christian Virtuoso: Shewing, That by Being Addicted to Experimental Philosophy, a Man is Rather Assisted, than Indisposed, to be a Good Christian* (1690)
Christian Virtuoso, I, Appendix	*Appendix* to *The Christian Virtuoso*, ed. Henry Miles, first published in Birch's 1744 edition of Boyle's *Works*, 5 vols (London, 1744), v, 665–83
Christian Virtuoso, II	*The Christian Virtuoso, Part II*, ed. Henry Miles, first published in Birch's 1744 edition of Boyle's *Works*, 5 vols (London, 1744), v, 684–736
Cold	*New Experiments and Observations Touching Cold, or an Experimental History of Cold, Begun* (1665)
Colours	*Experiments and Considerations Touching Colours* (1664)
Cosmical Qualities	*Of the Systematicall or Cosmicall Qualities of Things*, in *Tracts Written by the Honourable Robert Boyle* ([Oxford] 1671)
Cosmical Suspicions	*Cosmicall Suspitions (Subjoyned as an Appendix to the Discourse of the Cosmicall Qualities of Things)* in *Tracts Written by the Honourable Robert Boyle* ([Oxford] 1671)
CPE	*Certain Physiological Essays, Written at Distant Times, and on Several Occasions* (1661; 2nd edn, 1669)
Customary Swearing	*A Free Discourse against Customary Swearing* (1695)
Defence	*A Defence of the Doctrine Touching the Spring and Weight of the Air* (1662)
Degradation of Gold	*Of a Degradation of Gold Made by an Anti-Elixir: a Strange Chymical Narrative* (1678)
Effluviums	*Essays of the Strange Subtilty, Great Efficacy [and] Determinate Nature of Effluviums* (1673)
Examen	*An Examen of Mr. T. Hobbes his Dialogus Physicus De Natura Aeris* (1662)
Excellency of Theology	*The Excellency of Theology, Compar'd with Natural Philosophy . . . To which are annex'd Some Occasional Thoughts about the Excellency and Grounds of the Mechanical Hypothesis* (1674)
Exp. Obs. Physicae	*Experimenta & Observationes Physicae: Wherein are Briefly Treated of Several Subjects Relating to Natural Philosophy in an Experimental Way* (1691)
Final Causes	*A Disquisition about the Final Causes of Natural Things* (1688)
Forms and Qualities	*The Origine of Formes and Qualities, (According to the Corpuscular Philosophy) . . .* (1666)
Gems	*An Essay about the Origine & Virtues of Gems* (1672)
General History of Air	*The General History of the Air, Designed and Begun by the Honourable Robert Boyle Esq.* (1692)

Greatness of Mind	*Greatness of Mind, Promoted by Christianity*, appended to *Christian Virtuoso, I*, above
Hidden Qualities of Air	*Tracts: Containing Suspicions about Some Hidden Qualities of the Air* [etc.] (1674)
High Veneration	*Of the High Veneration Man's Intellect Owes to God; Peculiarly for his Wisedome and Power* (1685)
Human Blood	*Memoirs for the Natural History Of Humane Blood, Especially the Spirit of that Liquor* (1684)
Hydrostatical Discourse	*An Hydrostatical Discourse Occasioned by the Objections of the Learned Dr Henry More, against Some Explications of New Experiments made by Mr Boyle*, in *Tracts Written by the Honourable Robert Boyle* (1672)
Hydrostatical Paradoxes	*Hydrostatical Paradoxes, Made out by New Experiments, (For the Most Part Physical and Easie)* (1666)
Icy Noctiluca	*New Experiments, and Observations, Made upon the Icy Noctiluca* (1682)
Languid Motion	*An Essay Of the Great Effects of Even Languid and Unheeded Motion* (1685)
Mechanical Origin of Qualities	*Experiments, Notes, &c. about the Mechanical Origine or Production of Divers Particular Qualities* (1675)
Medicina Hydrostatica	*Medicina Hydrostatica: or, Hydrostaticks Applyed to the Materia Medica* (1690)
Medicinal Experiments	*Medicinal Experiments: Or, a Collection Of Choice and Safe Remedies, For the Most Part Simple, and Easily Prepared* (3 vols, 1692–4)
Notion of Nature	*A Free Enquiry Into the Vulgarly Receiv'd Notion of Nature; Made in an Essay, Address'd to a Friend* (1686)
Occasional Reflections	*Occasional Reflections Upon Several Subjects* (1665)
Porosity	*Experiments and Considerations About the Porosity of Bodies, in Two Essays* (1684)
Producibleness	*Experiments and Notes About the Producibleness Of Chymicall Principles*, appended to the second edition of *Sceptical Chymist* (1680)
Rarefaction of Air	*A Discovery of the Admirable Rarefaction of the Air*, English translation in *Tracts Written by the Honourable Robert Boyle* ([London] 1671)
Reason and Religion	*Some Considerations about the Reconcileableness of Reason and Religion* (1675)
Saltness of the Sea	*Tracts Consisting of Observations About the Saltness of the Sea* [etc.] (1674)
Salubrity of the Air	*An Experimental Discourse of Some Unheeded Causes of the Insalubrity and Salubrity of the Air*, appended to *Languid Motion*, above
Sceptical Chymist	*The Sceptical Chymist: or Chymico-Physical Doubts & Paradoxes . . .* (1661)

Seraphic Love	*Some Motives and Incentives to the Love of God. Pathetically Discours'd of, in a Letter to a Friend* (1659)
Specific Medicines	*Of the Reconcileableness of Specifick Medicines to the Corpuscular Philosophy* (1685)
Spring	*New Experiments Physico-Mechanicall, Touching the Spring of the Air and its Effects* (1660)
Spring, 1st Continuation	*A Continuation of New Experiments Physico-Mechanical Touching the Spring and Weight of the Air, and their Effects. The First Part* (1669)
Spring, 2nd Continuation	*A Continuation of New Experiments Physico-Mechanical Touching the Spring and Weight of the Air, and their Effects. The Second Part* (Eng. trans., 1682)
Style of the Scriptures	*Some Considerations Touching the Style of the Holy Scriptures* (1661)
Theodora	*The Martyrdom of Theodora, And of Didymus* (1687)
Theological Distinction	*Reflections upon a Theological Distinction*, appended to *Christian Virtuoso, I*, above
Things above Reason	*A Discourse of Things above Reason. Inquiring Whether a Philosopher Should Admit There Are Any Such* (1681)
Usefulness, I	*Some Considerations touching the Usefulnesse of Experimentall Naturall Philosophy. The First Part* (1663)
Usefulness, II, sect. 1	*Some Considerations of the Usefulnesse of Naturall Philosophy. The Second Part. The First Section. Of It's Usefulness to Physick* (1663)
Usefulness II, sect. 2	*Some Considerations Touching the Usefulnesse of Experimental Naturall Philosophy. The Second Tome, Containing the later Section Of the Second Part* (1671)

Preface

Robert Boyle died in the early hours of Thursday, 31 December 1691, and, though it may be more important to commemorate the anniversaries of great men's births than their deaths, it seemed a pity to allow this occasion to go unmarked. In fact, the 350th anniversary of Boyle's birth in 1977 seems to have passed virtually unnoticed: as the Bibliography appended to this volume illustrates, one has to go back to the 250th anniversary of Boyle's death in 1941 to find much commemorative activity, apart from a flurry inspired by the tercentenary of the publication of his best-known book, *The Sceptical Chymist*, in 1961. Hence it seemed appropriate to hold a gathering to mark the tercentenary of the great scientist's death, quite apart from the opportunity thereby provided to capitalise on the research on his writings and milieu which has proliferated in recent years.

It would have been possible for this event to have taken place in Oxford – where Boyle lived from 1655 to 1668 – or in London, where he shared a house in Pall Mall with his sister, Lady Ranelagh, from 1668 until their deaths within eight days of each other in 1691. Instead, it seemed better to escape from such predictable venues, and the symposium was convened near Stalbridge in Dorset; here, Boyle spent ten highly significant years of his life between 1645 and 1655, living on the estate which he had inherited from his father, the Great Earl of Cork, who had acquired it in 1636. The meeting took place between 14 and 16 December 1991 in the relaxed surroundings of the Horsington House Hotel, Horsington, Somerset, where participants were lavishly provided for through the generosity of the Foundation for Intellectual History.

Various people and organisations have assisted in making the symposium a success, and in bringing the book which has resulted from it to fruition. The indispensable support of the Foundation for Intellectual History was supplemented by that of those universities and other bodies which paid the transatlantic fares of participants. The conference was enriched by a visit to Stalbridge Park, site of the Boyles' mansion, where we were welcomed by Mr and Mrs Richard de Pelet and by Stalbridge's historian, Mrs Irene Jones. Subsequently, we drove to Yetminster to see Boyle's School, founded by John Warr with money left by Boyle, and Mrs Jenny King kindly showed us inside this attractive vernacular stone building. In the year that has since elapsed, the contributors have proved prompt and helpful in providing revised versions of the papers that they gave at the conference. In addition, they have usefully commented on the draft of my Introduction: particular thanks are due to John Henry in this connection, and helpful comments were also provided by Scott Mandelbrote. Lastly, various contributors have suggested additions to the Bibliography, which was circulated in draft form at the Stalbridge symposium.

Help with the Bibliography was also given by Roger Gaskell, who is preparing a revised edition of J. F. Fulton's *Bibliography of Boyle*, about which he briefly spoke at the conference, and who kindly made available for use here the entries relating to secondary literature on Boyle prepared by Graeme Fyffe of the Science Museum Library. The present listing aims to include all books and articles specifically devoted to Boyle published since 1940, together with more general works in which he figures prominently – though some discretion has had to be used in this, as it might otherwise have turned into a reading-list on seventeenth-century science as a whole. It should provide an illuminating chronicle of the scholarly attention which Boyle has received over the past half century, a story which is significantly advanced by the studies which this book comprises.

Hastings   Michael Hunter

Introduction

MICHAEL HUNTER

The state of Boyle studies

By any standards, Robert Boyle (1627–91) is one of the commanding figures of seventeenth-century thought. His writings are remarkable for their range, their significance and their sheer quantity: during his life he published over forty books, which between them will occupy twelve substantial volumes in a forthcoming new edition, about which more will be said later in this Introduction. Boyle achieved wide fame in the early 1660s through a series of experimental treatises in which he investigated the characteristics of air, outlined what he called 'corpuscularianism' and sought to extend its applicability to a wide range of natural phenomena. His early publications also included more programmatic statements about the principles of experimentation and about the great potential of the new philosophy for understanding and controlling the natural world. For the rest of his life he published various longer or shorter treatises about a wide range of natural phenomena, many of them purely descriptive, others more speculative in character. Equally important were a number of books in which he sought to define the relationship between God and the natural realm and the role of human understanding in assessing this; he also wrote a series of theological works, including some of a devotional character. Indeed, in his blending of a commitment to scientific work with deep piety, Boyle presented almost an ideal type of 'the Christian virtuoso', to quote the title of one of his own books – a great intellectual innovator who was at the same time a paragon of godliness and probity.

This image of Boyle was confirmed by the earliest and most influential account of him to appear in print, the sermon preached by Bishop Gilbert Burnet at Boyle's funeral at St Martin's in the Fields on 7 January 1692. This set the tone for most subsequent evaluation of Boyle, proving as influential as it did largely due to its brilliance in evoking a picture of him which was at one both with his view of himself and with the taste of his contemporaries. It was intended that Burnet would complement his sketch by a definitive life of Boyle, which would have given a more detailed and contextualised account of Boyle's achievement. When Burnet himself proved too busy to write this, the task was instead deputed to the scholar William Wotton: but he never completed it either, and, instead, Burnet's rather generalised eulogy held the field. Apart from Boyle's own writings, this sermon formed the sole source of the first book-length life of Boyle, that published by the miscellaneous writer Richard Boulton in 1715. Even the life published by the antiquary and divine Thomas Birch in 1744, though notable for the detail that it added concerning Boyle's career, was much under the shadow of Burnet when it came to the general evaluation of Boyle.

Birch's own principal contribution to Boyle's posthumous reputation – a highly significant one – was to publish a lavish collected edition of his works, which first appeared in 1744 and was reprinted in 1772. Already, epitomes of Boyle's writings produced by Boulton and by Peter Shaw had borne witness to an appetite for his works, which Birch's text seems to have succeeded in satisfying. Indeed, Boyle's posthumous reputation was largely subsumed in his writings, as if Birch's capacious volumes had provided all that needed to be known about him. Partly as a result of this, there is virtually no Boyle scholarship to speak of from the mid-eighteenth to the early twentieth century: evidently, there seemed no need for the kind of biographical labours that the likes of Brewster and De Morgan carried out for Newton. With one or two exceptions, Boyle was placed on a pedestal and left there – not ignored, but taken for granted.[1]

The rise of professional study of the history of science in the post-war years obviously changed this. Boyle claimed the attention of such luminaries of the history of science in this era as R. S. Westfall and T. S. Kuhn, but the scholar who devoted most effort to the elucidation of his ideas was Marie Boas, who later wrote under her married name, Marie Boas Hall. In two books and a series of articles, one of them itself of book length, Boas gave a detailed analysis of Boyle's aims and achievements, illustrating how he established the corpuscular philosophy on a foundation of brilliant experiment, and how he, more than anyone, sought to 'raise chemistry to a point where physicists recognised it as a real and important science, working in the same spirit as contemporary natural philosophy'.[2] Her writings form the starting point of a good deal of the investigation which is expounded in the following pages, even when the authors disagree with her conclusions.

But it might be felt that, for all this, Boyle did not receive as much attention in the post-war years as he might have done. To some extent this was because – despite his undeniable scientific significance – Boyle fitted badly into the view of the Scientific Revolution, which, as Roy Porter has perceptively pointed out, was typical of the prevailing historiography of this period.[3] Under the influence particularly of Alexandre Koyré, this stressed the role of heroic individual thinkers in revolutionising knowledge by achieving great syntheses of mathematicised physics, a vision epitomised by Newton and his *Principia*. This was a setting into which Boyle did not fit well. One reason for this was his own distrust of the higher reaches of mathematics because of the idealised view of the world which he associated with them.[4] Equally significant was the fact that his own opus comprised a variety of treatises, at best diffuse, at worst miscellaneous, bizarre and apparently trivial. How was a work like Boyle's *Experimenta & Observationes Physicae*, with its appendix of 'Strange Reports', supposed to fit into the heroic view of the Scientific Revolution? Undoubtedly embarrassment with aspects of Boyle which were at odds with the prevailing image of the rise of the new science helps to explain his relative neglect; it also accounts for a tendency in those who *did* study him, including Marie Boas, to place as 'rational' and mechanistic an interpretation as possible upon his views, and to neglect aspects of his ideas which fitted badly into this, including some – such as his alchemy – which will receive coverage in the pages below.

Since the 1970s, we have seen the growth of a scholarly tradition which, in reaction to the internalism of much post-war historiography, has sought to contextualise science in terms of social, economic and cultural change in its period; Boyle studies have undoubtedly benefited from this alternative approach. Here, the trail had been blazed in the 1930s by Robert K. Merton in his well-known 'Science, technology and society in seventeenth-century England', in which Boyle played a significant role: indeed, up to a point, the post-war historiography was itself reacting against views of this kind.[5] But Boyle has received particular attention in the context of the reformulation of a contextualist reading of early modern science over the past two decades.

One landmark was Charles Webster's *The Great Instauration* (1975), in which Boyle figures prominently – hardly surprisingly in view of his links during his formative years with the Hartlib Circle on which that book is focused. In addition, in various articles and a book published during the 1970s, J. R. Jacob made Boyle the subject of an attempt to forge a direct connection between the ethos of the new science and the ideological conflicts of the English Revolution. Jacob saw Boyle's personal values as redirected by his experience of the Civil War, and he claimed that, thereafter, Boyle's natural philosophy was formed at least in part to counter rival philosophies of nature. His views have had considerable influence over the past two decades.

On the other hand, though significant as a pioneering attempt at contextualisation, Jacob's views have not stood up to scrutiny in the light of subsequent research. Jacob's image of Boyle is in many ways rather partial and schematic, based on a selective reading of Boyle's writings, and often imputing ill-evidenced motives to him in his controversial works. In particular, the 'dialogue' that Jacob claimed to discern between Boyle as spokesman for an orthodox mechanistic worldview and a rival position which combined vitalism with politically subversive ideas is based on a mistaken presumption that natural philosophical positions were more clearly polarised than was actually the case; it also postulates a precise focus for the notions which Boyle attacked which is not warranted by the evidence.[6] In so far as Boyle was reacting to contemporary events at all, his position was more complicated than Jacob implied. This is illustrated in relation to the politics of the Civil War period by Malcolm Oster's essay below, while the same complexity in Boyle's attitudes is in evidence in other episodes of which Jacob's account presumed that certain motives were predominant, when in fact this is hard to substantiate; the need for a total revision of the view of Boyle's relationship with his milieu given by Jacob is acute.[7]

More recently, Boyle has been the focus of a different attempt to redefine the agenda of the history of science, in this case that of Steven Shapin and Simon Schaffer in *Leviathan and the Air-Pump* (1985), and a series of ancillary studies. In their book,

Shapin and Schaffer concentrate on Boyle's debate with Thomas Hobbes not only to argue for the self-consciously ideological role of science in the Restoration period, but also to offer a contextualised reading of the claims that Boyle made for his experimental findings and their role. Especially important is their attempt to problematise a phenomenon which earlier studies had taken for granted, and this is the issue of how knowledge claims were put forward and assessed, and the role of experiments and instruments in this connection. Thus they have actually investigated how and where Boyle carried out his experiments, illustrating difficulties about the replication of trials using the air-pump. They have also emphasised the significance of the way in which Boyle wrote about his experiments and the strategies he deployed in trying to convince others by invoking as value-free the publicly verified 'matters of fact' obtained by such means.

Undoubtedly, this approach has opened up new ways of looking at Boyle and his seventeenth-century milieu, stimulating fresh consideration of many important issues. But it, too, has its drawbacks. To some extent there is a problem about how the new science's ideological role is presented, overlapping with the shortcomings of the Jacob thesis which have already been noted. Thus both the unanimity of the 'Royal Society' position that Boyle is seen to speak for, and its influence in contemporary society, are exaggerated; the result is that the relationship of the new science to its milieu is misunderstood.[8] In addition, concentration on Boyle's exchange with Hobbes is unhelpful since it gives undue prominence to a position in natural philosophy which was in fact marginal: though 'Hobbism' may have been common in Restoration England, this seems to have represented a kind of eclectic cynicism rather than a systematic espousal of Hobbes' complete philosophical position, and protagonists of Hobbes' system of natural philosophy are almost impossible to find. Moreover concentration on this debate detracts attention from other significant threats with which Boyle had to deal, including the residual power of Aristotelian scholasticism. Shapin's and Schaffer's assertions about the context and significance of Boyle's methodological innovations, and particularly his preoccupation with 'matters of fact', have also been questioned, not least by Rose-Mary Sargent, who has shown that their arguments about the ideological implications of Boyle's claims concerning evidence and its proper interpretation are based on a somewhat distorted view of the precedents of such ideas in the legal tradition.[9]

There is a further, more fundamental, objection to the arguments of Shapin and Schaffer – which also applies to those of Jacob – and this is that they fail to do justice to Boyle because they offer too simplistically functionalist a reading of a man who in fact had a complex personality, which is easily distorted by an interpretation of this kind. As I have argued more fully elsewhere, we will never understand Boyle unless we take account of what might be termed the 'dysfunctional' as well as the 'functional' elements of his intellectual persona: the indefatigable experimenter was also the man who refused to become President of the Royal Society due to his 'great (and perhaps peculiar) tenderness in point of oaths', and arguably the two only make sense as complementary facets of the great 'scrupulosity' which Boyle showed in all areas of his life.[10] The need to link Boyle's personality with his scientific work has also been canvassed by J. J. MacIntosh, and Malcolm Oster has sought to throw light on the development of Boyle's personality by reexamining his upbringing and self-perception in his early years.[11] Sensitive study of this kind is likely to make sense of priorities and preoccupations on Boyle's part which cannot easily be accommodated in Shapin's and Schaffer's view of him.

Equally important is the risk that too much emphasis on Boyle's overall programme may detract from the *content* of his ideas and the extent to which the intellectual developments with which he was associated had an internal momentum of their own. Indeed, if it is possible to speak of a 'post-modernist' generation as far as approaches like those of Jacob, Shapin and Schaffer are concerned, it may be argued that the contributors to this book represent it. The scholars whose work appears here all take it for granted that ideas can only be properly understood through a full knowledge of their context, and they are quite receptive to certain aspects of the view of Boyle put forward by Shapin and Schaffer. Yet, at the same time, they are also sensitive to the power and complexity of intellectual traditions in their own right. Perhaps it is now possible to have the best of both worlds, with contextualism being combined with intellectual considerations which owe more to the post-war historiographical tradition, all in the setting of an understanding of Boyle's personality in its own right.

The present volume

Be that as it may, there is no doubt of the current vitality of Boyle studies: an unprecedented amount of work on him is currently being done, from a wide range of points of view. To some extent, as just noted, this marks a belated coming together of contrasting approaches to a figure of undeniable significance; it also reflects – and may in part be caused by – the fuller availability of resources for such study than ever before, the Boyle Papers having recently been catalogued for the first time and published in microfilm form. Whatever its explanation, it is this renaissance of Boyle scholarship which this book celebrates. Its immediate occasion was the symposium which, as explained in the preface, was held near Stalbridge in Dorset from 14 to 16 December 1991 to commemorate the tercentenary of Boyle's death. This provided an appropriate opportunity to bring together a group of mostly younger scholars working on Boyle, to offer a collective sample of their work, and to speculate a little on the direction in which studies of Boyle have developed in recent years and should develop in the future.[12]

The volume opens with Malcolm Oster's essay on Boyle's relationship to the turbulent events of the Civil War and its aftermath, in which, as already noted, he directly takes issue with J. R. Jacob's account of Boyle and the English Revolution. Oster indicates how Boyle's views during these decades were very much those which he had expounded in his earliest ethical writings, recently published for the first time by John Harwood, who has used them to condemn as schematic and misleading Jacob's account of the formative influences on the young Boyle.[13] Oster continues this analysis, providing a sensitive and helpful account of Boyle's position in the ideological and political cross-currents of the period; he also gives a fascinating view of the local conflicts by which Boyle was affected in Dorset and neighbouring counties. As he points out, Jacob's overall picture of Boyle depends on a serious distortion of the evidence, the essence of Boyle's position paradoxically being presented by Jacob as a concession to his claims as a whole. In fact, the young Boyle was prone to withdraw from political commitment, thus placing him in an intermediate position between royalists and parliamentarians. The same remained equally true in 1660: again, he was not particular about forms of government so long as they were effective. Throughout, he displayed a wish to be above politics, to transcend sectarianism. It was in this context that, from 1660 onwards, Boyle began to publish the writings which were in a sense to become programmatic of the new science.

John Harwood's essay is devoted to these books, looking in general terms at the published output which the remainder of the contributors to the volume analyse piecemeal. Drawing both on Boyle's works and on his correspondence, Harwood illustrates the extent to which Boyle consciously aspired to a public role through his activity as a writer, concerned as to the image of himself that his publications presented, and anxious about how his books were read and his ideas best communicated. Schooled in rhetoric from his early years, Boyle also sought to use rhetorical techniques to persuade his audience, attempting to develop an appropriate style for the presentation of findings about the natural world and of arguments about their philosophical and religious significance. Moreover – despite the hostility towards figurative language expressed by Thomas Sprat in his *History of the Royal Society* (1667), which Harwood sees as a polemical ploy in itself – Boyle and others in scientific circles were adept in their use of imagery. Appealing to experience, Boyle discovered how natural philosophy offered fresh opportunites for the interchange of the literal and the figural, which he sought to exploit. Though, as Harwood illustrates, Boyle often experienced difficulties in literary composition, this was balanced by a self-consciousness which has to be taken into account in order to understand him.

If Harwood's essay is indicative of a fresh dimension to our view of Boyle, with Rose-Mary Sargent's essay we come to the experimentalism for which he has long been celebrated. On the basis of a wide reading of his published and unpublished writings, she gives a helpful summary of Boyle's objections to Aristotelianism and a clearer account than hitherto of the relative importance of Baconian and Cartesian ideas in his thought. She also deals with his attitude to Galileo, Pascal and other experimental precursors, and surveys his indebtedness to the contemporary empirical traditions of chemistry, medicine and the practical arts. Above all, she emphasises the extent to which Boyle's views were formed by his actual experience of attempting to manipulate nature, and by his knowledge of the practices of artisans and tradesmen in direct contact with the natural world. Experimentation was a dynamic learning process, with a strong element of contingency which the more programmatic approach of Shapin and Schaffer obscures. Sargent also stresses that there was more to it than a simple empiricism: Boyle gave careful thought to the issue of how experiments should be interpreted, using his practical experience in the laboratory to provide a subtle strategy for the replication and verification of such findings and the assessment of their significance.

In Antonio Clericuzio's essay we turn to one of the specific themes dealt with by Sargent, Boyle's attitude to the chemists. Clericuzio focuses for this purpose on Boyle's most famous – if problematic – book, *The Sceptical Chymist* (1661), considering its intentions and impact, and clarifying its relationship both to an earlier treatise which he had written on related topics and to the lengthy appendix that Boyle added to the book when it was republished in 1680, *The Producibleness of Chymical Principles*. He takes issue with the view of *The Sceptical Chymist* which Jan Golinski has recently adopted, namely that earlier writers had failed to give the narrative detail that Boyle saw as crucial. Rather, Clericuzio argues that the problem with chemists was that neither the manner in which they expounded their findings nor the way in which they interpreted them seemed to Boyle properly philosophical; it was for this reason that their work could not be considered an integral part of natural philosophy. Here, the dialogue form which Boyle first adopted in this work was itself crucial, giving philosophical sophistication to a subject traditionally more practical. Clericuzio continues by clarifying Boyle's views on chemical compounds, and revaluing his impact on late seventeenth-century chemistry; in the course of this, sense is made of the subject matter of Boyle's 1680 annexe to the book, in which he dealt not least with the notion of 'spirit' and with the nature of metallic 'mercuries'.

Clericuzio ends by alluding to Boyle's alchemical interests, and these form the subject of the two subsequent essays, in which new light is shed on his activity in a field his participation in which would at one time have been dismissed by historians of science as an aberration on Boyle's part. First, Lawrence Principe provides a useful overview of Boyle's alchemical studies, emphasising not only the immense amount of time that he evidently devoted to them, but also the extent to which he used alchemical terminology and deployed the alchemical techniques of secrecy and the 'dispersal' of a message through a work. It is now clear that Boyle made many transmutation experiments of a fully alchemical kind, which he refers to in passing in his works; he clearly also sought contact with other alchemists who had carried out similar experiments, as seen particularly in the famous article by him published in *Philosophical Transactions* in 1676. Principe also probes at Boyle's motivation in his alchemical activities, and attempts to chronicle their progress during his career. He further speculates how – quite apart from specific concepts which may have come from the alchemical tradition (on which the next essay throws fuller light) – Boyle may have seen a broader significance in alchemy as illustrating the reality of spiritual activity in the world, and thus providing a potential bridge between natural philosophy and theology.

William Newman documents a more specific link between Boyle's alchemy and his concerns as a natural philosopher, claiming that many aspects of Boyle's corpuscularianism can be traced back to the 'minima' tradition of medieval alchemy, seen best in the pseudo-Geber's *Summa perfectionis*, of which Newman has himself recently

produced a definitive modern text. This entailed a particulate theory of matter, in terms
of which phenomena such as liquidity and volatility could be understood. Since 'Geber'
is known to have influenced all the major alchemical authors of the sixteenth and
seventeenth centuries, together with such mentors of Boyle's as Sir Kenelm Digby,
George Starkey and Frederick Clodius, it is not difficult to postulate routes by which
Geberian ideas might have reached Boyle. There are, of course, important aspects of
Boyle's corpuscularianism which are not paralleled in this source, since for 'Geber' size
was the only criterion, and no account was taken of either the movement of atoms or
their shape. But many Boylean concepts make better sense in this context than in that
of the revived atomism of the Epicurean school, such as his talk of the subtle and gross
parts of matter, or his conception of 'dregs' or 'denseness'. Hence we may well here
have an important source for Boyle's ideas, which has been obscured by an overstrong
contrast between 'mystical' and 'rational' traditions in seventeenth-century natural
philosophy, and by a mistaken presumption that Paracelsianism was more predominant
in the alchemy of Boyle's day than was in fact the case.

The need to abandon simplistic preconceptions about the mechanical philosophy in
its seventeenth-century context recurs in John Henry's essay, in which he concentrates
on the 'systematical or cosmical' qualities of things to which Boyle devoted a tract in
1671 – qualities of bodies which did not depend on the shape, size and motions of their
constituent corpuscles. Although such concepts would have been anathema to a strict
mechanist, Henry shows that Boyle always insisted on the possibility of unknown
corpuscles in nature which might have what he called 'peculiar faculties and ways of
working'. Indeed, Boyle apparently believed that a whole range of factors of this kind
must be at work in the world, even going so far as to hint that localised natural laws
might exist, by which the normal activities of matter may be modified. Boyle was
always somewhat defensive in his writings on such subjects, for reasons which will be
considered more fully later in this Introduction: but he seems to have been certain that
the principles of Cartesian mechanism were too limited, and that ultimately a true new
philosophy would emerge which would do justice to considerations of this kind.

Moving to Boyle's writings on theological and philosophical issues, we begin with an
essay by Jan Wojcik in which she convincingly anchors one of the most important,
Boyle's *Discourse Concerning Things above Reason* (1681), in the theological controversies of its day. One of the problems stemming from the image of Boyle that has come
down to us is that his writings often appear to be more timeless than they really are:
Wojcik's full and careful investigation, on the other hand, brings to light a context for
this work which has hitherto been entirely unsuspected. This stemmed from the fierce
debate over Socinianism which raged in England from the 1650s onwards, including
writings by nonconformists like Richard Baxter, John Owen and Robert Ferguson, and
Anglicans like Joseph Glanvill; it overlapped with a debate on predestination stimu-
lated by John Howe's *Reconcileableness of God's Prescience of the Sins of Men, with the
Wisdom and Sincerity of . . . whatsoever Means He uses to prevent them* (1677), a work
which Howe claimed that he had compiled at Boyle's request. Boyle's own book is to
be seen as an answer to the latter: in it, he provided a broad rationale for 'privileged things',

stressing God's inscrutability, warning against dogmatism, and making an implicit statement of his own belief in conditional predestination. Wojcik also stresses the parallel between Boyle's views on the limits of reason in religion and in natural philosophy.

An equally seminal work by Boyle was his *Disquisition about the Final Causes of Natural Things* (1688), to which two chapters are here devoted. First, Edward B. Davis is able to deploy an extraordinary and hitherto unknown survival among the Boyle Papers to throw light both on this book and on Boyle's relations with the author of the document in question, who turns out to be none other than Robert Hooke. This paper – printed as an appendix to Davis's chapter – comprises a series of notes in Hooke's hand on what was evidently a draft of *Final Causes*, dealing particularly with Descartes' views on the subject. Davis points out that it had in fact been Hooke who had initially introduced Boyle to Descartes in a systematic way, and he indicates how such evidence reveals Hooke as much more of an equal of Boyle's than accounts of the relations between the two men such as that of Steven Shapin would imply. Equally important is what can be learnt from the newly discovered paper about Boyle's rhetorical strategy in dealing with those with whom he disagreed: in his notes, Hooke urged Boyle to exercise charity in attacking such rival viewpoints, advice which Boyle was indeed to follow in the final version of the book, attempting to minimise disagreements among Christians and hence achieve a united front against the over-riding infidel threat that so preoccupied him.

In his chapter, Timothy Shanahan looks at the *Disquisition* from a somewhat different perspective, arguing for the originality and importance of Boyle's treatment of final causes and the novelty of his conceptual apparatus for dealing with them. Elucidating Boyle's concern about the position both of the Epicureans and the Cartesians, Shanahan goes on to illustrate how he sought to eliminate ambiguity by clarifying both the terminology of types of ends and the modes of reasoning used to ascertain them. On this basis, Shanahan systematically reexamines Boyle's argument, showing how powerfully he vindicated final causes, and claiming a subtlety in Boyle's analysis which previous commentators have overlooked. Shanahan also delves more deeply than Davis into the grounds on which Boyle disagreed with Descartes, and the type of material that he felt it most appropriate to use to illustrate final causes. More-over, in taking issue with the recent account of the subject by James Lennox, who saw Boyle as primarily concerned with final causes in science, Shanahan points out that in fact most of Boyle's examples are taken from natural history: his primary concern was a more general philosophical one, of strengthening the case for acknowledging the existence of a divine author of nature and stimulating devotion to him.

Implicit in Shanahan's paper is the more general case that Boyle is easily under-valued as a philosopher. It has often been presumed that, philosophically, Boyle had little that was original to say, or, in so far as he had, that Locke said it better. Just how wrong this is is illustrated by J. J. MacIntosh's chapter. On the basis not least of extensive use of material to be found among Boyle's unpublished manuscripts, MacIntosh explicitly compares Boyle's views on God's existence and on the argument from miracles with Locke's, and finds them superior. Thus he offers a rather critical reading of Locke, whose arguments for God's existence are seen as 'almost startlingly weak'. Boyle, on the other hand, comes across as possessed of a subtle view on such matters, due not least to his awareness of the complexity of natural phenomena and hence the extent to which events were rarely as straightforward as might have been predicted. The views of the two men on miracles show a similar contrast. Locke's are seen as ambitious but not self-sustaining; Boyle, on the other hand, put forward a more limited but more subtle account of supernatural events as a whole, and the way in which decisions should be made as to what should be deemed miraculous. Again, a sophistication appears with which at one time Boyle might not have been credited.

The new Boyle

What is the image of Boyle which cumulatively emerges from these essays? The intention here is not to try to impose a synthesis on a thinker who was himself always reluctant to synthesise, who deliberately contrasted his manner of writing discrete expositions of specific themes to the systematising tendency which he saw as the bane of earlier philosophical traditions. But it is appropriate to end this Introduction by reflecting a little more broadly on Boyle's methods and preoccupations, and the way in which these affect the manner in which he should be interpreted. In doing so, I will draw partly on the findings of the essays that follow, and partly on my own recent work, published and unpublished, much of it related to the cataloguing of the Boyle Papers at the Royal Society and to the preparation of a new edition of Boyle which is to appear in the 'Pickering Masters' series, and which will finally supersede Thomas Birch's edition after two and a half centuries.[14]

One thing that is clear is that detailed study of the kind exemplified by the essays in this volume tends to enhance one's respect for Boyle's sophistication as a thinker – whether it be in his complex and novel typology of final causes, or his attempt to avoid some of the pitfalls of arguments about reason and its limitations. The same subtlety is in evidence in his attitude towards experiment and the means by which knowledge claims might be verified. Indeed, as both Rose-Mary Sargent and Antonio Clericuzio make clear, recent emphasis on 'matters of fact' as the basis of Boyle's scientific strategy has tended to obscure the true sophistication of Boyle's stance. 'Our way is neither short nor easy', as Boyle put it in the latter connection (see below, p. 73), and it may be argued that whatever Boyle applied himself to, he did so with an intensity which is significant in itself and which frequently took him beyond the position of his more superficial contemporaries.

This is as true of Boyle's life as of his thought. As I have argued elsewhere, there is a clear parallel between the indefatigability that Boyle displayed in the laboratory and the assiduity with which he sought to salve his conscience through casuistry.[15] The same sedulity is also seen in the care with which he supervised the administration of his landed estates, as shown, for instance, by a paper of twenty numbered points outlining the 'Method' to be observed in managing his Irish estates in 1683.[16] Nor is it fanciful to recognise the same prudent ratiocination in his attitude to his health, as evidenced by Sir Hans Sloane's story that 'he had divers sorts of cloaks to put on when he went abroad, according to the temperature of the air; and in this he governed himself by his thermometer', or by the memorialist, Roger North's, account of his 'chemical cordials calculated to the nature of all vapours that the several winds bring; and [Boyle] used to observe his ceiling compass every morning that he might know how the wind was, and meet the malignity it brought by a proper antidotal cordial.'[17]

Linked to this general assiduity was an acute concern on Boyle's part about the fairness of the decisions he came to and the steps he took, and a wish to avoid divisive and morally reprehensible actions. In his personal life, Boyle was ceaselessly concerned about the rectitude of his role in financial and other transactions, and the dilemmas which this caused him are well illustrated by the surviving notes on his interviews with his confessors, Edward Stillingfleet and Gilbert Burnet, in the last months of his life.[18] In politics, Boyle's objective was to avoid partisanship, aware as he was of the short-comings of the more vigorous protagonists of either side: yet he could not entirely escape the political world, and hence, as Malcolm Oster indicates below, difficult decisions about allegiance were presented to him which he was reluctantly forced to confront.

In his intellectual life, Boyle's fastidiousness was juxtaposed with the need to balance various objectives in a milieu which was a veritable minefield. Here the problem was that, though Boyle might be convinced that his aims and priorities as a thinker and writer were interrelated, they sometimes pulled in different directions. The difficulties involved may be illustrated by briefly summarising his principal ends. First and foremost, there was his hope to achieve a reformed natural philosophy which would also form part of a properly defined relationship between God and man. To this end, one priority was to challenge the scholastic modes of thought which were still much in evidence in his day, including such unnecessary obfuscations as forms and qualities, together with the engrained and commonplace habits of mind which scholasticism inculcated in related spheres. Equally, there was the threat of 'atheism', generally associated with a thorough-going materialism, and personified for many of Boyle's contemporaries by the philosophical and political views of Thomas Hobbes: this was of particular concern to Boyle because of the extent to which the study of nature was traditionally, if in Boyle's view quite wrongly, suspected of encouraging atheistic attitudes.[19] In addition, there were the dangers which Boyle perceived in the other alternatives to Aristotelianism which thrived in his period, whether they be Paracelsianism, the various vitalist and magical traditions of the day, or the ideas about a 'spirit of nature' espoused by such thinkers as the Cambridge Platonists.

Balancing these priorities was not easy, particularly for one as fastidious as Boyle, and the result is to explain some of the tensions and difficulties – as well as some of the profundities – which are revealed in the course of this book. Thus, as John Henry points out, while too austere a mechanism might seem to smack of materialism and hence make the idea of admitting non-mechanical attributes in matter attractive, this had the drawback of making it seem as if Boyle was faltering in his commitment to a mechanical worldview and succumbing to the temptation to readmit scholastic occult qualities to natural philosophy. There was also the problem that even thinkers with whose views and objectives Boyle substantially agreed – such as Descartes or Henry More – might adopt questionable positions on key issues. Such dilemmas constantly faced Boyle, and arguably this accounts for various of his traits as a writer and thinker, including some that seem initially puzzling.

One might start with Boyle's characteristics as a writer, since all who have read his works will be familiar with the sheer convolution of his style. Indeed, there is an immediate tension here, between Boyle's aspiration to clarity and accessibility, which John Harwood illustrates below, and his parallel need to do justice to the complexity of his thoughts on many difficult issues. Often, it is the latter rather than the former which triumphs on the printed page. In part, this reflected a longstanding difficulty in composition to which Boyle often refers in his prefaces, as Harwood indicates. One symptom of this was a constant tension between fact and argument, between the provision of information and the deployment of it to prove a general point, a difficulty which is particularly marked in Boyle's *Usefulness of Natural Philosophy*, which grew longer and more shapeless in the course of composition due to the repeated insertion of extra data.[20] More often, however, Boyle's convolution was due to a need to reconcile conflicting arguments and counter-arguments.

Draft material for Boyle's writings in the Boyle Papers reveals revisions made during composition which range from parentheses clarifying or modifying his position to the insertion of whole sections of text to deal with some new difficulty which had occurred to him; the latter are sometimes not even inserted in the most appropriate place in the original draft, and the result is rarely to add to a work's clarity. Indeed, where draft versions of a text survive which are significantly different from the published version, the former frequently have a directness which the latter has lost.[21] It is presumably due to this process of accretion that Boyle's printed works are often hard to follow, as with the *Disquisition about Final Causes*, to which, as Timothy Shanahan points out, his exposition gives an architectonic form which is missing from Boyle's somewhat shapeless original. Hence we have to thank the assiduity with which Boyle applied himself to issues for the sprawling and often rather unreadable opus that he has bequeathed us.

Similar factors may also help to account for one of the vexing features of Boyle's writings to which more than one contributor draws attention below, and this is Boyle's vagueness in referring to the adversaries against whom his polemical works are aimed. A few such antagonists are specified – notably Hobbes, Francis Linus and Henry More – but in general Boyle is prone to retreat into an unhelpful obscurity about his enemies, speaking only in rather generalised terms about such groups as 'Peripateticks' or 'Chymists' as his target. This has led to some egregious misinterpretations, as with J. R. Jacob's postulate of Boyle's 'dialogue' with a putative but unsubstantiated group of politically subversive 'pagan naturalists', which has already been alluded to. In other cases (as with Jan Wojcik's study below), sensitive scrutiny of the context from which a work by Boyle emanated can sometimes fill in a background which has been overlooked due to Boyle's failure to make explicit reference to it. But in many instances it has to be admitted that it is far from clear who Boyle was getting at.

With some of those whom Boyle criticised in such a circumspect way, this may reflect the policy of 'Parcere nominibus' advocated by Hooke in his advice to Boyle, as discussed in Davis's essay below; even for more obvious adversaries, it could be linked to the general strictures on debate in *Certain Physiological Essays* on which Steven Shapin and Simon Schaffer have partly built their case.[22] But an alternative reading of it is as a kind of subterfuge on Boyle's part, a ducking out of the added complications inherent in direct confrontation with a real antagonist – which he had experienced in the Hobbes dispute – by referring in a generalised way to adversaries who would hardly thus be encouraged to respond. Indeed, one often has the suspicion that Boyle would himself have been hard put to cite chapter and verse for the antagonists' views that he attacked, as against having a general sense that these were the kind of views which scholastics, for instance, adopted. In other cases, these objections may even have reflected doubts of his own, as with those on miracles which he confided to his episcopal confessors.[23] Either way, this vagueness seems to me best interpreted, less as a matter of strategy, than as a symptom of Boyle's ambivalent response to the complex state of affairs by which he was presented.

A similar approach helps us better to understand those aspects of Boyle's ideas which comprise one of the most striking findings of this book – namely the extent to which he dabbled in notions which seem incompatible with a strict interpretation of the mechanical philosophy. Both the alchemical study adumbrated here and the strange effluvialist cosmology to which John Henry draws attention show a degree of eclecticism on Boyle's part which is far greater than traditional accounts either of him, or of the Scientific Revolution, would have allowed. In part, this eclecticism is itself a symptom of Boyle's assiduity in pursuing any line of enquiry which seemed likely to enhance his understanding of the workings of the natural world. But in the case both of alchemy and of 'cosmical qualities', what is hardly less significant than Boyle's interest in such ideas is his peculiar practice in relation to them. Attractive as such notions may have been for Boyle, they presented him with problems in terms of his conflicting priorities, and his response was to take refuge in obscurity or even secrecy.

Thus 'cosmical qualities' raised the dilemma that has already been alluded to, their apparent advantage in alleviating the rigours of Cartesian mechanism being offset by the drawback of reintroducing qualities suspiciously like scholastic ones. The result – apart from encouraging 'a typically Boylean piece of trying to have one's cake and eat it', in John Henry's words – was to impel Boyle to a degree of obscurity in presenting his ideas on such matters which has hitherto largely prevented commentators from noticing their existence at all, and which restricts the extent to which they can be fully reconstructed even now.

Boyle's attitude to alchemy is even more revealing. Here, too, there were incentives to its study, partly the possibility that there might be truth in its adepts' claims about the power over and insight into nature to which they aspired, and partly, as Principe intriguingly speculates, because it appeared to offer an empirically verifiable spiritualist realm bridging the natural and the supernatural. It might thus provide a weapon against the threat of 'atheism' comparable to that which his contemporaries More and Joseph Glanvill found in witchcraft, and Principe further suggests that this facet of Boyle's interest in alchemy may have intensified in his later years as the godless threat increasingly worried him.

Yet here, too, Boyle was presented with a dilemma, and this explains his ambivalence about the way he expressed his ideas, which is itself almost as revealing as the fact that he indulged in such pursuits at all. Of course, alchemy brought with it traditions of secrecy, and Boyle was under some obligation to partake of these. But equally significant is the extent to which such enquiries brought into play various of the tensions in Boyle's intellectual persona which have already been referred to. In part, the pursuit of alchemy presented him with a dilemma because of moralistic scruples to which I have drawn attention elsewhere.[24] If, as Boyle believed, alchemical insight was linked to access to the spirit realm, it was problematic whether it was legitimate to achieve this. Thus the spirits to which access was vouchsafed might be evil ones rather than angels, and even the frequently mysterious origin of the powder of transmutation through which alchemical transformation was achieved was a potential cause for disquiet. Such considerations were clearly significant for as scrupulous a man as Boyle, and they can only have had the effect of inhibiting him in enquiries of this kind.

In addition, there was the view of alchemy as obscure and hence implicitly fraudulent to which Boyle gave expression in *The Sceptical Chymist*. This overlapped with the view of alchemists – together with devotees of such practices as astrology – as potentially subversive 'enthusiasts', which clearly reflected a significant prejudice in the dominant culture of Restoration England.[25] Of course, for those who accepted the power of supernatural forces in the world, such criticism was at worst hypocritical and at best beside the point: it was perfectly possible to believe that pursuits like alchemy possessed a core of truth even if they had been abused by unworthy practitioners, and it seems likely that Boyle's characteristically careful consideration of all issues would have prevented him from any hasty reaction in this connection. But – as his own contribution to the tradition of denigration illustrates – the extent to which alchemy offended against the appeal to clarity and reasonableness which was such a rallying-call in Restoration England was a further factor making it difficult for Boyle to be too overt in divulging interests of this kind.

The result is that Boyle's writings on alchemy have a complex history of publication and non-publication, which has to be taken into account fully to understand them. Relatively little was openly published about his alchemical activities – of his *Dialogue on Transmutation*, for instance, only one section appeared – and even what he divulged was oddly presented, as Principe shows in his chapter. Moreover, the same is also true of Boyle's writings on subjects which might seem more harmless, which he either wholly or partially suppressed because he apparently could not resolve the conflicts of values to which they gave expression.

An example is Boyle's writings on medicine, which clearly presented him with comparable dilemmas. Boyle evidently drafted a treatise in which he developed the critical view of orthodox medical practice expressed in parts of his *Usefulness of Natural Philosophy*; synopses of this work survive among the Boyle Papers, together with one section of its text.[26] Though Boyle clearly tried to be judicious and even-handed in his criticism, however, this did nothing to mitigate the hostility with which his efforts were received by an embattled medical profession, which caused him considerable unease. As a result, he almost entirely suppressed his critique of orthodox therapy, instead publishing in his last years works comprising piecemeal suggestions as to how it might be supplemented by the use of easily prepared remedies. Moreover, the diffidence he felt even in presenting these is reflected by the abnormal number of manuscript versions that his introductory comments went through before they appeared in print.[27] Again, one sees Boyle's acute scrupulosity as a writer and thinker, which in many ways is the obverse of that philosophical sophistication, that ability to tease through an argument in the light of all relevant considerations, which is so much in evidence in this volume.

Fortunately, a number of such suppressed compositions survive in the Boyle Papers, now all the more accessible than hitherto due to their availability in catalogued and microfilmed form. Important conclusions may therefore be drawn as to just what Boyle put into print, and what he did not, and why. Indeed, in this respect, that component of the Boyle Papers which comprises draft treatises or fragments of them needs to be treated as a part of Boyle's opus as a whole alongside his published works: the two often only make sense through their mutual relationship, which requires careful elucidation. For all his apologies, Boyle's prolific output shows that he was not loath to publish: if sections of text were never printed, there may well be good reasons for this, probably along lines similar to those indicated here. Moreover, it is important to stress that this needs to be taken into account in interpreting such unpublished material. The Boyle Papers are not to be treated as a kind of lucky dip from which Boylean *bon mots* can be randomly selected, a tendency which the contributors to this volume have fortunately largely eschewed. Rather, they provide evidence of Boyle's fastidious, painstaking and convoluted intellectual persona.

What the Boyle Papers also provide is the best basis for a full understanding of Boyle's evolution as a thinker and writer, many aspects of which remain distressingly obscure. Even in the case of his major works, our current knowledge of when they were written, when they reached the form in which they were published, and what previous stages they had been through often depends largely on brief asides by Boyle in the course of his prefatory or other remarks. Though some such questions will probably always remain hard to resolve, the Papers have far more potential to throw light on them than has yet been realised, not least through the use of handwriting for dating purposes. As I have indicated in the introduction to my catalogue, careful scrutiny of the content of the archive has resulted in the recognition of distinctive handwritings which can often be dated: this makes it possible to identify discrete 'strata' of material of different dates within the Papers, dating from the 1650s, the 1660s and from later in Boyle's life.[28] The potential of this will be exemplified by the study which Edward Davis and I are completing of Boyle's *Free Enquiry into the Vulgarly Receiv'd*

Notion of Nature, in which we will bring together handwriting and other evidence to indicate which parts of the book were written in the 1660s and which were added when Boyle revised the work *c.* 1680, as well as outlining the stages by which the process of revision occurred and the way in which these altered the work's character and content. The impending new edition of Boyle will provide further information of this kind for all of Boyle's published writings.

In this way – and also by taking account of material in the archive which was never published at all – we should acquire a much clearer view than hitherto of how Boyle's interests developed during the course of his life, and how and when his leading preoccupations changed. Moreover, this should help in the interpretation of Boyle's ideas, not least by making it possible to avoid egregious errors due to ignorance of such data. Instead, researchers will be directed more precisely to the exact chronological context in which Boyle's ideas on different subjects took shape. A new age of Boyle scholarship should result.

Hence there remains much to be done. But the agenda of such future studies will undoubtedly owe a great deal to the work represented in this volume. What we have gained is a rather different Boyle from the one we had a few years ago, more nuanced, more detailed and more true to life, his positive and more negative qualities coming together to make sense of each other and of the man as a whole. What is more, this is a picture which will undoubtedly be further refined by the work which all those contributing to this book have in hand. We leave it to others to subordinate Boyle to some great enterprise of historical reconstruction or sociological theorisation. We consider it sufficiently important to do justice to his ideas, their evolution and their legacy in the context of a sensitive understanding of the mind of their maker.

NOTES

1 For a survey of writings on Boyle before 1940 see J. F. Fulton, *A Bibliography of the Honourable Robert Boyle*, 2nd edn (Oxford: Clarendon Press, 1961), though this has a few omissions, such as the valuable account of Boyle in E. A. Burtt, *Metaphysical Foundations of Modern Physical Science* (1924; rev. edn, London: Kegan Paul, 1932). From 1940 onwards, Fulton's listing of secondary works is superseded by the Bibliography appended to this volume.

2 Marie Boas [Hall], *Robert Boyle and Seventeenth-Century Chemistry* (Cambridge University Press, 1958), p. 232.

3 Roy Porter, 'The Scientific Revolution: a spoke in the wheel?' in *Revolutions in History*, eds. Roy Porter and Mikuláš Teich (Cambridge University Press, 1986), pp. 290–316.

4 See Boas [Hall], *Boyle and Seventeenth-Century Chemistry* (n. 2), p. 216; Steven Shapin, 'Robert Boyle and mathematics: reality, representation and experimental practice', *Science in Context* 2 (1988): 23–58.

5 Porter, 'Scientific Revolution' (n. 3), p. 295; R. K. Merton, 'Science, technology and society in seventeenth-century England', *Osiris* 4 (1938): 360–632 (reprinted in book form, New York: Howard Fertig, 1970).

6 Jacob's writings are listed in the Bibliography. For a critique of his account of the background to Boyle's *Free Enquiry into the Vulgarly Receiv'd Notion of Nature*, see the study of the making of this work which Edward B. Davis and I are currently preparing; see also N. H. Steneck, 'Greatrakes the stroker: the interpretations of historians', *Isis* 73 (1982): 161–77 (many of the points in which still stand, despite Jacob's response in his *Henry Stubbe, Radical Protestantism and the Early Enlightenment* (Cambridge University Press, 1983), pp. 164–74). On the falseness of Jacob's polarisation, see

particularly John Henry, 'Occult qualities and the experimental philosophy: active principles in pre-Newtonian matter theory', *Hist. Sci.* 24 (1986): 335–81, p. 357 and n. 80. On the position of J. R. (and M. C.) Jacob more generally, see Michael Hunter, 'Science and heterodoxy: an early modern problem reconsidered', in *Reappraisals of the Scientific Revolution*, eds. D. C. Lindberg and R. S. Westman (Cambridge University Press, 1990), pp. 437–60.

7   See Michael Hunter, 'The conscience of Robert Boyle: functionalism, "dysfunctionalism" and the task of historical understanding', in *Renaissance and Revolution: Humanists, Scholars, Craftsmen and Natural Philosophers in Early Modern Europe*, eds. J. V. Field and F. A. J. L. James (Cambridge University Press, 1993), pp. 147–59. For further criticism of Jacob's views, see Michael Hunter, *Establishing the New Science. The Experience of the Early Royal Society* (Woodbridge: The Boydell Press, 1989), ch. 2; Nicholas Canny, *The Upstart Earl: A Study of the Social and Mental World of Richard Boyle, First Earl of Cork, 1566–1643* (Cambridge University Press, 1982), pp. 146ff.; and Harwood, *Essays*, Introduction, esp. pp. xxiii–iv, xli.

8   Hunter, *Establishing the New Science* (n. 7), pp. 13–14, 67.

9   Rose-Mary Sargent, 'Scientific experiment and legal expertise: the way of experience in seventeenth-century England', *Studies in History and Philosophy of Science* 20 (1989): 19–45.

10 Hunter, 'Conscience of Robert Boyle' (n. 7). See further below, 'The new Boyle'.

11 See J. J. MacIntosh, 'Robert Boyle's epistemology: the interaction between scientific and religious knowledge', *International Studies in the Philosophy of Science* 6 (1992): 91–121; Malcolm Oster, 'Biography, culture and science: the formative years of Robert Boyle', *Hist. Sci.* 31 (1993): 177–226.

12 One component of the symposium is not included here, a paper by Steven Shapin, which will form part of his forthcoming book on Boyle and the constitution of trustworthiness in the seventeenth century. Readers will have the opportunity to compare the approaches of his volume and the present one in due course.

13 See above, n. 7.

14 Edited by Edward B. Davis, Antonio Clericuzio and myself, this will include both Boyle's published works (in twelve volumes) and his correspondence (in four volumes); it will be based on an exhaustive examination of all relevant materials, and will include a full critical apparatus.

15 See Hunter, 'Conscience of Robert Boyle' (n. 7).

16 Chatsworth House Lismore MS 33, no. 124.

17 *Works*, i, cxxxvi; Roger North, *Lives of the Norths*, ed. A. Jessopp, 3 vols. (London, 1890), iii, 146. North added facetiously: 'So that if the wind shifted often he was in danger of being drunk.'

18 See Michael Hunter, 'Casuistry in action: Robert Boyle's confessional interviews with Gilbert Burnet and Edward Stillingfleet, 1691', *Journal of Ecclesiastical History* 44 (1993): 80–98.

19 For further elaboration of these points, see Hunter, 'Science and heterodoxy' (n. 6).

20 Compare the draft version of *Usefulness, Part II, sect. 1* surviving as BP 8, fols 3–28, with the published text in *Works*, ii, 64ff. This difficulty was probably one of the factors which led Boyle to curtail the second section of Part II of the work: compare the work as actually published, *Works*, iii, 392–494, with the synopsis of its intended contents in BP 8, fol. 1, communicated by Boyle to Oldenburg on 25 April 1666 and printed by Joseph Glanvill in *Plus Ultra* (London, 1668), pp. 104–6. This matter will be explored more fully in conjunction with a tabulation of all the extensive surviving draft material for the work in my forthcoming study of the making of *Usefulness*.

21 Compare, for instance, Boyle's published essay 'Of the Usefulness of Mathematicks to Natural Philosophy', *Works*, iii, 425–34, with the draft material relating to this in BP 17, fols 71–4.

22 Shapin and Schaffer, *Leviathan and the Air-Pump*, pp. 72ff.

23 See Hunter, 'Casuistry in action' (n. 18), pp. 90, 95, 97.

24 Michael Hunter, 'Alchemy, magic and moralism in the thought of Robert Boyle', *BJHS* 23 (1990): 387–410.

25 See Brian Vickers, 'The Royal Society and English prose style: a reassessment', in B. Vickers and N. S. Struever, *Rhetoric and the Pursuit of Truth* (Los Angeles: William Andrews Clark Library, 1985), pp. 1–76.

26 BP 18, fols 133–4; R.S. MS 199, fols 114v–20. See also *ibid.*, fol. 44.

27 For the published version, see *Medicinal Experiments*, *Works*, v, 314–16. Draft material relating to this survives in R.S. MSS 186, 187 and 189. In addition, a Latin version of a longer prefatory text survives in BP 17, fols 1–36. See also Boyle's *Specific Medicines* and *Medicina Hydrostatica*, *Works*, v, 74–129,

453–507, and R. E. W. Maddison, 'The first edition of Robert Boyle's *Medicinal Experiments*', *Ann. Sci.* 18 (1962): 43–7. I am currently preparing a study of the material referred to here and in note 26, and its significance.

28 Michael Hunter, *Letters and Papers of Robert Boyle: A Guide to the Manuscripts and Microfilm* (Bethesda, Md.: University Publications of America, 1992), pp. xxix–xxxix.

Virtue, providence and political neutralism: Boyle and Interregnum politics

MALCOLM OSTER

Over the last two decades, a developing sociology of science has placed Boyle within the specific contention that the Scientific Revolution rested essentially on political, social and material foundations. A key writer in this emerging orthodoxy was James Jacob, who argued that Boyle's science, politics and religion were direct ideological responses to the wider context of political instability in the Interregnum and Restoration. On this reading, Boyle's material circumstances during the English Civil War and its aftermath produced a social ethic of expediency, which, in turn, was transformed at the Restoration into the ideological advancement of the mechanical philosophy on behalf of the Royal Society and the Anglican Church.[1]

Although this work proved a timely corrective to internalist studies that gave little attention to social context, and thus opened up new fields of enquiry into the social relations of science, Jacob's treatment of both Boyle and the Royal Society was open to the charge of being excessively monocausal. In so far as the Society was supposed to have had a specific corporate ideology contained and expressed in Thomas Sprat's *History*, Michael Hunter convincingly showed that though the *History* was certainly commissioned by the Society, it was not very closely supervised because a specific public role for science representing the Society's ethos remained very imprecise beyond a broad Baconianism.[2] Nevertheless, Jacob had argued that 'the leading Fellows who preached a social message in the Society's name did so with a view to building a particular kind of kingdom that would serve particular ends', self-consciously advocating a harmonious alliance between mechanism, monarchy and moderate episcopacy, to secure social order and material advance.[3] In doing so, he clearly parted company from Barbara Shapiro's earlier insistence that the Society's proclaimed withdrawal from the political and religious arena characterised a genuine and pervasive latitudinarian moderation found among those occupying the middle ground of Anglicanism.[4]

While there is no question that the Society felt it required a reconcilable public stance to secure the ambition of institutional permanence within a potentially hostile

landscape of court wits and High Churchmen, no greater role of political calculation
has been laid at the door than that of Boyle, one of the Society's key founders. This has
been given added impetus through Schaffer and Shapin's *Leviathan and the Air-Pump*,
which, though successful in showing how Boyle and some of his associates created
a distinct social and intellectual 'space' for their experimental community, is less
than convincing in its broader depiction of general unanimity within the Society's
leadership for a distinct strategy to secure political and social stability for experimentalism.[5]

I want to argue that by briefly examining the political circumstances of Boyle's
father, the Earl of Cork, and the New English in Ireland, and, more centrally, Boyle's
own reflections and experiences during the Civil War, Interregnum and transition to
Restoration, when he was a far less public figure, one can arrive at a different view
regarding Boyle's politics, despite the fragmentary nature of the evidence.[6]

Virtue and riches

Cork's interest in the themes of virtue, fortune and providence, which were to
preoccupy Boyle in the 1640s, needs to be seen against Cork's broader political stance.
Politically, with the outbreak of the Irish Rebellion in October 1641 and his supposed
leanings towards royalism in the Civil War now unleashed on the mainland, the reality
was a militant neutrality explicable within the fragility of the New English situation in
Ireland, and he advised his sons accordingly. For example, he advised his sons Richard,
who succeeded him as the Second Earl of Cork in 1643, and Roger, Baron of Broghill,
then seeking aid in England to quell the revolt in Munster, 'To be extreame[ly] carefull
how you carry [your]selves & what company you keep, for the tymes are so full of
danger that no honest man knowes whom to trust'. As I hope to show, the similarity
with Boyle's political attitudes during the Interregnum is unmistakable.[7] It had always
been a primary objective of Cork and his contemporaries to populate the Munster
plantation with Protestant settlers to both withstand pressure from the wider Catholic
community, and to secure a working Protestant majority in the Irish House of
Commons. The most likely estimate of the plantation's English population by 1641,
including those outside seignories, was 22 000. A Protestant working majority was only
attained in Dublin in 1634, though by 1640 Protestants had a comfortable majority in
Munster of twenty-six to fourteen in representation.[8] Broghill's taking up the mantle
from Cork, in order to continue the defence of Munster, gave later allegiance to whatever force represented stability and order. It might, therefore, be suggested that the
Civil War in England hopelessly complicated what was essentially a pragmatic policy of
loyalty to the crown.[9]

Cork's prudent caution in the display of political postures was largely grounded in
the fragility of both his own personal position and the New English experience in
Ireland. In similar fashion, in order to appreciate the impulse behind Boyle's social and
political ethic from the mid-1640s, it is necessary to recognise his predicament in war-

torn Dorset, while comprehending the personal, intellectual and religious resources that he had ready to hand to understand that predicament. In examining those resources, it is important to go beyond Boyle's more immediate anchoring in Protestantism to the Renaissance heritage that informed it, Christian humanism, which had as its ultimate objective the reconstruction of the social order. Boyle was almost certainly inculcated with the Renaissance view that 'both Roman stoics and primitive Christians evinced more purity of life and a more godly social and political ethic than the despised medieval church had been able to achieve . . . '.[10] The aspiration of northern humanists from the late fifteenth century to reshape individual, social and political behaviour was rooted in the Bible, the works of the Church Fathers, and Greek and Roman moralists.[11] The writings and example of Erasmus were central: whereas his *Praise of Folly* attacked scholastic theologians, the aim of his *Enchiridion* was to commend reformed character which could pursue what the ancients understood as an ethical life. The wider belief that truths contained in pagan philosophy were necessary for the Christian believer was an abiding feature of humanism and Protestant moral philosophy in the sixteenth and seventeenth centuries. John Colet's statutes for St Paul's School in London, which Milton attended, outlined a classical, humanist Christian education very similar to that which the young Boyle experienced at Eton from 1635 to 1638 and then on the continent.[12]

The syncretistic character of writers such as Cicero and Seneca were appealing precisely because they could provide humanist theoreticians with a new vocabulary of virtue, imbued with reformist optimism, while retaining a Christian notion of sin. The godly layman, as active participant in the Commonwealth, could come to embody new descriptive concepts such as gravity, temperance and prudence; in addition Cicero's belief that man's rational soul would help subdue passion and work for the public good had a central place in humanist manuals of behaviour. Similarly, Erasmus's writings advised a tight rein on the passions and disclaimed all the Protestant sins (that Boyle was to make so much of in his early writings), such as gluttony, gambling, blasphemy and idleness. Although humanists saw social ills rooted in individual sin, they insisted that the reformed individual change his corrupted society.[13] Boyle, in fact, pointed to the ideal social type of Christian humanist – the lay intellectual, educated in the classics and exercising virtue in practical social activity. Boyle's positive view of human potential achieving virtue, expressed throughout his ethical treatise of 1645/6, the *Aretology*, rested, as it did for humanists, on the centrality of learning and the belief that moral instruction had to be primarily practical and applied.[14]

Though there are few overtly political postures expressed in *Aretology*, the range of concerns Boyle hoped to cover in the work indicates early on his priorities for the individual in pursuing felicity, that individual's implied relationship with civil society and the centrality of knowledge linked to virtue:

> For according to the Seueral humors of men, and the Diuers Courses of life, that
> they affect: som place it in the Erthly and obuius Goods; either of the body, as

Plesure, Helth, Buty, and the like; or of fortune, namely, Riches; or of Opinion, to wit, Honor: others in the more inward and solid Goods of the Mind; either Theoreticall, namely Knoledg; or Practicall, to wit, Vertu.[15]

The principal dangers impeding virtue come from the passions and affections, but after working his way through them Boyle is able to affirm, using scriptural example and vivid metaphor, their value, evoking a possible personal response to the political turmoil all around him in Dorset:

Thus we see Anger and Shame, the Whetstones of Courage; Loue and Compassion the Exciters of Bounty; Feare the Sharpner of Industry, and caution against [In] Rashnesse and Indiscretion: as also Hope the Anchor of Patience that keepes her from that Shipwrac in the midst of the Tempestuus frownes of aduerse Fortune.[16]

Though I have elsewhere detailed Boyle's estimate of the gentleman or aristocrat in society, and have argued that Boyle's broad intent is captured in his claim, 'My Dessein is, to turn both the Greatnesse & the Obscurenesse of men's Birth, into Motives unto Vertu', Boyle's own reduced circumstances naturally come into view.[17] In a short second section of 'The Gentleman', Boyle considers the question of fortune, and insists that it is impracticable to assess what the appropriate size of a fortune should be in each case, but as a rule, 'his [Fortune] <Estate> ought to be Suitable to his Condition'. Perhaps thinking of his own case, he realises that it is far better if an estate can generate new titles than one that is hardly able to support old ones. 'Especially in Familys whose Honors are of a Late <fresh> Creation: where greater Portions are requisite, to purchase Alliances equall/proportionable/ to their new Titles'. In these cases men are obliged to live beyond their true means in order to disguise to others the meanness of their recent ancestry. One can only speculate on whether Boyle indeed had his own family in mind, particularly in view of the high cost of Cork's marriage alliances for his sons and daughters to secure the lustre of the line.[18]

In answering the charge that the pursuit of riches through virtuous means entails poverty, Boyle employs arguments that were for the most part familiar to both Puritan and Anglican theology and ethics down to the Civil War: 'Pouerty in it's own Nature, is no necessary attendant on Vertu . . . Tis tru, that it is vsually longer in building vp a fortune then is Couvetousness but that Fortune, tho slowly, is both more certin in the building, and more [secure] <lasting> when it is built.'[19] Boyle expresses the widespread Protestant view that poverty, like wealth, was neither good nor evil in itself, but that it could be used as a purifying agency. It is likely that the perception he had of his own 'calling' to some destiny was complemented by the wider Protestant belief in God placing each individual within the social order; this was as much believed by the prominent Anglican bishop and later close friend of Boyle, Robert Sanderson, as it was by Puritans like William Perkins.

Just as Cork had believed that his rise to fortune had come about through the providence of God, so Boyle had no strong motive for presuming that his predicament had taken on permanence. He had already experienced the good fortune of finding his elder sister, Katherine, in London when returning from the continent in 1644, as also the manor in Stalbridge on the Dorset borders which his father had left for him. He would have been urged by Katherine, and probably other family members in the first months after returning, to recover the fortune that was clearly seen as his inheritance and part of God's scheme for his own future. That fortune was part of Cork's legacy to be carried on into future generations, and, as long as Boyle's attitude to recovering his fortune was pursued with that remit, he could utilise his thoughts on ethical instruction for others. It is, of course, perfectly possible that Boyle suffered extensive self-delusion on this question, but the rigorous and searching daily regime he followed in self-examination, meditation and prayer makes Boyle a singularly implausible candidate for self-delusion.[20] Boyle would have been all too aware, as his comments in the *Aretology* demonstrate, that earthly treasure was in constant danger of being corrupted by sin and avarice, and this awareness was shared almost without exception by all Anglican and Puritan writers, who warned men that abundance of temporal blessings was no definitive sign of God's approval. To work simply for riches as an end in itself was condemned as a grievous sin, and Protestant ministers preached continually on the damnation awaiting those who would not heed that warning.

However, it is worth noting in regard to Boyle's views that Anglicans did differ from Puritans, particularly during the 1630s, on two points – they rarely specified how much wealth was adequate, and they saw ambition and acquisition as less serious sins. Henry Hammond and Robert Sanderson both exemplify this attitude, and Boyle follows suit in his openness towards building a fortune and in his refusal to fix a limit to what it should be in favour of depending where a man finds himself in society.[21] Boyle's rejoinder to the charge that virtue can lead to poverty is to ask in a marginal insertion, 'What pray you is the End for which we desire Riches, but Contentment; which if Vertu can afford vs in the Absence of our Riches; why shud we not beleeue that we may be welthy enuf without them?'. Elsewhere Boyle brings into play the scriptural link between virtue and riches in II Chronicles i, 10–11; the reward Solomon enjoyed for not seeking riches was to receive them with honour.[22]

In describing Boyle's motives in writing the *Aretology*, Jacob argued that 'Boyle's own displacement as a result of events in England and Ireland acted as a catalyst, triggering and sustaining his rejection of an inherited ethic'. This was linked to the need for Boyle to prove to himself the credibility of his faith, so that 'His piety thus drove him to make his life a model of Christian virtue'.[23] At issue here is a question of handling: there is no question of the turmoil Boyle was plunged into after his father's death, and it would have been curiously out of character if he had not drawn upon the resources of his pietistic faith. Yet, not only does Jacob never seriously consider the influence of Boyle's upbringing and education, but the 'ambiguity' and 'tension' that Jacob perceived 'between virtue as felicity itself and virtue as a mere expedient to temporal gain' was well understood by Boyle and most of his Protestant contemporaries.[24] To depict Boyle as an innovator of a new social ethic in which fortune becomes a means to virtue was to seriously misunderstand both Boyle and the broad heritage he drew upon.

Boyle's politics in Dorset

At sixteen years old, Boyle suddenly found himself stranded on the continent in the summer of 1643 with the news of his father's death, his inherited fortune seemingly gone, and even his physical safety in some doubt. Cork had well understood the increasing gravity of his situation and had written to Marcombes, Boyle's tutor, in March 1642, indicating that £230 was being sent and that the two brothers should either return to Ireland or get to Protestant Holland and place themselves under the Prince of Orange.[25] Cork was understandably uneasy about Robert and his brother Francis going to England until the political and financial situation of the family in Ireland had recovered, which would otherwise 'draw contempt upon us all'.[26] Boyle had written back to Cork in May from Lyons, suggesting that even if he could reach Holland his youth would give him little reason to be received among the troops, and instead had decided to accept Marcombes's invitation to stay at his residence in Geneva, continue his studies and 'gather some force and vigour to serve and defend my Relig[ion] . . . my King and my Country according to my little power . . . '.[27] Boyle may well have had doubts about his suitability for army service, but, in any event, Marcombes probably pressured Boyle and his brother to return to Geneva as the most prudent course of action in the short term. As the situation continued to deteriorate in Ireland, Cork had been happy to instruct Boyle to stay in Geneva at the end of the summer until the following spring. Finally, in 1644, Boyle decided to leave Geneva because 'being unable to subsist any longer, I was forced to remove thence . . . '.[28]

There is no evidence ready to hand that he knew about the Stalbridge manor and estate before finding Katherine in London, though her circle clearly interpreted his change of fortune as providential, as Boyle himself confirmed. Boyle's increasing sense of difference from his peers and unease in 'bad' company, something he would soon write on at length, goes some way to explain why he decided not to enlist on the royalist side. Boyle writes that though he would have had 'the excellent king himself, divers eminent divines, and many worthy persons of several ranks; yet the generality of those he would have been obliged to converse with, were very debauched, and apt, as well as inclinable, to make others so . . . '.[29] It is important to note Boyle's assumption that he would enlist on the king's side, despite the fact that the Anglo–Irish circle around Lady Ranelagh were solidly for the parliamentary camp. The reason is likely to be that on returning to England at seventeen years old, and largely insulated from the political debate in Geneva, fighting for the king would have simply been a continuation of the noble's sense of duty, as his earlier letters indicate. There might have also been an awareness through Marcombes of how much Cork relied on the monarchy for political survival, whatever private sympathies for parliamentarianism Cork may have had. Jacob only entertains cowardice as Boyle's motive for not joining the Dutch army, whereas for not enlisting in England, 'The evidence suggests that he gave up the idea as soon as he learned that he was heir to Stalbridge'. Nevertheless, Jacob goes on to pointedly concede what is being argued for here, which is that Boyle 'tends not to read events from either the royalist or the parliamentary point of view but to invest them with a significance that transcends politics and is ultimately spiritual'.[30]

It has been increasingly recognised in historiography over the last three decades that four types of explanation have emerged from studies of allegiance in the Civil War. There is the 'deference' thesis that the English common people had no real allegiance to either side, and thus analysis of gentry politics is crucial. A second and recently popular view has been that of neutralism, in which it has been asserted that the priority of both the lower orders and large portions of the gentry was to protect home, family and community. There is the equally popular class interpretation, upheld by Civil War propagandists themselves, that 'the people' did largely take sides and that Parliament's cause of reason and religion was supported by the 'middling sort', such as yeomen, independent craftsmen and freeholders. Finally, there is a fourth thesis that contrasts in popular allegiance had a regional basis, related to local differences in economic complexion, social structure and culture, which particularly explains the existence of a 'popular royalism' in some parts of the country.[31] Of course, in reality no sharp distinctions can easily be made between these interpretations, and frequently they complement each other. The complexity of explanation and motivation in the period is well expressed by B. L. Manning's early characterisation of neutralism: 'Neutralism lay somewhere between moderate Parliamentarianism and moderate Royalism. For many it was merely a resting place in the passage from one side to the other. In the realm of thought neutralism did not represent a third force . . . '.[32] This early view has been enlarged by the perceptive survey of Aylmer, in which he concurs with the view that the war was initially brought about by very small, determined minorities on both sides, but that 'once begun it was sustained, with varying degrees of enthusiasm, by a much wider range of people – positive neutralism and side-changing as well as sheer escapism being most strongly marked at the top; indifference, apathy, and involuntary side-changing at the bottom levels of society'. Though Aylmer takes due account of the work that has been done at county level, he concludes that even the local parish or manor cannot really explain why 'individual members of different social groups and classes divided as they did. What we have to account for is a huge number of personal and group decisions'.[33]

Be that as it may, research that has been done at local level has been inclined to find Dorset among the quietest counties in the south, with a fair division of opinion among the gentry and a clear absence of ideological conflict. The general tendency, once the siege of the royalist fortress of Sherborne was over in September 1642, was to protect estates. In March 1643 a neutralist agreement was negotiated with royalist MPs Rogers and Strangeways in order to prevent forces from either side entering the county. More-over, the personal correspondence of Dorset gentry reveals that even those hitherto partisan in the war now yearned only for peace and order.[34] Hutton found that in Dorset, Somerset and Devon, not only did the rural populations forcibly stop royalists raising soldiers and aid parliamentarians in expelling royalist troops; they subsequently made clear their antipathy to Parliament's cause.[35] Despite the King's nephew, Prince Maurice, reducing many of the parliamentarian strongholds in Dorset and Devon after mid-1643, with the arrival in early June 1644 of the Earl of Essex, the region was to be repeatedly overrun for the next year by marauding armies and multiplying garrisons on both sides. By the early summer of 1645, the endless fighting in which Wiltshire, Somerset and Dorset had become embroiled since the start of Goring's campaigns for the King, resulted in the emergence of 'Peace-keeping' associations of Clubmen, a movement of farmers and small labourers determined to protect homes and communities. When Cromwell's New Model Army destroyed Goring at Langport, the mid-Somerset Clubmen declared for Parliament, but the Wiltshire–Dorset Clubmen, in contrast, attacked the New Model for wrecking local peace and were consequently crushed by Cromwell in Gillingham Forest, only a short distance from Stalbridge, in August.[36] Dorset's strategic location as a bridge between London and the South-west, including royalist Cornwall, meant constant military activity and passage of armies, no matter what the local inhabitants desired. As a result, there was a substantial fall in the average proportion of total wealth held by Dorset farmers in the form of household goods in the 1640s; it was down to 34% from 41% in the previous decade, and was to remain at that level into the 1670s.[37]

Boyle's Stalbridge estate was situated in the pasture area known as the Blackmore Vale, a region where copyholders seemed to have had unusual security of tenure and where there was no clear correspondence between wealth and status before the war. The Vale had a less nucleated residential pattern than that found in villages on the chalk, yet resembled them in being less socially polarised, less affected by market economy pressures, or influxes of landless poor, than that experienced in clothing parishes and towns. As an area that maintained a distinct cultural conservatism and a resistance to Puritan moral reformation, its politics could be characterised as one of 'popular royalism'. Although its royalism was partly the product of peer and gentry influence, Underdown persuasively argues that this was only a minor ingredient. Its royalism was mainly a product of local culture in which traditional assumptions about church, state and society, and the reciprocal rights of King and Parliament, went alongside a probable covert toleration of Laudian ritual.[38] A few miles away from Stalbridge was the royalist fortress of Sherborne, unique in having been garrisoned for long periods by the King, and the presence of the Digby family at the castle ensured that every able-bodied male had taken up arms at some point.[39]

Travelling constantly between Stalbridge and London in the 1640s, Boyle was exposed both to the dislocation the war caused, as well as the bewildering range of political and religious opinions it unleashed. The first intimation by Boyle of feeling somewhat unsafe in the Stalbridge area because of the Dorset committee comes in a letter to Katherine, in March 1646. Frustrated that writing his ethical work is taking so much time, he hopes he will not be staying in Stalbridge for too long given 'the plague, which now begins to revive again at Bristol and Yeovil six miles off, fits of the committee, and consumption of the purse . . . '.[40] The likelihood is that Boyle was learning his politics very quickly in this period, and, when he writes to Marcombes from London in October 1646, a mature grasp of political developments on the mainland is evident.[41] He opens the letter with a belief that things are now very finely balanced; whereas he correctly views the position of the King in England as virtually at an end militarily, he appreciates that the measures taken by Parliament to encourage the Scots to return home and cease occupation in Newcastle, where the King is still held, are by no means resolved. He voices doubt that a speedy settlement is in view, and probably with his correspondent in mind disclaims, 'but for my part, that have always looked upon sin as the chief incendiary of the war . . . I cannot without presumption expect a recovery in that body, where the physic that should cure, but augments the disease'.[42]

An undated letter from Boyle to Benjamin Worsley, his colleague in the Invisible College, probably written soon after 21 November 1646 because of the reference to the saltpetre project, provides further evidence of Boyle's disquiet in the Stalbridge locality. Boyle writes that his hope of composing discourses in solitude have failed to materialise, given his endless financial troubles:

> and the plain truth of it is, that (betwixt the injustice of those, that hold the civil sword, and the unruliness of those, that draw the martial) things are carried in so strange a way in the West, that these parts can afford little content or safety to any man, that is not either a soldier, a sequestrator, or a committee man.[43]

As probably the largest landowner in the Stalbridge locality, Boyle may well have been under suspicion from the Dorset committee as a possible royalist sympathiser, as other large landowners were, but he would at least have been able to point to his parliamentary connections in London. Boyle had enough financial problems of his own without incurring the fate that befell most wealthy royalists at the end of the war of paying a penal fine graded according to the extent of their delinquency and net landed wealth. Evidence of continued pressure from the Dorset committee on the Stalbridge area comes from the measures taken against the Reverend William Douch, who had been instituted as Rector of Stalbridge in 1621, had tutored Robert and Francis Boyle in 1638–9 before they were sent to the continent, and was now 'ousted for scandal and delinquency' in January 1647/8.[44] Boyle is almost certain to have been the most socially eminent communicant at the local church, and given his earlier association with Douch as a boy, through Cork's direction, the moves against Douch for his supposed Laudian or royalist sympathies must have placed Boyle in a very difficult position. We simply have no evidence whether he made any advances to the committee on Douch's behalf, but the likelihood is that he would have kept out of the affair for fear of compromising his own delicate position.

It has already been conceded that there are few overtly political remarks made in his ethical writing of this period and that what there is tends to cohere with the tone taken in his letters that sin, greed and ambition are the stock-in-trade of many of those who climb to positions of political leadership. Though the knowledge of natural philosophy and the furtherance of learning are usually cited as spurs to virtuous behaviour, this can hardly be said for some practitioners of statecraft. For how can a man 'consider the vast Immensity of the heu'nly Orbs, without lafing at the meane Ambition of our Lofty Statesmen, that can part with their Assurances of all this to purchase an Inconsiderable share of a molehill?'.[45] It is probably fruitless to speculate on specific individuals that Boyle may have had in mind, and the same can be said for a further vivid use of metaphor, which, if it has any proximity to actual events, may simply indicate frustration at the political conflict between factions at the top which were shared by most of his countrymen:

> Neither ar we bound to beleeue all thes Macchiavillians to be such Incomparable Politicians as they ar generally giuen out for. Our own times will be able to afford both recent and very pregnant Examples, that for the most part this <so> admired Subtility is [a] but a cunning way of vndoing themselues, at the last: and many by their heaping Projects vpon Proiects haue made and climbd a Ladder, at the top of which, when they thouht to haue seated themselues vpon a Throne, they haue <but> found themselues vpon a Scaffold.[46]

It has been implied so far that to try and establish whether Boyle was a royalist or a parliamentarian is to somewhat pursue the wrong question. Boyle does not seem convinced of the intrinsic rightness of either position and is rather concerned with what kind of regime conduces to the wider objectives he has in mind for the individual and society. His short treatise, 'The Gentleman', illustrated, for example, his desire to see virtue advanced rather than blocked by rankings in traditional society. In considering in similar fashion the possibility of 'Heroick Virtu' expressing itself in individuals, Boyle is frank in having a less than enthusiastic conviction that monarchy can deliver what is required:

> These Heroical Men ar more frequently <obseru'd> (for the most part) in Common-welths then in Monarchys: not that they ar more frequently born there; but partly because that in [Common] Republickes the way to honour and prefer-ment lys more open to desert, which is a quickning Spur and a great incitement to Noble Spirits: and Partly too, because the lesser Inequality of Men's Conditions in Common-welths, renders these Heroick Spirits more conspicuus: which in Monarchys wud be swallow'd by the Glory of the King or Princes; to whom for the most part ar attributed the most Glorius Action of their Subjects.[47]

This should not be quickly interpreted, however, as a clear indication that he has been converted to the parliamentarian perspective without reservations; only that advance-ment through merit is one dimension where commonwealths have the edge, but which may be nullified by, for example, unbridled ambition in the headstrong and untutored. Therefore, the most instructive lesson is 'that Vertu is the Cement of humane society', and whatever society can hasten its realisation is to be preferred.[48] Boyle's enthusiasm for commonwealths that promote talent, a work ethos and public service is echoed again in a letter to Samuel Hartlib in May 1647, where, in the context of discussion of the 'Invisible College' and Hartlib's utopian schemes, Boyle points to examples in Europe he is familiar with that possess the appropriate virtues:

> Certainly the taking notice of, and countenancing men of rare industry and publick spirit, is a piece of policy as vastly advantageous to all states, as it is ruinously neglected by the most. And therefore we may evidently observe those common-

wealths (as the Hollanders and the Venetian) to be the most happy and the most flourishing, where ingenuity is courted with the greatest encouragements.[49]

The provisional conclusion one might arrive at regarding Boyle's exposure to the war and its aftermath, particularly in the Stalbridge locality, is that the popular, though generally qualified, forms of royalism and parliamentarianism he encountered, together with the pervasive neutralism in the region, would have reinforced his own partially formed stance of non-partisanship. The pragmatic political caution of his father was an instructive guide for Boyle's own survival. Although Boyle shared much common ground with the Puritan desire for moral reformation, his own approach drew upon a rather broader Erasmian temper, and he had little enthusiasm for the Puritan strategies often employed against opponents, as in the case of Douch. However, his sympathy towards the parliamentary regime probably rested upon the conviction that virtue and personal industry would more easily flourish in a commonwealth than a monarchy. The clear support for the regime by other members of his family, as well as the Hartlib circle he came into contact with by March 1647, no doubt helped confirm his perception of states, such as Holland and Venice, as providential allies in the wider Protestant cause.

From Republic to Restoration

Turning now to the political crisis of the late 1650s, which ultimately led to the collapse of the Republic, it is reasonable to suggest that right up to the last days of the Cromwellian regime, there was no sense of inevitability among those of Boyle's circle that a new monarchical regime was in view. Hartlib, for example, slowly gave up his aim of establishing an Office of Address in 1659, but only against the background of unceasing petitioning for the project. No clear conscious alternative was being actively promoted, and, indeed, only a few weeks before the dissolution of the Republic, Henry Oldenburg, close friend of Boyle and future secretary of the Royal Society, remained optimistic in correspondence that a further republican settlement was on the cards. Notions of providence also placed constraints on conscious political designs as they could work against God's intentions. For example, Oldenburg wrote to Boyle on 4 July 1657 from Saumer, excited about the rumour of English victory over the Spanish fleet in the Canary Isles, and commented, 'If this is true . . . one might say (it seems to me) that the continued success of the Protector is an evident sign that Heaven has some extraordinary task for him'.[50]

Nevertheless, Hartlib's group, including Boyle who had been largely immersed in Oxford science after 1656, probably recognised that their aspirations could be halted or perhaps significantly undermined by the ending of the Protectorate, as the role of the state had been central to their projects, if not their successes. Indeed, their very lack of success in promoting schemes in the Republic pointed to the need for continued vigilance and confidence in God's historical design, whatever configuration it took.

This included responding positively to the exigencies of any new political regime. As late as November 1659, one finds Hartlib's desire for a new fraternity that would maintain the flame of reformation, embodied in 'Antilia', a refashioned Macaria, drawing Boyle into the subsequent correspondence.[51]

In the event, it was not to be a revived Macaria that came to embody hopes for a reformation in science, morals and learning, but a Royal Society quickly consolidated within the first three years of the restored monarchy. Clearly, all of the older sources have serious deficiencies in helping to determine who were the active nucleus. Webster was able, however, to arrive at a reasonably accurate picture of the twelve most active members in the first two-and-a-half years' life of the Society: Boyle, Brouncker, Charleton, Croone, Evelyn, Goddard, Moray, Oldenburg, Petty, Rooke, Wilkins and Wren – all of whom were at the inaugural meeting. Significantly, in reconstructing the number of references of participants in meetings between December 1660 and June 1663, Boyle came third with 116, only overtaken by Moray (172) and Goddard (139).[52]

Yet Boyle's scientific affiliations through the Interregnum from the Invisible College and Hartlib's Agency to the Oxford Experimental Club and Gresham College only underline his membership of a broad Protestantism with a correspondingly rich Baconian and utilitarian conception of the social role of experimental science. The Oxford Club, in particular, provided Boyle and his associates with an arena for formulating practices of organised, collective experimental discipline, so crucial to the Royal Society. Although Boyle had remained in close contact with the millenarian-inspired pansophic strain of applied science exemplified by Puritans like Hartlib, John Dury and Theodore Haak, the Oxford science he came to dominate after the mid-1650s was distinguished by its highly specialised experimental work and political heterogeneity. The Oxford community was concerned on the one hand with abstract mathematics, astronomy and the classic problems of natural philosophy, and on the other with practical chemistry, physiology and mechanical instruments. In effect, the foundation and institutionalisation of the Royal Society reflected a very broad set of intellectual forces which Boyle himself, as a gentleman naturalist interested in the relationship between theoretical abstraction and practical application in experimental philosophy and the manual arts, could embody.[53]

By 1660, quite apart from the lack of headway of Hartlib's Agency, of which Boyle had been a trustee, Boyle may have seen the need to reappraise the whole practicality of income funded by the state. In theory, subscriptions from the Society's Fellows presumed an organisation freed from state regulation and the vagaries of national policy, which was complemented by Boyle's own financial independence from the chemical/medical market. However, Hunter's research has shown clearly enough that the Society experienced a very precarious existence without a permanent and substantial transfusion of funds from the monarchy, whatever prestige it gained from royal patronage.[54] Of greater importance was the political implication of Boyle's lifelong concern for individual self-discipline and industry, a tight rein on the passions, the construction of axioms, a sceptical temper and a higher premium placed on the moral ends of faith rather than the speculative content of theology. These traits of Boyle not only found expression in the experimental philosophy, but also pointed to order and rationalisation in civil society, whether expressed in technological, educational and economic reform or, for example, projects for a new 'universal' language. Although Boyle could share with figures like Petty a strong belief in the efficacy of reform, the efficient husbanding of resources and the banishing of idleness (particularly as it applied to that arena of experiment, Ireland), it is also worth noting that Boyle moved away from many of the social engineering schemes associated with Hartlib. Here Boyle's Interregnum past may have left him with divided loyalties, and provides some clue as to why he virtually never mentions Hartlib by name in his published work, even though correspondence between them continued regularly until Hartlib's death in 1662. Did the public Boyle now also think it politically prudent to eradicate in print his fifteen-year association with the Puritan emigré and virtuoso? What is probable is that Boyle, like most of his associates, continued to believe in the necessity of strong government from the centre, whatever its precise political complexion, in order that the new philosophy could more easily advance. In this sense, it had been as appropriate to provide the republican government with qualified support as it was now equally appropriate to ally with monarchy – ideally, limited monarchy.

Conclusions

It has been argued by Schaffer and Shapin that, following the experience of the Civil War and Interregnum, it became axiomatic for Boyle and his experimental community to claim that their knowledge alone could ensure social harmony. While Hobbes remained preoccupied with the problem of uncontrolled private belief, made worse in his view by the independent claims of clergymen and experimental philosophers to competence in disseminating knowledge and policing consciences, Boyle decided both to provide new apologetic weapons for weakened churchmen and to present his community as an ideal society of social harmony and right knowledge. Pursuing a social contextualist thesis similar to that of the Jacobs, Schaffer and Shapin suggest the 'rhetoric' of limited toleration was set against legal coercion and an irenic space was allowed for all those committed to the necessity of good order, an ideal that was promoted by Hartlib, Dury, Ussher, Broghill and, of course, Boyle through the 1650s.[55] Similarly, experimental knowledge would be generated by the authority gained from communal discipline and free judgement. A 'godly' community emerged based upon inductive proofs and Boylean 'matters of fact' alongside a latitudinarian, doctrinally imprecise religion. In sum, the mechanical philosophy became a self-conscious political weapon marrying experimentation with political conservatism to secure the Restoration settlement and the long-term success of experimental science.[56]

While this thesis contains valuable insight, there is an unwarranted assumption about the unity of the experimental community and its transparent political objectives, which is tightly linked to the deliberate promotion by Boyle of a rudimentary empiricism, largely devoid of both theoretical sophistication and causal hypothesis.[57]

The concentration on 'matters of fact' as an ideological gambit not only misconstrues the serious role it had for Boyle in determining experimental effects, as Sargent has pointed out, but also shifts the focus away from other equally, if not more, important motifs of Boyle, such as intelligibility. Moreover, though Boyle was represented by Hall and Kuhn as a strict adherent of the mechanical philosophy, other commentators have shown how qualified Boyle was in his scientific writings regarding mechanism.[58] Clericuzio has pointed out that it was only at the start of the eighteenth century that Boyle became a symbol of the mechanical philosophy, an image particularly drawn by Leibniz, who posited a sharp line between Boyle's and Newton's theories of matter. Yet many of Boyle's contemporaries understood that he was reluctant to explain chemical phenomena by simply having recourse to the mechanical properties of particles, and that his chemistry was corpuscular before it was mechanical.[59] Boyle, therefore, was an advocate of the mechanical philosophy only at the level of general principles of intelligibility over other rival philosophies. This was very carefully arrived at and was not accepted by Boyle because it could, as the Jacobs have argued, be used as the Society's ideological stance against occultist sectaries and radicals who saw science as a 'powerful tool for promoting religious, political, and social revolution'.[60] That Boyle (and his Oxford colleagues) in the 1650s consciously developed a new 'Anglican' synthesis crystallised around 'a metaphysics of God and matter that authorized a conservative interpretation of the social hierarchy', erodes the epistemological, philo-sophical and theological roots of Boyle's philosophy of nature for a more fundamental, and unwarranted, recourse to socio–political explanation.[61]

Boyle was undoubtedly a key figure in articulating an intellectual space through the rhetorical distancing of science from religion and politics. In response to those critics who charged the Society with providing ammunition for materialists and atheists, Boyle helped formulate a rationale in functional differentiation, with reason and faith occupying different dimensions and ends of enquiry. In this sense, the industrious pursuit of moral and natural philosophy begun in the 1640s helped decipher a world of physical and moral order in the midst of war and political disorder. Both realms pointed to an appropriate and providential accommodation of politics, experimentalism and moral knowledge in the altered circumstances of the Restoration.

NOTES

1   J. R. Jacob, 'The ideological origins of Robert Boyle's natural philosophy', *Journal of European Studies* 2 (1972): 1–22; 'Robert Boyle and subversive religion in the early Restoration', *Albion* 6 (1974): 275–93; 'Restoration, reformation and the origins of the Royal Society', *Hist. Sci.* 13 (1975): 155–76; 'The New England Company, the Royal Society and the Indians', *Social Studies of Science* 5 (1975): 450–5; 'Boyle's circle in the Protectorate: revelation, politics and the millennium', *JHI* 38 (1977): 131–40. These articles were incorporated into Jacob, *Robert Boyle and the English Revolution: A Study in Social and Intellectual Change* (New York: Burt Franklin, 1977). See also his 'Boyle's atomism and the Restoration assault on pagan naturalism', *Social Studies of Science* 8 (1978): 211–33; 'Restoration ideologies and the Royal Society', *Hist. Sci.* 18 (1980): 25–38; and, with M. C. Jacob, 'The Anglican origins of modern science: the metaphysical foundations of the Whig constitution', *Isis* 71 (1980): 251–67.

2 Michael Hunter, 'Latitudinarianism and the "ideology" of the early Royal Society: Thomas Sprat's *History of the Royal Society* (1667) reconsidered', in *Establishing the New Science. The Experience of the Early Royal Society* (Woodbridge: The Boydell Press, 1989), pp. 45–71, on pp. 46–7. Jacob's view of Sprat's *History* as the official manifesto of the Society essentially followed that of Margery Purver in her *The Royal Society: Concept and Creation* (London, 1967). Purver's views had already been subjected to an important critique by Charles Webster in 'The origins of the Royal Society', *Hist. Sci.* 6 (1967): 106–28.

3 Jacob, 'Origins of the Royal Society' (n. 1), pp. 155, 169.

4 Barbara Shapiro, 'Latitudinarianism and science in seventeenth-century England', *Past and Present* 40 (1968): 156–41, reprinted in *The Intellectual Revolution of the Seventeenth Century*, ed. Charles Webster (London: Routledge, Kegan & Paul, 1974), pp. 286–316; also *John Wilkins 1614–72: an Intellectual Biography* (Berkeley: University of California Press, 1969), ch. 8.

5 Shapin and Schaffer, *Leviathan and the Air-Pump*, esp. ch. 7.

6 The following symbols are used in transcriptions from manuscripts: [ ] deletion; < > interlinear; { } marginal insertion; ( ), // Boyle's own marks, unless otherwise indicated.

7 Cork to Dungarvon, 21 November 1642. British Library MS, Egerton 80, fol. 14r. The most important study of Cork remains Nicholas Canny, *The Upstart Earl: A Study of the Social and Mental World of Richard Boyle, First Earl of Cork, 1566–1643* (Cambridge University Press, 1982).

8 M. MacCarthy-Morrough, *The Munster Plantation. English Migration to Southern Ireland 1583–1641* (Oxford University Press, 1986), pp. 260–1, 278.

9 Broghill's plan to travel to France after the execution of Charles I in January 1649, in order to help the King's son, should be seen in this light. Broghill was subsequently forced by Cromwell to change allegiance. At the same time, Boyle's attempt to aid Broghill in acquiring a pass from the French ambassador indicates the unease Boyle must have felt after Charles's execution. See Thomas Morrice, 'The life of the Earl of Orrery', in *A Collection of the State Letters of the Right Honourable Roger Boyle, the First Earl of Orrery . . .* (London, 1742), pp. 9–11; Boyle to 'Madam' (unidentified), 26 March 1649 from Marston Bigot. Draft letter, Yale University Medical Library, Conn., USA. I am grateful to Antonio Clericuzio for a copy of the letter.

10 Margo Todd, *Christian Humanism and the Puritan Social Order* (Cambridge University Press, 1987), p. 22. This is now the most exhaustive survey of the transmission of humanism into Protestant social ethics.

11 *Ibid.* p. 23.

12 *Ibid.* pp. 24–5. See also P. Duhamel, 'The Oxford lectures of John Colet', *Journal of the History of Ideas* 14 (1953): 493–510. For an example of Boyle's continental education, see *The Lismore Papers*, ed. Alexander B. Grosart, 5 vols. (First Series, privately printed, 1886) and 5 vols. (Second Series, privately printed, 1887–8; hereafter *II Lismore*), iv, 100–1.

13 Todd, *Christian Humanism* (n. 10), pp. 28–9, 32–3. See also her 'Seneca and the Protestant mind: the influence of Stoicism on Puritan ethics', *Archiv für Reformationsgeschichte* 74 (1983): 182–99.

14 Todd, *Christian Humanism* (n. 10), pp. 40, 46. In so far as Todd argues that the university curricula at Oxford and Cambridge were heavily permeated with humanist assumptions, this suggests that the complexion of Boyle's private education was a lot closer to those formal curricula than may have been supposed. Boyle's major ethical treatise is in R.S. MS 195, of which MS 192 appears to be a draft, titled *The Aretology or Ethicall Elements of Robert Boyle. Begun at Stalbridge, The [Blank] of 1645. That's the tru Good that makes the owner so.* For the dating of MS 195 note the title page and *Works*, i, xxx, xxxiv. It has now been published in Harwood, *Essays*, pp. 3–141.

15 Harwood, *Essays*, p. 5. A detailed analysis of Boyle's principal arguments in the *Aretology* is not appropriate here but can be found in Malcolm Oster, 'Nature, ethics and divinity: the early thought of Robert Boyle' (unpublished D.Phil. thesis, University of Oxford, 1990), pp. 86–97.

16 Harwood, *Essays*, p. 17. Boyle goes on to affirm that the true temper of the wise man consists 'not in the being Without Passions, but in the being Aboue them', p. 18.

17 BP 37, fols. 160–3. Quoted from fol. 161v. Titled 'The Gentleman', it clearly dates from the 1640s, though its relationship to the *Aretology* is unclear. Jacob interprets Boyle's understanding of the aristocratic code simply as a matter of peer group assessment, quite beyond issues of private conscience; Jacob, *Robert Boyle* (n. 1), p. 47. Boyle's 'The Gentleman' is read as a study in the 'expediency of descent', a device that plays, for Jacob, a 'curious' part in the apparent reversal of this

tradition worked out in the *Aretology* where temporal success is predicated upon virtue and expediency 'is swept under the carpet'; Jacob, *Robert Boyle* (n. 1), pp. 48–9. For a broader view of Boyle's understanding of social rank as it pertained to experimentalism and the manual arts, see Malcolm Oster, 'The scholar and the craftsman revisited: Robert Boyle as aristocrat and artisan', *Ann. Sci.* 49 (1992): 255–76 (Boyle quotation from p. 259).

18 BP 37, fols. 162v–163r.

19 Harwood, *Essays*, p. 70. Wide-ranging discussion on Protestant perceptions of wealth is in Timothy Breen, 'The non-existent controversy: Puritan and Anglican attitudes on work and wealth, 1600–1640', *Church History* 35 (1966): 273–87.

20 See Boyle's *The Dayly Reflection* in Harwood, *Essays*, pp. 203–35.

21 Breen, 'Non-existent controversy' (n. 19), p. 286.

22 Harwood, *Essays*, pp. 71, 113.

23 Jacob, *Robert Boyle* (n. 1), pp. 50–1.

24 *Ibid.* p. 66.

25 Grosart, *II Lismore* (n. 12), v, 22.

26 *Ibid.* p. 23.

27 *Ibid.* p. 72, Boyle to Cork, 15 May 1642.

28 *Works*, i, xxvii.

29 *Ibid.* Boyle wrote that finding Lady Ranelagh was 'acknowledg'd as a seasonable Providence', as was being detained 'constantly in a Religious family' where 'gracious Providence' prepared him for his future. British Library MS Sloane 4229, fol. 68. For details on the Anglo–Irish parliamentarian circle Boyle encountered in London, see Charles Webster, *The Great Instauration* (London: Duckworth, 1975), pp. 57–67. For Ireland, T. C. Barnard, *Cromwellian Ireland, English Government and Reform in Ireland 1649–1660* (Oxford, 1975), esp. ch. 8. I show the significance of Boyle's developing personality in 'Biography, culture, and science: the formative years of Robert Boyle', *Hist. Sci.* 31 (1993): 177–226.

30 Jacob, *Robert Boyle* (n. 1), pp. 13–14.

31 Examples of the 'deference' view include Lawrence Stone, *The Causes of the English Revolution* (London: Routledge & Kegan Paul, 1972), p. 145, and Robert Ashton, *The English Civil War*, 2nd edn (London: Weidenfeld & Nicolson, 1989), p. 174. Neutralism and county loyalty is covered by B. L. Manning, 'Neutrals and neutralism in the English Civil War 1642–1646' (unpublished D.Phil. thesis, University of Oxford, 1957); Alan Everitt, *The Community of Kent and the Great Rebellion* (Leicester University Press, 1966); J. S. Morrill, *The Revolt of the Provinces: Conservatives and Radicals in the English Civil War 1630–1650* (London: Allen & Unwin, 1976). The class thesis is strongly evident in Christopher Hill, *Puritanism and Revolution in Pre-Revolutionary England* (London: Secker & Warburg, 1964), pp. 199–214; *The World Turned Upside Down* (London: Penguin, 1972); *Milton and the English Revolution* (London: Faber, 1978); and by Brian Manning, *The English People and the English Revolution* (London: Penguin, 1976). The regionalist interpretation is shown with regard to Dorset, Somerset and Wiltshire in David Underdown, *Revel, Riot, Rebellion, Popular Politics and Culture in England 1603–1660* (Oxford University Press, 1985).

32 Manning, 'Neutrals and neutralism' (n. 31), p. 144.

33 G. E. Aylmer, *Rebellion or Revolution? England from Civil War to Restoration* (Oxford University Press, 1987), pp. 42–3.

34 Anthony Fletcher, 'The coming of war', in *Reactions to the English Civil War 1642–1649*, ed. John Morrill (London: Macmillan, 1982), pp. 35, 39.

35 Ronald Hutton, 'The royalist war effort', in Morrill, *Reactions* (n. 34), p. 53.

36 Underdown, *Revel, Riot, and Rebellion* (n. 31), pp. 147–8. Hutton, 'Royalist war effort' (n. 35), p. 64. Hutton believes that the main purpose of the Dorset Clubmen was to stop plunder by recently established local garrisons on both sides through ensuring a proper levy of the local rate for local soldiers. For plundering in Dorset, see F. J. Pope, 'Sidelights on the Civil War in Dorset', *Somerset and Dorset Notes and Queries* 12 (1910–11): 52–5.

37 Hutton, 'Royalist war effort' (n. 35), p. 64; J. H. Bettey, 'Agriculture and rural society in Dorset 1570–1670' (unpublished Ph.D. thesis, University of Bristol, 1977), p. 339. Bettey maintains that the 'great majority of large owners were sympathetic to the royalist cause', *ibid.*, p. 351.

38 Underdown, *Revel, Riot, and Rebellion* (n. 31), pp. 25, 96–7, 104–5, 140, 173–4. For the Vale's economic and agricultural complexion, see Bettey, 'Agriculture' (n. 37), pp. 3, 5, 88, 306, 327.

39 Underdown, *Revel, Riot, and Rebellion* (n. 31), p. 197. Nowhere were celebrations more festive for the return of the King than at Sherborne, *ibid.*, p. 271.

40 *Works*, i, xxx. It remains unclear whether Boyle is referring to a local committee. In the absence of other evidence, the county committee is a more suitable candidate.

41 *Ibid.* pp. xxx–xxxiv: Boyle to Marcombes, 22 October 1646.

42 *Ibid.* pp. xxx–xxi. The letter indicates the general prudence he had to exercise in this period: 'I was once a prisoner here upon some groundless suspicions, but quickly got off with advantage . . . I have been forced to observe a very great caution, and exact evenness in my carriage, since I saw you last, it being absolutely necessary for the preservation of a person, whom the unfortunate situation of his fortune made obnoxious to the injuries of both parties, and the protection of neither', *ibid.*, p. xxxiii.

43 *Works*, vi, 40.

44 C. H. Mayo, *The Minute Books of the Dorset Standing Committee, 23rd Sept. 1646 to 8th May 1650* (Exeter: William Pollard & Co., 1902), p. 317. Douch was to subsequently die 11 June 1648, *ibid.* Boyle's continued interest in the parish in later life, despite his residence in London after 1668, is shown by the possession of a copy of the petition for the removal of Samuel Rich, instituted in 1675, for immorality in BP 40, fols. 10–11. For useful information on this area, see I. M. Green, 'The Persecution of "scandalous" and "malignant" parish clergy during the English Civil War', *English Historical Review* 94 (1979): 507–31. Generally, the committees were the architects of their own unpopularity, and in the post-war period began to recognise this. For example, in December 1646 the Dorset committee began to reduce the number of sequestration officials. See Underdown, *Revel, Riot, and Rebellion* (n. 31), pp. 225, 244–6.

45 Harwood, *Essays*, p. 56.

46 *Ibid.* p. 67.

47 *Ibid.* p. 132. Monarchy does have the virtue, however, of maintaining order as he points out elsewhere: 'And tis a Maxime amongst Statesmen, that <a> Monarchy, (tho de[ne] generated into Tyranny) is yet les Intollerable then an Anarchy', *'Of Sin'*, *ibid.*, p. 145.

48 *Ibid.* p. 146.

49 *Works*, i, xl, 8 May 1647. Boyle spent considerable time in Italy on the Grand Tour and visited the Netherlands from the end of February to the beginning of April 1648. See Maddison, *Life*, p. 72. For further comment on possible meanings of Boyle's use of the term 'ingenuity' see Shapin and Shaffer, *Leviathan and the Air-Pump*, pp. 130–1, and my own response to their appraisal in Oster, 'Scholar and the craftsman' (n. 17), pp. 275–6.

50 Oldenburg to Becher, 2 March 1660, *Oldenburg*, i, 359. The anticipation of a further republican settlement is also pointed out by Charles Webster, 'Puritanism, separatism and science', in *God and Nature: Historical Essays on the Encounter between Christianity and Science*, eds. David Lindberg and Ronald Numbers (Berkeley: University of California Press, 1986), p. 212.

51 *Works*, vi, 132. Hartlib to Boyle, 15 November 1659. The initial interest soon dissipated, and by October 1660 Hartlib was apologetic for the inflated hopes raised over the colony. See *The Diary and Correspondence of Dr John Worthington*, ed. J. Crossley, 2 vols. (Manchester, 1847–86), i, 210–15. Hartlib to Worthington, 15 October 1660. For information on the early project of 'Macaria', see Charles Webster, 'Macaria: Samuel Hartlib and the Great Reformation', *Acta Comeniana* 26 (1970): 147–64.

52 Webster, *Instauration* (n. 29), pp. 90–2. Webster points out that this active nucleus was split fairly evenly between those who had been in collaboration with the parliamentary regime, such as Petty, Wilkins and Goddard, and those who had opposed it, such as Moray, Wren and Evelyn: *ibid.*, p. 95.

53 For details on the political and philosophical composition of the Oxford Club, see Webster, *Instauration* (n. 29), pp. 153–78. For Boyle and Oxford science, see Robert Frank, *Harvey and the Oxford Physiologists* (Berkeley and Los Angeles: University of California Press, 1980).

54 For the Society's financial difficulties, see Hunter, *Establishing the New Science* (n. 2), ch. 1.

55 Jacob, 'Boyle's circle' (n. 1), pp. 134–5, and with M. C. Jacob, 'Anglican origins' (n. 1), pp. 265–7; Shapin and Schaffer, *Leviathan and the Air-Pump*, p. 300.

56 For Shapin and Schaffer's analysis of Boyle's use of 'matters of fact', see *Leviathan and the Air-Pump*, pp. 22–6, 43, 139, 162, 166, 175, 222. The long tradition of legal usage involving the term is documented in Rose-Mary Sargent, 'Scientific experiment and legal expertise: the way of experience

in seventeenth-century England', *Studies in History and Philosophy of Science* 20 (1989): 19–45, esp. pp. 42–5.

57 Shapin and Schaffer, *Leviathan and the Air-Pump*, p. 341, and Hunter's response, *Establishing the New Science* (n. 2), p. 14.

58 For the centrality of intelligibility for Boyle, see Antonio Perez-Ramos, *Francis Bacon's Idea of Science and the Maker's Knowledge Tradition* (Oxford: Clarendon Press, 1988), pp. 166–79; Sargent, 'Scientific experiment' (n. 56), p. 43. For Boyle as strict mechanist, see M. B. Hall, *Robert Boyle and Seventeenth-Century Chemistry* (Cambridge University Press, 1958) and T. S. Kuhn, 'Robert Boyle and structural chemistry', *Isis* 43 (1952): 12–36. For a different view, see A. Clericuzio, 'A redefinition of Boyle's chemistry and corpuscular philosophy', *Ann. Sci.* 47 (1990): 561–89; Charles Webster, *From Paracelsus to Newton: Magic and the Making of Modern Science* (Cambridge University Press, 1982), p. 69; Yung Sik Kim, 'Another look at Robert Boyle's acceptance of the mechanical philosophy: its limits and its chemical and social contexts', *Ambix* 38 (1991): 1–10.

59 Clericuzio, 'A redefinition' (n. 58), p. 561; Sargent, 'Scientific experiment' (n. 56), pp. 34–5.

60 Jacob and Jacob, 'Anglican origins' (n. 1), p. 253. Similarly, Hill, *World Turned Upside Down* (n. 31), ch. 14.

61 Jacob and Jacob, 'Anglican origins' (n. 1), pp. 253–4. See Timothy Shanahan, 'God and nature in the thought of Robert Boyle', *Journal of the History of Philosophy* 26 (1988): 547–69, esp. pp. 548–9.

Science writing and writing science: Boyle and rhetorical theory

JOHN T. HARWOOD

> But I think it will be convenient to make the very first words of the *English* account itself I sent you, to be preceded by a long stroke in the same line, to intimate that the letter did not begin there; lest, without such intimation, the beginning may appear either too abrupt, or not so civil to you, as I desire my writings should be thought, as well as be.[1]

> But besides the unintentional deficiencies of my style, I have knowingly and purposely transgressed the laws of oratory in one particular, namely in making sometimes my periods or parentheses over-long: . . . I chose rather to neglect the precepts of rhetoricians, than the mention of those things, which I thought pertinent to my subject, and useful to you, my reader.[2]

The tercentenary of Robert Boyle's death was commemorated by historians and philosophers of science, not by literary scholars. Since an important dimension of Boyle's career was literary, such neglect is surprising. Part of the reason for this neglect may be the textual and bibliographical limitations of the Birch edition, which makes Boyle seem like a writer of the mid–eighteenth century rather than a contemporary of Sir Thomas Browne, John Milton and John Dryden.[3] Given the volume of his publications during a thirty-year career, his achievement as a writer can hardly be questioned. When we consider the range and variety of these publications, we should not be surprised by his publisher's extraordinary claim in *Specifick Medicines* (1685) that Boyle had become known 'both at home and abroad . . . [as] *the English Philosopher*'.[4] Yet Boyle never defined himself solely as a natural philosopher or solely as a moral philosopher, having noticed that religious books written by laymen, 'especially gentlemen, have (*caeteris paribus*) been better entertained, and more effectual, than those of ecclesiasticks . . . [He] is the fittest to commend divinity, whose profession it is not'.[5] Instead, he established and exploited personae that enabled him to address both natural and moral philosophy.

We should recall that Boyle emerged from the late 1640s not as a natural philosopher but as an essayist and a moral philosopher. He was a writer with a passionate interest in contemporary issues, especially social reform. As a novice moralist, he was particularly critical of the failure of divines and contemporary moral philosophers to persuade. 'Too many of our moralists', he complained, spend

> more time in asserting their method, than the prerogatives of virtue above vice; they seem more sollicitous, how to order their chapters than their readers actions; and are more industrious to impress their doctrine on our memories than our affections, and teach us better to dispute of our passions than with them.[6]

If a reformer writes but does not persuade, he has failed. For someone so passionate about reform, rhetorical effectiveness was crucial.

From Boyle's prefaces and letters to his friends – especially to Lady Ranelagh, Samuel Hartlib, Robert Hooke, John Beale and Henry Oldenburg – we learn how Boyle read other writers, how some readers read him, how he responded to some of his readers and how he wanted his works to be read. His self-consciousness about writing emerged from anxieties about the dangers of misreading and the importance of correct reading. What has not been satisfactorily recognised is the role of writing and publication in defining his career within a highly literary culture. To make this case, I first survey Boyle's sense of authorship, showing that, despite his personal modesty, he pursued a highly ambitious literary program. Next I examine some of Boyle's remarks on rhetoric, suggesting his lifelong interest in the ways that rhetorical practices affected readers' responses and examining some of his strategies for achieving his rhetorical aims. Finally, I consider Boyle's quest for eloquence and a style that would create an audience for natural philosophy, arguing that metaphor was central not only to his sense of style but also to understanding nature.

Fashioning a literary life

To describe Boyle as a major writer of the Restoration is hardly an understatement. He was the most prolific and most prominent writer among the Fellows of the early Royal Society. Between 1659 and 1700, more than eighty editions of Boyle's writings appeared in English and more than one hundred translations were published in Latin. Of this total, more than one hundred editions appeared in England, and the remainder were published on the continent, most in Latin but nearly a dozen in French, German, and Dutch.[7] In only three years – 1677, 1679 and 1689 – did an original work, a new edition or a translation of an earlier work by Boyle not appear in England. Since several books (e.g., *Occasional Reflections* [1665] and *Martyrdom of Theodora* [1687]) were written several decades before they were published, his literary career was far more complicated than his contemporaries realised. Indeed, a complete account of his publishing history, when available, will shed important light on his career as a writer.

Even if we acknowledge the sincerity of Boyle's personal modesty, we should also recognise his sensitivity to his status as a writer and his ambition to be an influence in

Europe as well as in England.[8] Ambition and modesty are not contradictory attributes; Boyle had both traits, though he chose to emphasise only his modesty. In the most famous portraits and engravings of himself, he pointed away from himself and toward his writings, thus deflecting attention toward his work or toward Nature, the subject of his writings.[9] When William Faithorne was engraving Boyle's portrait, Hooke asked Boyle for detailed advice, enclosing Faithorne's sketch 'to see whether you approve of the dress, the frame, and the bigness; what motto or writing you will have on the pedestal, and whether you will have any books, or mathematical, or chemical instruments, or such like, inserted in the corners, without the oval frame, or what other alterations or additions you desire'.[10] Hooke would not have raised such questions unless Boyle cared about the visual details. When it came to the rituals and conventions of print culture, Boyle mastered at least four strategies for fashioning a literary identity. I shall refer to these strategies as (1) embedding self-references in his writings, (2) encouraging publishers' catalogues, (3) securing Latin translations for his major works and (4) adopting a variety of social roles and personae on his title pages. Let me touch briefly on each strategy.

Boyle was more than willing to make himself a character in his own writings, thereby shaping a public image of himself. Consider, for example, the numerous references to 'Boyle' as a character who was defended in a dialogue, the image of 'Boyle' in the *Sceptical Chymist*, or even a self-reference to *Seraphic Love* in *Occasional Reflections*.[11] Boyle frequently cited his own publications and alluded to forthcoming publications, some of which never appeared. When we also consider the extent of his involvement in the early issues of the *Philosophical Transactions* – as an author or as the subject of book reviews or other accolades – we see how print rapidly made him a public figure.[12] Anyone who read all of his *oeuvre* would have a rich store of images and associations with the character of 'Robert Boyle'.

Even more striking evidence of his literary self-fashioning are the catalogues of his publications. Full analysis of Boyle's publication lists must await the forthcoming Hunter and Davis edition, which will deal with archival and bibliographical evidence too complex to be summarised briefly, but what is worth emphasising here is the role of these catalogues in signalling, preserving and enhancing his status as an author. I argue that as early as the 1660s Boyle understood that he could use print culture to shape his career. Authorship had its privileges.

Seventeenth-century printers often used the end-papers of their volumes to list other books that they had published. The printers' motives for such advertising are easily understood, but Boyle's catalogues are different for two reasons: (1) he worked with a variety of printers, thereby reducing their immediate self-interest; and (2) the catalogues were so frequent – at least ten were produced during his lifetime. One can hardly question Boyle's participation in these publications: the earliest ones took the form of a publisher's note to the reader and were signed by 'H.O.' (Henry Oldenburg).[13] On 17 October 1667, Boyle wrote to Oldenburg: 'I think you have pitched upon a good course about the catalogue of my writings; only if you mention it at all, you may be pleased to remember to add to the reference, that the two tracts of the

Origin of Forms, and of *Hydrostatical Paradoxes*, mentioned in the catalogue you speak of, annexed at the end of the *History of Cold*, came forth since that history'.[14] It was unlikely that Oldenburg would have initiated such ventures without Boyle's approval and cooperation. In addition, the catalogues continued long after Oldenburg's death. The final catalogues included Boyle's last publications, indicating that, as Boyle approached the end of his life, he was concerned about the accuracy and completeness of the record.

The earliest catalogue of Boyle's publications accompanied his *Experimental History of Cold* (1665). In it, Oldenburg listed writings already published by Boyle as well as 'such Philosophical Writings of the same Author, as being occasionally mention'd (here and there) in the above-named Books, are not yet publish'd, but (though not absolutely promis'd) by divers of the Curious expected'.[15] As the convoluted syntax and teasing tone suggest, Boyle was aware of readers' expectations and sought to attract and to please the 'Ingenious Reader'.

From the outset of his career, Boyle wanted to ensure that both English and Latin editions were available. As early as 1667, he asked Oldenburg to monitor how his works were being received on the continent.[16] For almost all of the major works of the 1660s and 1670s, the Latin translation appeared either in the same year as the English edition or in the next year; in some cases, the Latin edition appeared first or included material not printed in the English editions, as the forthcoming edition will indicate. These editions are noted in the catalogues, which thus enabled Boyle to signal that he was more than an 'English' author and more than a writer of natural philosophy or theology. Through the Latin editions published in England and on the continent, he addressed a European audience.

Steven Shapin and Simon Schaffer have properly emphasised that Boyle's social status gave him special influence and authority in the early Royal Society. But we should not underestimate the significance of an even simpler fact: Boyle's wealth. The same wealth that allowed him to build his own vacuum pump and hire an operator also allowed him to pay his Latin translators and, if necessary, his printers. As a result, his translators and printers were prompt. By contrast, the proposed Latin translations of Hooke's *Micrographia* and Thomas Sprat's *History of the Royal Society* were never completed.[17] But promoting his translations was not his only strategy. During the 1660s, Boyle experimented with several other conventions of publication, including anonymous and semi-anonymous attribution on his title pages.

How should a gentleman identify himself on the title page of a book? This question assumes, of course, that a gentleman has already agreed to publish his writings and has thereby disregarded the vague sense of impropriety associated with publication. We should recall that some of Boyle's friends believed that a gentleman who was also a philosopher should not have published *Occasional Reflections*.[18] Boyle identified himself in several ways: as an anonymous author, as a semi-anonymous author, as an author-as-gentleman and as an author as Fellow of the Royal Society. Each choice has implications for readers' expectations and for Boyle's unfolding career.

Boyle used some of his title pages either to signal or partially to mask his identity, playing off against, perhaps arousing and certainly confounding readers' expectations about 'Robert Boyle'. Even when he *seemed* to publish anonymously, he ensured that readers of *The Excellency of Theology* (1674) would recognise his authorship through acronyms like 'T.H. R.B.E.'. As readers decoded the acronym and articulated the message one letter/word at a time ('The Honourable Robert Boyle, Esquire'), it accentuated Boyle's status far more prominently than if the words had been printed on the title page. We should contrast the modest attribution of the dedication of the English translation of de Bils' *Anatomy* (1659) to 'R.B.' with the authorial charade of 'T.H. R.B.E.'. Boyle was aware of the change in his status as a public figure between 1659 and 1674, and his title pages signalled it. Moreover, in *Excellency of Theology* the marginal references to Boyle's earlier publications are so copious that his identity could almost be determined by this feature alone. We should consider his several variations on the claims of authorship.

Boyle published anonymously *Things above Reason* (1681), though he acknowledged the work in his catalogues soon thereafter. In 'Some Considerations about the Possibility of the Resurrection', annexed to *The Reconcileableness of Reason and Religion* (1675), the author was 'T.E. A Layman'. The authorship of *A Free Enquiry into the Vulgarly Receiv'd Notion of Nature* (1686) and of *The Christian Virtuoso* (1690) was assigned to 'R.B. Fellow of the Royal Society', an identification that would have mystified nobody; *The Martyrdom of Theodora* (1687) was ascribed to a 'Person of Honour' but was immediately attributed to Boyle in the catalogues. Since Boyle did not use such ruses for his works of natural philosophy – the title pages indicated his social status and, after 1663, his affiliation with the Royal Society – we may infer that the author did not wish to implicate directly the Society in discourse unrelated to its charter. At the same time, the transparent mask of anonymity simultaneously pointed away from and directly toward the author; it provoked and resolved suspense in the same gesture, insisting on the importance of the author's identity even while thinly concealing it. If readers in the 1680s wondered who 'Robert Boyle' was, then he had successfully established a range of personae suitable for his two favourite subjects, theology and natural philosophy.

Above all else, he sought to influence the *manners* and *minds* of his readers.[19] Regardless of genre, Boyle's primary rhetorical purpose was not to provide information; it was to *persuade*. To do this, he adjusted his presentation for different kinds of audiences, both popular and learned.[20] If he had been addressing atheists or sceptics, he would have used different arguments.[21] Audiences determined both the kinds of arguments and the kinds of style. One strategy for dealing with the audience was to control the characteristics of his persona.

The social status of the writer, Boyle knew, shaped readers' expectations. But status was not the only element. He could control other attributes (e.g., good manners in dealing with adversaries, avoidance of dogmatism, modesty, self-deprecation, etc.) and create an attractive persona. If his social standing would help to persuade others to write occasional meditations, the same ethos would also encourage gentlemen to do experiments.[22] The most important implication of his social standing is that it enabled him to pursue moral and natural philosophy without being suspected of ulterior motives or of partisan interests.

> [It] will somewhat add to the reputation of almost any study, and consequently to that of things divine, that it is praised and preferred by those, whose condition and course of life exempting them from being of any particular calling in the common-wealth of learning, frees them from the usual temptations to partiality to this or that sort of study, which others may be engaged to magnify, because it is their trade or interest, or because it is expected from them; whereas these gentlemen are obliged to commend it, only because they really love and value it.[23]

Thus, he wrote *Occasional Reflections* 'not to get reputation, but company, [so he is] unwilling to travel alone; and had rather be out-gone than not at all followed, and surpassed than not imitated'.[24] By systematically cultivating a persona that included the attributes of 'gentleman', 'layman', and 'virtuoso', he could justify his writing on theological topics (though not a divine) and on medical subjects (though not a physician). The fluidity of his personae generated questions that he felt obliged to answer: Why had he made natural philosophy a 'secondary' topic of study after he had written so many books to persuade readers to value it highly? Why had he never made divinity, philosophy or physic his profession after writing so extensively about the subjects? Why did he write as a layman and not a cleric?[25] Besides complicating readers' views of his identity, he also gained stylistic latitude.

Boyle's manuscripts and prefaces richly document his struggles with writing. Taken collectively, his publications are festooned with apologies, complaints, excuses and self-justifications. In various works, he apologised for beginning so abruptly, for not making transitions between sections, for not being able (because of illness) to re-read what he had written, for making the parts of the discourse of unequal size, for unintended repetitions caused by poor eyesight, for leaving dialogues unconnected and for presenting an incomplete argument. He chronicled a litany of problems (illness, haste, lack of method) that undermined the revision of his *Free Enquiry into the Vulgarly Receiv'd Notion of Nature*. He justified why he had included quotation in an experimental work, why he had not cited other writers in a work on style and why he had excluded references to the Church Fathers and school philosophers in a theological work.[26] He even digressed about the need for digressions.[27] Though all of these gestures may be as sincere as they are tiresome to read, they also affect Boyle's persona.

Such elaborate meta-discourse has at least one important rhetorical effect. It heightens the reader's awareness of Boyle's efforts to produce rhetorically effective prose and thus strengthens the author's ethos, his earnestness, his moral credibility. As Boyle commented in his *Notion of Nature*, his method of presentation was proof that he 'pretends not to dogmatize, but only to make an inquiry'.[28] His meta-discourse sends a clear message about the persona and his programme. Boyle did not present his finished thoughts or unalterable conclusions: he offered readers his working drafts and thus invited them to imitate his practice.

Through a wide network of friends and correspondents, Boyle created a literary circle that provided encouragement, advice and detailed responses to his writing. Their correspondence with him creates a rich secondary literature about reading, writing and publishing in the late seventeenth century. Whereas the title pages and catalogues were public documents, his letters privately discussed other dimensions of authorship. Because his correspondence has been purged twice – an action that John Beale recommended – we should be especially attentive to manuscripts and letters that have survived.[29] These letters shed new light on Boyle's vocation as an author.

Rhetoric and reading

Boyle considered the problems of communication from the perspective of the reader as well as the writer, providing abundant evidence of how he performed both roles. By all accounts, he was a sensitive reader, responding strongly – perhaps viscerally – to what he read and believing that other readers were like him. Whenever he read two particular verses of Lucan, for example, he felt a chill.[30] He was opposed to the reading of romances for a deeply personal reason: they mesmerised his imagination.[31] He carefully recorded how he read Scripture and how he interpreted it; we learn how he read books of devotion.[32]

Even reading wise authors could be problematic, he warned, because the dangers of misreading were so numerous and so grave:

> A wary person reads the wisest authors, with a reflection, that they may deceive him by being themselves deceived; and undergoes a double labour, the one in investigating the meaning, and the other in examining the truth of what they deliver: but in the Bible, we are eased of the latter of these troubles.[33]

For Boyle, reading Scripture was a matter of listening either to God or to Satan;[34] it was a matter of mining underground.[35] He was attracted to eloquence because he understood its power, although he also recognised that persons who read only for figures were like those who heard 'the most pathetick sermons, not as Christians but orators'.[36] Clearly, *how* one read determined *what* one read: readers were active participants in determining the meaning of texts. Because many readers required some of the rhetorical figurations that misled other readers, Boyle needed to accommodate them:

> There being a nicer sort of readers, which need instruction (and to whom it is therefore a charity to give it) who are so far from being likely to be prevailed on by discourses not tricked up with flowers of rhetorick, that they would scarce be drawn so much as to cast their eyes on them.[37]

If reading was a complex act, writing could hardly be easy. His awareness of and sensitivity to readers no doubt contributed to his anxiety about writing. Certainly, none of his contemporaries left more evidence of the aspirations to conduct both an 'experimental life' *and* a literary life. His remarks about rhetoric therefore have particular relevance.

Writing autobiographically about his education in Geneva in the late 1640s, Boyle adopted the pseudonym 'Philaretus'. He recalled that Philaretus studied 'both Rhethoricke & Logicke, whose Elements (not the Expositions) Philaretus wrote out with his owne hand; tho afterwards he esteem'd both those Arts (as they are vulgarly handled) not only vnseasonably taught, but obnoxious to those (other) Inconveniences & guilty of those Defects, he dos fully particularize in his Essays'.[38] The parts of rhetorical theory, he knew, included 'embellishments of our conceptions and . . . the congruity of them to our design and method, and the suitable accommodation of them to the various circumstances considerable in the matter, the speaker, and the hearers'.[39] This definition is hardly surprising, for it was a commonplace of Renaissance rhetorical training. But like much of Boyle's learning in other areas, his knowledge of rhetoric was both self-taught and remarkably extensive. He knew far more about the subject than we might have expected from the dismissive remarks of Philaretus.

Like alchemy or astronomy, rhetoric has specialised terms and concepts that Boyle had to master. In several works, he discussed – and assumed his audience understood – such technical terms as protasis, apadosis and reddition as well as the principles of decorum associated with them.[40] His highly original analysis of the styles found in Scripture, first published in *Some Considerations Touching the Style of the Holy Scriptures* (1661) but written a decade earlier, deepened his knowledge of classical rhetoric and the importance of accommodating style to one's audience. As Boyle was to demonstrate, natural philosophy was a subject for investigation, not a fixed literary genre.

In the seventeenth century, Richard Kroll argues, 'all forms of knowledge (including natural philosophy) were commonly known and confessed to be rhetorical'.[41] Although Boyle never developed an original theory of rhetoric, his concerns about reading and misreading were certainly rhetorical and were shared by the larger culture: the

> mid-seventeenth century could comprehensively revise its view of itself by rereading classical culture; it subsequently adapted a method – in part from ancient culture – that could reveal and discuss that very process of rereading; it argued for a general practice of reading governed by such methodical self-consciousness; and . . . it treated as tantamount to an ethical or political failure the failure to read with sufficient empirical rigor.[42]

Indeed, because Boyle worried a great deal about how readers read and misread texts, he richly documented how he read and wrote about texts as diverse as Scripture and Nature.

Boyle's correspondence contains abundant references to readers and writers, authorship and readership, the fame and the influence of successful writers.[43] In *Science and Society in Restoration England*, Michael Hunter has noted that John Beale's concern with fame is a frequent topic in his correspondence with Hartlib, Evelyn and Boyle.[44] Beale encouraged Boyle to model his career after Erasmus's, whose fame was secure because he published in short works and because he carefully preserved his letters for publication.[45] Beale urged him to collect and preserve his letters during his lifetime. 'Your own amanuensis, and your own directions, are the best conservatories of these; and you may then expunge and enlarge, as you see it of public conducement'.[46] (We know that Boyle closely read Descartes' correspondence.)[47] At stake, of course, was more than fame: it was the survival of one's ideals and one's works. All authors, even Bacon, were at risk of being forgotten, Beale believed. For instance, if Boyle's experiments appeared to continue the *Novum Organum*, he would save Bacon from oblivion.[48] Only authors whose works continued to be printed could avoid 'the *Lethean lake*', a sobering thought for any writer.

Beale's critiques of his friend's writings raise the same issues that Boyle addressed so often in his prefaces. In 1661, for example, Beale described how he reacted to the language and style of Boyle's *Certain Physiological Essays*, topics about which the author cared deeply.[49] We should observe how directly Boyle justifies the styles of *Certain Physiological Essays* (1661) and describes the uses of each:

> I have endeavoured to write rather in a philosophical than a rhetorical strain, as desiring, that my expressions should be rather clear and significant, than curiously adorned . . . For though a philosopher need not be solicitous, that his style should delight its reader with his floridness, yet I think he may very well be allowed to take a care, that it disgust not his reader by its flatness, especially when he does not so much deliver experiments or explicate them, as make reflections or discourses on them: for on such occasions he may be allowed the liberty of recreating his reader and himself, and manifesting, that he declined the ornaments of language, not out of necessity, but descretion, which forbids them to be used, where they may darken as well as adorn the subject they are applied to.[50]

We should notice the explicit attention to figurative language, which Boyle regards as quite appropriate when the philosopher reflects on experiments. The writer's discretion is critical. Later in his career, Boyle asked Beale for specific criticisms of a theological work,[51] and Beale responded with highly detailed remarks.[52] From such comments, Boyle could learn how his readers understood and reacted to his writings.

Beale was not the only correspondent who shared reactions to Boyle's writings and who offered advice about the pursuit of a literary life. John Evelyn's reaction to *Seraphic Love* was nothing short of rapturous, at least as far as its style was concerned.[53] In 1665, Lady Ranelagh, whose encouragement and support continued for forty-five years, told Boyle that many people wished to borrow her copies of his works.[54] In the same year, when Londoners were terrified by the plague, she urged him to publish a second edition of *Occasional Reflections* to teach people how to 'read' the world.[55] In 1665, Boyle informed Richard Baxter that his *Occasional Reflections* had been more fortunate than he had expected 'even from divers poets and wits, who were indisposed to any literary work that promoted piety'.[56] In 1666, Boyle told Oldenburg that he was gratified by the reception of his *Usefulness of Experimental Philosophy*.[57] A surgeon wrote to Boyle about a specific page in a work and then discussed the case with the author.[58] From praise and criticism, queries and suggestions, he learned to anticipate or forestall criticism of his style, argument or method. Again, we must allow his correspondents to suggest the range of rhetorical interests: many of Boyle's letters have not survived. One such topic deals with visual techniques for improving the clarity of presentation.

Boyle was interested in how the layout of material on the page affected the reception of content, as we know from his extensive correspondence with Hooke and his prefatory remarks about 'reading' the plates in the first continuation of *New Experiments Physico-Mechanical, Touching the Spring and Weight of the Air*.[59] He was acutely aware of the importance of illustrations in helping his audience understand and accept natural philosophy.[60] The illustrations needed to be so lavish that even novice experimentalists could carry out their own experiments.[61] As in natural philosophy, no detail was too small to ignore. Even the marginal references on the page were important to the correct reading of natural philosophy. Beale claimed that

> every marginal reference might advertise of a fresh method by union, parallel, graduation, limitation, and thousands of other inferences, whereof some might prove of comprehensive and pregnant importances. And a list of experiments well chosen will hold out against all times, as firmly as any of Euclid's propositions. Sir, the marginal remark is to signify, how you may make an hundred experiments serve for a thousand uses, and escape all oppositions.[62]

If illustrations and marginal references were important cues to readers, so too were citations.[63] Boyle was even willing to indicate passages that his readers could skip, using brackets to mark material that could be safely ignored.[64] To state the matter more simply, Boyle took great pains to make his texts readable. Conscious of the rhetorical decorum expected of a philosopher and of his readers' expectations and tastes, he sought to make difficult texts accessible. The combination of high ideals and problematic approaches to composition, however, made it very difficult for him to complete his works satisfactorily.

What was his method of composing? His prefaces and the Boyle Papers provide a rich collection of explanatory illustrations. The following account is representative of his general approach to writing, which I characterise as an 'additive' or 'incremental' strategy:

> And, that I might not exclude, by too early a method, those things, that, for aught I knew, might hereafter be pertinent and useful, I threw my reflections into one book, as into a repository, to be kept there only as a heap of differing materials, that, if they appeared worth it, they might be afterwards reviewed, and sorted, and drawn into an orderly discourse . . . [though he was unable to] give them the finishing stroke.[65]

Because Boyle was attracted to 'heaps' of material, he assiduously gathered and preserved information on an astonishing range of topics. Like his contemporaries, he kept commonplace books and *adversaria*, and he generally composed 'heads' of an argument as an aid to planning.[66] Indefatigable at identifying topics to discuss, problems to investigate and materials to assemble, he was far less successful at finding principles of order or organisation for the materials he collected. The hope of 'an orderly discourse' was all too often only a hope. The 'finishing stroke' often eluded him.

In reflecting on the proper way to write a natural history, for example, he described three preliminary steps: reading, conference and meditation. He then identified various kinds of trials to be made.[67] But to move from individual trials to a coherent discourse was troublesome, and he experimented with various techniques to produce an orderly draft. One approach was to change the kind of paper on which he composed his material. Because Boyle had lost some of his early papers and even whole treatises, he told Oldenburg, he resolved to write in loose and unpaged sheets:

> and sometimes (when [Boyle] had only short memorials and other Notes, to set downe) even in lesser papers. And though [he] knew it probable that by this course, [he] might loose some Papers, yet supposing that, however, [he] should by that course keep some men from the Temptation they had before, [he] presumed that by the help of the remaining Papers and [his] memory, [he] should quickly be able to supply the Loss of a sheet or two, in case it should happen.[68]

Loose pages did not, however, solve his problem with 'method', a word that recurs frequently when he described how he performed experiments or meditations.[69] He later composed only on one side of the leaves so that he would have room to insert later experiments.[70] He was quite willing both to 'pad' and to insert materials originally written for one occasion into quite different kinds of works, a tactic that made coherence even more challenging.[71] As a result, he preferred loosely structured discourses to tightly structured arguments: his points accumulate, sometimes even cascading. A loose structure allowed for subsequent additions. In retrospect, the large number of publications entitled 'continuations' or 'second part' capitalise on this approach.

Whether he wrote systematically or haphazardly, in short tracts or long books, Boyle found writing very difficult.[72] If drafting was hard, revision was excruciating, often more a matter of the scissors than the pen. This lengthy account from *Some Considerations Touching the Style of the Holy Scriptures* is typical of his approach:

> because that in divers of the other parts of my essay I had here and there, frequently enough, occasion to say something of the same theme, I have been obliged, that I might obey you, not only to dismember, but to mangle the treatise you perused, cutting out with a pair of scissars here a whole side, there half, and in another place perhaps, a quarter of one; as I found, in the other parts of my discourse, longer or shorter passages, that appeared to relate to the style of the scripture, that I might give you at once all those parts of my essay, which seemed to concern that subject. And though I have, here and there, by dictating to an amanuensis, inserted some lines or words, to make the loose papers less incoherent, where I thought it easy to

be done; yet in many others I have only prefixed a short black line to the incoherent passages, if I could they could not be connected with those, whereunto I have joined them, without such circumlocution, as either the narrowness of the paper would not permit, or my present distractions . . . and the weakness of my eyes, would not allow of.[73]

Not surprisingly, neither his amanuenses nor his printers were always successful in producing the document that Boyle had wanted, a fact for which many 'advertisements' and publishers' notes apologised.[74] Such problems only compounded his concerns about style. A nagging problem was the style in which he was to present his material and, by implication, to present 'Robert Boyle' to his audience.[75]

Pleasing the 'Ingenious Readers'

When we consider the rhetorical variety of his *oeuvre* – letters, dialogues, reports, descriptions, sermons, meditations and oratory – I doubt that Boyle would accept Vickers's dichotomy between 'rhetorical' and 'philosophical' writing.[76] Vickers assumes that because Boyle wrote *New Experiments Physico-Mechanical* in the form of a letter, he 'had not yet moved from the "literary" to the "scientific" mode'.[77] I do not think that the modes were so firmly determined or that Boyle was constrained by rigid generic expectations. It is true that in the early 1660s Boyle distinguished between the 'rhetorical strain' and the 'philosophical strain' in the preface to *Certain Physiological Essays*, but neither his discussion nor his practice in that work wholly rejected the 'rhetorical strain'.[78] Unlike Sprat or Hooke, whose early work was supervised by more senior members of the Royal Society, he was always free to experiment with rhetorical forms and a range of styles.[79] Indeed, he was never satisfied that he had found the best way to *persuade* his intended audience.

His letters provide numerous details about the production and distribution of books, his own and others'. We learn from Hartlib, for example, that Boyle bought forty copies of de Bils' *Anatomy* (1659) for distribution to gentlemen.[80] This gesture was a shrewd tactic to reach (or help create) an audience of gentlemen and to render it sympathetic to his outlook and method on a subject not in itself likely to attract them. To reach a larger audience of gentlemen after the Restoration, Boyle would need to present his material attractively. Style was crucial to this goal.

Boyle defended his style of writing as one driven by the needs of readers who were novice experimentalists or who were relatively unskilled in practical divinity, casuistry, or theology.[81] For *rhetorical* reasons he demonstrated how he made reflections and experiments, how he extracted significance from them and how he integrated personal experience and biblicism. In each mode, eyewitness testimony was critical to his ethos, his credibility. And in each mode, he assumed that his audience would be responsive both to evidence and to style.

While he would never have attended the London theatres after the Restoration –

both Hooke and Beale were shocked at the content of the plays and the conduct of the audience – he certainly was aware that contemporary audiences were expected to recognise and evaluate subtle shifts of style in poetry and drama. In his *History of His Own Time*, Gilbert Burnet, Boyle's close friend, frequently commented on their contemporaries' rhetorical style as speakers and as writers.[82] Both in speaking and writing, style was as prominent a trait as a person's appearance or character. Just as Sprat worried about the derision of the 'Wits' in his *History*, Boyle's sensitivity to the outlook of the 'Wits' or 'Gallants' was evident even in the 1640s and endured through-out his career.[83] If audiences were so closely attuned to stylistic nuance, how was he to proceed?

Boyle had an acute sense of the need to accommodate his material to his audience. The issue was important because strategies of accommodation were shaped by assumptions about readers – their knowledge, outlook and interests. Boyle made quite similar assumptions about the audiences for his works of natural philosophy and his works of moral philosophy. For natural philosophy, his readers were defined as potential experimentalists whose lives would be sharply improved by pursuing what Steven Shapin and Simon Schaffer call 'the experimental life'.[84] His nephew Richard Jones was addressed as 'Pyrophilus', thus representing a fictionalised 'Every Reader'. As such, his fictional attributes are important.

Pyrophilus belonged to 'this sort of virtuosi' among the nobility and gentry who were disgusted by what the schools taught about forms, and generation, and corruption, a reaction that was 'usual among ingenious readers'.[85] Boyle's rhetorical task was to show such readers precisely what the experimental life involved while depicting its rewards in such a way that gentlemen would want to embrace it. He thought that gentlemen were attracted to eloquently presented arguments and that figurative language was crucial to eloquence.[86]

Why was figurative language so important? Boyle believed that powerful metaphors made a lasting impression on the memory; they were aids to seeing and remembering. Those 'truths and notions, that are dressed up in apt similitudes, pertinently applied, are wont to make durable impressions on [the memory]', he noted in *Occasional Reflections*, and he exemplified this by a repeated rhetorical pattern of a concrete observation leading to a spiritual application.[87] Moreover, the same literary itch appears as a subtext in natural philosophy as well: after describing quite graphically the experiment of the bee in the air pump in *New Experiments Physico-Mechanical* – a familiar kind of experiment in which the creature almost dies and then is revived – Boyle comments thus: 'This invited us thankfully to reflect upon the wise goodness of the creator, who, by giving the air a spring, hath made it so very difficult, as men find it, to exclude a thing [i.e., air] so necessary to animals'.[88] The movement between literal and figurative, physical and spiritual, is swift and revealing. Clearly, he found similitudes powerful and hoped that they would be as memorable for readers as they were for him.

Indeed, in *The Christian Virtuoso*, Boyle described his efforts to please contemporary taste through his choice of similitudes:

> And (that the stile might not be unsuitable to the writer, and the design) I thought
> fit, in my arguments and illustrations, both to employ comparisons drawn from
> telescopes, microscopes &c. and to make frequent use of notions, hypotheses, and
> observations, in request among those, that are called the new philosophers. Which
> I rather did, because some experience has taught me, that such a way of proposing
> and elucidating things, is either more clear, or, upon account of its novelty, wont to
> be more acceptable, than any other to our modern virtuosi; whom thus to gratify, is
> a good step towards persuading of them.[89]

Boyle's fascination with metaphoric language was widely shared by his contem-
poraries, but our understanding of this important issue has been badly complicated by
Thomas Sprat's presentation of the topic in his *History of the Royal Society*.[90] The early
Royal Society both feared and courted the 'Wits', nowhere more clearly than in
Sprat's *History*. I argue that Sprat's claims about language were designed to appease the
Wits, not to describe how natural philosophy should be written. Despite his well-
known account of the 'plain style' followed by the Royal Society – 'And to accomplish
[their aims, the Fellows] have indeavor'd, to separate the knowledge of *Nature*, from
the colours of *Rhetorick*, the devices of *Fancy*, or the delightful deceit of *Fables*'[91] –
both Boyle and Hooke made extensive use of figurative language. Even 'that severe
philosopher, monsieur *Des Cartes*,' Boyle observed, 'somewhere says, that he scarce
thought, that he understood any thing in physics, but what he could declare by some
apt similitude'.[92] Though Hooke is less explicit than Boyle in justifying his style, we
should not overlook the important similarities between them.

In *Micrographia* (1665), for example, Hooke wondered whether a correct way to
understand Nature might involve decoding the shapes and properties of bodies and
thus to 'read' the text that God has written:

> Who knows but *Adam* might from some such contemplation, give names to all
> creatures? If at least his names had any significancy in them of the creature's nature
> on which he impos'd it; . . . And who knows, but the Creator may, in those charac-
> ters, have written and engraven many of his most mysterious designs and counsels,
> and given man a capacity, which, assisted with diligence and industry, may be able
> to read and understand them.[93]

This master trope of 'reading Nature' – alluding to the metaphor of the 'Two Books' –
was invoked by many writers in the early Royal Society and is, in my judgement, the
most important metaphor in Boyle's writings.[94] By considering Nature as a text, Boyle
defines his role as a reader. Just as he had needed to learn ancient languages in order
to understand Scripture, he needed to acquire new skills to read Nature. The text
of Nature, as he called it, was 'God's stenography', hieroglyphics, or God's epistle;[95]
God was the writing master.[96] Perceiving metaphors in written texts was expected of
good readers; discovering metaphoric relationships in Nature enabled discovery
and then communication. If Nature is an encrypted text, the natural philosopher
needs to find the correct keys to unlock it. Metaphor was crucial to seeing and
communicating.

In *The Excellency of Theology*, Boyle observed that 'naturalists are as much obliged to God for their knowledge, as we are for our intelligence to those, that writ us secrets in cyphers, and teach us the skill of decyphering things so written; or to those, who writ what would fill a page in the compass of a single penny, and present us to boot a microscope to read it'.[97] This view of metaphor and the master trope of 'reading the world' applies as easily to Boyle's natural philosophy as it does to his moral philosophy. In the second part of *The Christian Virtuoso*, Boyle again stresses the metaphor of texts, locks, and keys, comparing the object of theology

> to a book of excellent notions and secrets, but written in cyphers; and the knowledge of divine things, conferred by bare philosophy, to a very imperfect key; consisting of not half the letters of the alphabet: for though, by the help of this, a man may make a shift to understand, here and there, many whole words, that consist of such letters, as his key answers to, and may perchance, on the same account, understand some few short and scattered sentences, and other periods; yet the whole scope, and contrivance of the book, and many of the principal things, contained in it, he must be ignorant of: but revelation is like the remaining part of the key, which compleats the other; and which, when he is once possessed of, he will not only learn, many new truths, and understand satisfactorily, many passages, which he but doubtingly, guessed at before; but, which is the main thing in our comparison, he will clearly perceive, that what he has last discovered, both perfectly agrees with what he had learned before, and compleats what is wanted, to make up a symmetrical piece, worthy of the author.[98]

Notwithstanding Sprat's strictures on figurative language, many of Hooke's descriptions are quite as figurative and almost as contemplative as those in *Occasional Reflections*.[99] Neither was restricted to the stylistic conventions of the 'plain style'.

In cultivating an audience of 'Ingenious Readers', Boyle could draw on and impart his knowledge of natural philosophy. He had access to a rich, novel vocabulary for description and explanation, one in which objects were worthy of study both in themselves and as part of a richly symbolic universe of meaning. Natural philosophy provided new occasions for pushing literal details into figurative significance; figurative language illuminated natural phenomena. Rightly used, metaphors were to the mind what microscopes were to the eye.[100] For readers, metaphors aided comprehension; for writers, metaphors enabled perceptions and understanding.

The Royal Society needed both microscopes and metaphors, and Boyle was eager to promote the uses of both instruments. Indeed, both 'mechanical philosophy' and the 'corpuscularian hypothesis' are heavily dependent on figurative language for their meaning. Boyle employs the metaphor of body-as-engine in a half dozen works of natural philosophy and moral philosophy.[101] Part of his task in promoting the 'New Philosophy' was to find the best possible metaphors for unobservable subjects.[102]

In summary, the movement between literal and figurative is more than an important aspect of Boyle's sense of style. It is also a link between his moral philosophy and his natural philosophy. His early writings demonstrate an interest in natural philosophy as a source of figurative language and an incitement to piety. Later, he was eager to find metaphors capable of explaining the natural world. His abiding assumption was that metaphors *affected* an audience as they *affected* him. His rhetorical self-consciousness helps to explain his abiding interest in how he responded to texts as well as his obsessive worries about his own success. Boyle's conviction that writing was a powerful agent of change enabled him to triumph over the frustrations and anxieties that characterised a long career as an author. If he had truly become the '*English Philosopher*', it was because he had learned to exploit both the vacuum pump *and* the printing press.

NOTES

1 *Oldenburg* iv, 93. Boyle to Oldenburg, 29 December 1667.

2 *CPE, Works*, i, 305.

3 In *Selected Philosophical Papers of Robert Boyle* (Manchester University Press, 1979), M. A. Stewart laments that 'Boyle scholarship is at present dominated by the use of the verbally corrupt 1772 edition of the *Collected Works*, . . . which has acquired an authority it never deserved, partly because commentators have mistaken the eighteenth-century archaism of its typography, spelling and punctuation for the authentic convention of a century before, and partly because they have been over-ready to take nonsense for period quaintness' (p. viii). Besides establishing reliable texts of his publications, the forthcoming edition of Boyle's works by Michael Hunter and Edward B. Davis will address at least three other critical issues: Boyle's publication history in English and Latin, his involvement with successive editions of individual works and his complicated relations with amanuenses, translators and publishers. Hunter's guide to the MSS in *Letters and Papers of Robert Boyle: A Guide to the Manuscripts and Microfilm* (Bethesda, Md., University Publications of America, 1990) is invaluable.

4 *Works*, v, 76.

5 *Excellency of Theology, Works*, iv, 2–3.

6 *Style of the Scriptures, Works*, ii, 287.

7 My figures are based on material found in J. F. Fulton's *A Bibliography of the Honourable Robert Boyle*, 2nd edn (Oxford: Clarendon Press, 1961). I have not attempted to determine whether a volume printed in (say) 1680 was a re-issue of a previous edition or a new edition: publishers were quite entrepreneurial about printing a new title page and declaring a 'new' edition.

8 His reputation in Italy is well described in Clelia Pighetti, *L'Influsso Scientifico di Robert Boyle nel Tardo '600 Italiano* (Milan: F. Angeli, 1988).

9 See Steven Shapin, 'Pump and circumstance: Robert Boyle's literary technology', *Social Studies of Science*, 14 (1984): 481–520; and Shapin and Schaffer, *Leviathan and the Air-Pump*, pp. 256–60.

10 *Works*, vi, 488.

11 See *Saltness of the Sea, Works*, iii, 734; *Sceptical Chymist, Works*, i, 459; and *Occasional Reflections, Works*, ii, 395. For other self-references, see *Sceptical Chymist, Works*, i, 556; *Excellency of Theology, Works*, iv, 1, 24, 59; *Cold, Works*, ii, 468, 666; and *Hydrostatical Paradoxes, Works*, ii, 739. Such examples could be easily multiplied.

12 For instance, see Oldenburg's review of *Cold* in *Phil. Trans.* 1 (1666): 46–52, in which he wrote: 'the inquiring World [may] take notice, that this subject, as it hath hitherto bin almost totally neglected, so it is now, by this Excellent Author, in such a manner handled, and improved by near *Two hundred* choice *Experiments* and *Observations*, that certainly the *Curious* and *Intelligent* Reader will in the perusal thereof find cause to admire both the Fertility of a Subject, seemingly so barren, and the Author's Abilities of improving the same to so high a Degree' (p. 47).

13 A draft of a publisher's note to accompany *Degradation of Gold* survives: BP 8, fol. 171; cf. *Works*, iv, 371.

14 *Oldenburg*, iii, 532.

15 Robert Boyle, *New Experiments and Observations Touching Cold* (London, 1665), pp. 846–7.

16 *Oldenburg*, iv, 93–4. In *Boyle: la vita, il pensiero, le opere* (Milan: Accademia, 1978), Clelia Pighetti has

shown that Boyle had a powerful effect on his Italian contemporaries, serving Malpighi not just as a scientific exemplar but also as a literary model.

17 John Wilkins made very slow progress on a Latin translation of *Micrographia*, alarming Oldenburg (Oldenburg, iii, 31) and depriving Hooke of some recognition that his pioneering work deserved. One can speculate whether this episode could be related to Hooke's later bitterness about being deprived of credit for other work, a topic most recently considered by Rob Iliffe, '"In the warehouse": privacy, property and priority in the early Royal Society', *Hist. Sci.* 30 (1992): 29–68. Sprat's *History* suffered a similar fate, with neither a Latin nor a French translation ever being completed (*Oldenburg*, iii, 485, 487, 515–16).

18 *Works*, vi, 525.

19 *Style of the Scriptures*, *Works*, ii, 287.

20 *Ibid.* p. 274.

21 *Excellency of Theology*, *Works*, iv, 65.

22 *Occasional Reflections*, *Works*, ii, 326.

23 *Excellency of Theology*, *Works*, iv, 2–3.

24 *Works*, ii, 335.

25 See *Excellency of Theology*, *Works*, iv, 3–4 and *Usefulness*, II, *sect. 1*, *Works*, ii, 62; *Notion of Nature*, *Works*, v, 160; and *Christian Virtuoso*, *Works*, v, 509.

26 *Rarefaction of Air*, *Works*, iii, 495; *Final Causes*, *Works*, v, 394; *Spring, First Continuation*, *Works*, iii, 178; *Things above Reason*, *Works*, iv, 406; *Reason and Religion*, *Works*, iv, 191; *Notion of Nature*, *Works*, v, 254; *Cold*, *Works*, ii, 475; *Style of the Scriptures*, *Works*, ii, 254; and *Reason and Religion*, *Works*, iv, 153.

27 *Style of the Scriptures* was 'designed to serve particular acquaintances of mine, [so] it was fit it should insist on those points they were concerned in; and that . . . much of the seeming desultoriness of my method, and frequency of my rambling excursions, have been but intentional and charitable digressions out of my way, to bring some wandering friends into theirs, and may closely enough pursue my intentions, even when they seem most to deviate from my theme' (*Works*, ii, 254).

28 *Works*, v, 161.

29 Of Boyle's letters to Hartlib, for example, not one survives among the Hartlib Papers at Sheffield University; likewise, few of Boyle's letters to John Beale are extant. For discussion of the survival of Boyle's correspondence, see Harwood, *Essays*, pp. xx–xxi.

30 *Languid Motion*, *Works*, v, 20.

31 Boyle's attraction to romances is a frequent concern in the early writings (Harwood, *Essays*, pp. xlviii, 116, 187, 192, 196, 239). In 'Robert Boyle and the epistemology of the Novel', *Eighteenth-Century Fiction* 2 (July 1990): 275–82 and *Before Novels: The Cultural Contexts of Eighteenth-Century English Fiction* (New York: Norton, 1990), J. Paul Hunter attributes to Boyle an important role in creating an audience for fiction, a remarkable irony given Boyle's profound distrust of fictional forms.

32 See *Reason and Religion*, *Works*, iv, 172; *Excellency of Theology*, *Works*, iv, 17; and *Style of the Scriptures*, *Works*, ii, 289.

33 *Ibid.*

34 *Ibid.* p. 269.

35 *Ibid.* p. 279.

36 *Occasional Reflections*, *Works*, ii, 449.

37 *Ibid.* p. 434.

38 Maddison, *Life*, p. 30. For a broad overview of how schoolboys learned logic and rhetoric, see Wilbur Samuel Howell, *Logic and Rhetoric in England, 1500–1700* (Princeton University Press, 1956).

39 *Style of the Scriptures*, *Works*, ii, 301.

40 *Occasional Reflections*, *Works*, ii, 328–30. He defined and illustrated many rhetorical schemes in *Style of the Scriptures*, trying to convince Orrery that the Bible achieved sublime eloquence.

41 Richard W. F. Kroll, *The Material Word: Literate Culture in the Restoration and Early Eighteenth Century* (Baltimore: Johns Hopkins University Press, 1991), p. 22.

42 *Ibid.*

43 In his *Aretology*, a sustained work of ethics written in the late 1640s, Boyle stressed that honour was more important than fame: Harwood, *Essays*, p. 72.

44 Michael Hunter, *Science and Society in Restoration England* (Cambridge University Press, 1981), pp. 194–7.

45 *Works*, vi, 414. For Erasmus's manipulation of print culture, see Lisa Jardine, *Erasmus, Man of Letters* (Princeton University Press, 1993).

46 *Ibid.* p. 417.

47 *Reason and Religion, Works*, iv, 163.

48 *Works*, vi, 417; for Beale's commitment to protecting Bacon's reputation, see Hunter, *Science and Society* (n. 44), p. 195. On the importance of experiment, see Marie Boas Hall, *Promoting Experimental Learning: Experiment and the Royal Society, 1660–1727* (Cambridge University Press, 1991).

49 *Works*, i, lxiii.

50 *CPE, Works*, i, 304.

51 *Works*, vi, 341.

52 *Ibid.* pp. 393–5.

53 *Ibid.* pp. 291–4.

54 *Ibid.* p. 528.

55 *Ibid.* pp. 525–6.

56 *Ibid.* p. 520.

57 *Oldenburg*, iii, 65. Oldenburg discussed the sales of Boyle's works (*Oldenburg*, ii, 653).

58 *Works*, vi, 648.

59 *Ibid.* pp. 500ff; *Works*, iii, 178.

60 Beale counselled that the schemes should indicate how they are to be read or used (*Works*, vi, 404); for their importance to Boyle's publications, see *Works*, vi, 393.

61 *Usefulness, II, sect. 2, Works*, iii, 30, and *Hydrostatical Paradoxes, Works*, ii, 738.

62 *Works*, vi, 404–5.

63 Boyle analysed the 'politics' of citation in *Excellency of Theology, Works*, iv, 56; he justified his quotations, which were inserted after he had written his epistle, in *Reason and Religion, Works*, iv, 156; he explained his reasons for quoting authors in their original language in *CPE, Works*, i, 313; and he asserted the need to check others' quotations carefully in *Cold, Works*, ii, 475.

64 'Those, that do not relish the knowledge of the opinions and rites of the ancient . . . may skip': *High Veneration, Works*, v, 138–9.

65 *Reason and Religion, Works*, iv, 153.

66 For examples, see *Works*, v, 562, 732–3. In *Cold* (*Works*, ii, 469) he developed comprehensive titles (or headings) even if he could not adequately complete them; sections were of unequal length; he was obliged to present much circumstantial detail, even if it weakened his style, because of his regard for experimental facts.

67 *Works*, iv, 598. He listed the heads or titles in *Works*, iv, 599, but see also p. 618 for his heads for human blood, p. 799 for mineral waters, or pp. 601–2 for urine.

68 *Oldenburg*, iv, 98–9.

69 Regarding Boyle's early discussion of method, see Harwood, *Essays*, pp. lviii, 75, 171, 203, 216.

70 The footnote says that he left the MS with Oldenburg to show the intention of the author: *Exp. Obs. Physicae, Works*, v, 569. For problems with loose sheets of paper, see *General History of Air, Works*, v, 611; *Notion of Nature, Works*, v, 159; and *Oldenburg*, iv, 98. In addition, loose sheets were penned over time and not available when he wanted to revise; some sheets were lost: *High Veneration, Works*, v, 130.

71 When he found himself a little short on material, he drew on an epistolary discourse and inserted it, or at least the first fifteen or twenty pages: *Christian Virtuoso, Works*, v, 557.

72 *Excellency of Theology, Works*, iv, 55–6.

73 *Style of the Scriptures, Works*, ii, 252.

74 The Advertisement to *Notion of Nature* (London, 1686; omitted by Birch) contained this painful admission: 'The Reader is here to be advertis'd of a great Oversight that happen'd to be made by several Transpositions of the loose Sheets, wherein (and not in a Book), the Copy was sent to the Press. For the Discourse beginning at the sole Break that is to be met with in the Hundred and Fiftieth Page, and ending with another Break at the second Line of the Hundred Fifty and Sixth Page, ought to have been plac'd at the sole Break that is to be met with in the Hundred Sixty and Second Page. . . . And though also some Connections and Transitions, relating to the Transpos'd Papers, be not such as

they should be, yet 'tis not judg'd fit, that the Reader be troubled with long Advertisements about them; because his Discretion may easily correct them, and the Incongruities are not of Moment enough to spoil the Discourses they relate to' (sig. A, italics reversed).

75 See also Tony Davies, 'The ark in flames: science, language and education in seventeenth-century England', in *The Figural and the Literal: Problems of Language in the History of Science and Philosophy, 1630–1800*, eds. Andrew E. Benjamin, Geoffrey N. Cantor and John R. R. Christie (Manchester University Press, 1987), pp. 83–102; *The Literary Structure of Scientific Arguments*, ed. Peter Dear (Philadelphia: University of Pennsylvania Press, 1991). Dear's chapter, 'Narratives, anecdotes, and experiments: turning experience into science in the seventeenth century' (pp. 135–63), is particularly valuable.

76 Brian Vickers, *English Science, Bacon to Newton* (Cambridge University Press, 1987), p. 46. Vickers has supplanted Richard Foster Jones's view of Sprat in 'The Royal Society and English prose style: a reassessment' in B. Vickers and N. S. Struever, *Rhetoric and the Pursuit of Truth: Language Change in the Seventeenth and Eighteenth Centuries* (Los Angeles: William Andrews Clark Library, 1985), pp. 3–76; see also B. Vickers, *In Defense of Rhetoric* (Oxford University Press, 1988).

77 Vickers, *English Science* (n. 76), p. 46.

78 *CPE, Works*, i, 304.

79 On Sprat's supervision, see Michael Hunter, 'Latitudinarianism and the "ideology" of the early Royal Society: Thomas Sprat's *History of the Royal Society* (1667) reconsidered', in *Establishing the New Science: The Experience of the Early Royal Society* (Woodbridge: The Boydell Press, 1989); on Hooke's, see John T. Harwood, 'Rhetoric and graphics in *Micrographia*' in *Robert Hooke: New Studies*, eds. Michael Hunter and Simon Schaffer (Woodbridge: The Boydell Press, 1989), pp. 131–3.

80 *Works*, vi, 130.

81 For recognition that few readers would actually do experiments, see *Cold, Works*, ii, 471; *Medicina Hydrostatica, Works*, v, 454; and *Final Causes, Works*, v, 393.

82 See Gilbert Burnet, *History of his Own Time*, ed. Osmund Airy, 2 vols (Oxford, 1897–1900), *passim*.

83 For Boyle's fears of the Wits, see Harwood, *Essays*, pp. 69, 83, 214, 242, 246. 'Wits' was generally a codeword for irreligion: see *Christian Virtuoso, Works*, v, 508. For discussions of Sprat's *History*, see Paul B. Wood, 'Methodology and apologetics: Thomas Sprat's *History of the Royal Society*, *BJHS* 13 (1980): 1–26; Peter Dear, '*Totius in Verba*: rhetoric and authority in the early Royal Society', *Isis* 76 (1985): 145–61; and Hunter, 'Latitudinarianism' (n. 79).

84 Shapin and Schaffer, *Leviathan and the Air-Pump*, ch. 2.

85 *Forms and Qualities, Works*, iii, 5.

86 Vickers praises Boyle's use of metaphors in *English Science* (n. 76), p. 15. A splendid account of ideas about eloquence is provided in Debora K. Shuger, *Sacred Rhetoric: The Christian Grand Style in the English Renaissance* (Princeton University Press, 1988), ch. 2.

87 *Works*, ii, 333, and 359ff., *passim*. Similar comments are found in *Style of the Scriptures, Works*, ii, 299–305, and many other places.

88 *Works*, i, 109, 111.

89 *Works*, v, 509.

90 For similar issues in Hooke, see Harwood, 'Rhetoric and graphics in *Micrographia*' (n. 79). Vickers' *Rhetoric and the Pursuit of Truth* (n. 76) clarifies and corrects the pervasive misunderstanding of Sprat's views of rhetoric (see esp. pp. 20–5).

91 Thomas Sprat, *History of the Royal Society* (London, 1667), p. 62.

92 *Christian Virtuoso, Works*, v, 511. Boyle then cited several of Descartes' metaphors.

93 Robert Hooke, *Micrographia* (London, 1665), p. 154.

94 Kroll considers the analogy between atoms and letters to be Boyle's most pervasive simile in *The Material Word* (n. 41), p. 107. On the 'Two Books', see E. M. Klaaren, *Religious Origins of Modern Science* (Grand Rapids, Mich.: W. B. Eerdmans, 1977), p. 62 and ch. 4, *passim*.

95 *Usefulness I, Works*, ii, 63; *Occasional Reflections, Works*, ii, 349; and *Excellency of Theology, Works*, iv, 77. For similar views, see *Usefulness I, Works*, ii, 19–20, and *Occasional Reflections, Works*, ii, 340. The same metaphor can be found in the writings of Sir Thomas Browne: 'Thus there are two bookes from whence I collect my Divinity; besides that written one of God, another of his servant Nature, that universall and publik Manuscript, that lies expans'd unto the eyes of all; those that never saw him in the one, have discovered him in the other' (*Religio Medici* i. § 16).

96   *Final Causes, Works*, v, 438.
97   *Works*, iv, 24.
98   *Works*, vi, 788.
99   In *Micrographia* (n. 93), see Hooke's discussion of the kettering-stone (p. 95), rose-leaves (p. 124), moss (p. 134), the seeds of thyme (p. 153) and poppy (p. 155), the feet and eyes of flies (pp. 171–2, 179–80), the gnat (pp. 193–4), the wandering mite (p. 207) and the book-worm (p. 210). Perhaps because of the similarities to Boyle's style, Beale thought that Boyle had significantly influenced Hooke's work (*Works*, vi, 338). Hooke's later writings do not feature such extended meditations or figurative language.
100 *Christian Virtuoso, Works*, v, 511. Though dealing primarily with Descartes and Newton, Colin Turbayne's argument about metaphor and science also applies to Boyle (*The Myth of Metaphor* [Columbia, S.C.: University of South Carolina Press, 1970]).
101 See *Effluviums, Works*, v, 241; *Specific Medicines, Works*, v, 84; *Occasional Reflections, Works*, ii, 367; *Final Causes, Works*, v, 442; *Usefulness II, sect. 1, Works*, ii, 175; and *Notion of Nature, Works*, v, 235.
102 Turbayne, *The Myth of Metaphor* (n. 100), p. 96.

Learning from experience: Boyle's construction of an experimental philosophy

ROSE-MARY SARGENT

In the 'Introductory Preface' to an otherwise lost work which Boyle apparently wrote in the early 1650s, he claimed that 'ever since I first addicted my Time and Thoughts to the more serious parts of Learning [I] found my selfe by a secret but strong Propensity, inclined to the Study of Naturall Philosophy'.[1] This evidently occurred before the three-year tour of the continent that he undertook under the tutelage of Isaac Marcombes from 1639 onwards, since he states that, prior to 'the commands of my Parents engaging me to visit divers forreigne Countrys', 'the first course I tooke to satisfy my Curiosity, was to instruct my selfe in the Aristotelian Doctrine, as that whose Principles I found generally acquiesced in by the Universitys and Schools, and by numbers of celebrated Writers, celebrated for little lesse, then Oraculous'.[2]

According to Boyle, it was while he was on the continent that he was first 'strongly tempted to doubt' the 'solidity' of Aristotelianism, because he 'met with many things' both in reading and in observation, that were 'capable to make me distrust the Doctrine'. In time, he formulated three general objections to it. First, he found that Aristotle's principles had not been 'proved by his admirers' but had been convincingly 'opposed not only by the Chymists in generall and great store of Moderne Physitians, but [by many] acute and famed Philosop[hers]'. Secondly, among the wonders of the continent, Boyle noted that he had 'observed many things' that 'were wholly unintelligible' from the Aristotelian standpoint because the doctrine was 'much too narrow and slight to reach either all or the more abstruse effects of Nature'. Finally, in addition to its lack of proof and its explanatory failures, Boyle found Aristotelianism 'altogether barren as to useful productions and that with the assistance of it I could do no more than I could have done when I was a stranger to it.'[3]

Embedded within these objections, one can see the beginning of Boyle's desire to devise a new way of doing natural philosophy that would extend the bounds of knowledge beyond the superficial appearances of things and lead to the discovery of truths that would have practical applications. Also, one can see that his objections to Aristotelianism were based in part upon his knowledge of the work that had been performed by his immediate predecessors in philosophy, mechanics, chemistry and medicine, and in part upon his own observations of nature, to which he would shortly add considerations drawn from his extensive investigations into the practices employed by various tradesmen and artisans. In order to understand his rejection of Aristotelianism, therefore, it is necessary to take into account not only the influence of his intellectual predecessors, but also what he learned from his actual experience of the circumstances surrounding practical attempts to manipulate nature.

As an alternative to Aristotelianism, Boyle was to advocate the new 'corpuscularian philosophy' that he described as consisting of two elements: (1) the Baconian programme that sought 'to restore the more modest and useful way practiced by the antients, of enquiring into particular bodies, without hastening to make systems'; and (2) the Cartesian programme that advocated 'the application of geometrical theorems, for the explication of physical problems'. The 'experimental and mathematical way of enquiring into nature' was once again gaining ascendancy because of the work of these 'restorers of natural philosophy', and their direct influence upon Boyle's early objections to Aristotelianism is undeniable.[4] Descartes' influence can be seen in Boyle's acceptance of the idea that explanations in terms of occult forms and qualities should be replaced by hypotheses concerning the motions of the least parts of matter. It was the set of precepts suggested by Bacon for the discovery and justification of such hypotheses, however, that significantly affected the way in which Boyle would formulate the details of his experimental method.

Most importantly, Boyle followed Bacon's new 'physical logic' that inverted the order of discovery and proof.[5] Instead of beginning with speculations about the universal causes that may be operative in nature, philosophers should first compile a vast amount of information about natural effects in order to discover 'how things have been or are really produced'.[6] Boyle saw himself as an 'under-builder'.[7] His work was not designed to establish principles so much as 'to devise experiments to enrich the history of nature with observations faithfully made and delivered' so that 'men may in time be furnished with a sufficient stock of experiments to ground hypotheses and theories on'.[8] The results of experimental trials had to be included within natural histories, because it was by their help that 'nature comes to be much more diligently and industriously studied' and numerous processes may be 'discovered and observed' that 'would not be heeded' by 'the lazy Aristotelian way' that depended upon the superficial appearances of things presented to us by simple sense perceptions.[9]

Boyle also accepted the link made by Bacon between knowledge and power – those 'twin objects' of *The New Organon*.[10] He was careful to stress, however, that the 'famous distinction, introduced by the Lord Verulam, whereby experiments are sorted into luciferous and fructiferous' had to be 'rightly understood'. It would be a mistake to think that although experiments could be sorted as to their dominant end, there was no commerce between the two types. Luciferous experiments 'whose more obvious use is to detect to us the nature or causes of things' may be 'exceeding[ly] fructiferous', because 'man's power over the creatures consists in his knowledge of them; whatever does increase his knowledge, does proportionately increase his power'. Fructiferous experiments, on the other hand, could be of use not merely to 'advantage our interests' but also 'to promote our knowledge', because by giving us the ability 'to produce effects, they either hint to us the causes of them, or at least acquaint us with some of the properties or qualities of the things concurring to the production of such effects'.[11]

As Bacon had called for a marriage of the rational and empirical, so Boyle sought a 'happy marriage' of the practical and the speculative parts of learning.[12] Not only had the order of inquiry and demonstration been inverted, but the very meaning of knowledge had been altered. No longer was the sign of knowledge to be the deductive certainty of the logical or metaphysical systems of classical philosophy. Knowledge was now that which has 'a tendency to use'.[13] The barren speculations of system-builders were to be replaced by the concrete accomplishments of experimenters actively intervening in nature. The Baconian programme provided Boyle with the general epistemological outline of his experimental philosophy. He did not slavishly follow Bacon's precepts, however. Rather, Boyle's broad knowledge of the work, performed in a number of areas of learning in the years following Bacon's death, led him to expand upon and modify the Lord Chancellor's vision of philosophy and to formulate a more comprehensive and sophisticated account of experimental inquiry.

Boyle's eclecticism was more than a somewhat idiosyncratic personality trait. He constructed, and subsequently justified, his experimental philosophy by incorporating within it what he found to be the most successful aspects of various intellectual and practical disciplines. The purpose of the following is to provide a summary account of how Boyle's almost inordinate curiosity about all areas of learning led him to develop the details of a complex epistemological alternative to classical forms of natural philosophy. The first section examines how Boyle compared and contrasted experimentalism with the work performed by his predecessors in three general areas of learning: physico-mechanics, chemistry and medicine. The second section examines the way in which Boyle developed methodological strategies for the new philosophy by using what he learned about experimental activity from his acquaintance with numerous tradesmen and artisans.

Boyle's experimental predecessors

Among the new 'mechanists', Boyle admired Mersenne's work on sound and Bacon's on heat, but it was Galileo who remained for Boyle 'the great master of mechanics' and whose work provided him with most of the concrete examples that he would use for the promise of an experimental method in the physico-mechanical realm.[14] He first encountered Galileo's work while on his tour of the continent, when, in 1642, upon his arrival in Florence, he became interested in the new telescopic observations that had been made in favour of Copernicanism, and he taught himself Italian in order to read 'the new paradoxes of the great star-gazer *Galileo*; whose ingenious books, perhaps because they could not be so otherwise, were confuted by a decree from *Rome*'.[15]

Galileo's telescopic observations had significantly contributed to the successful overthrow of Aristotelian cosmology, and his experiments in mechanics were to change the science of motion. Galileo did not attempt to set up a grand philosophical system, but focused upon particular aspects of mechanics, and, as Boyle did later, wrote that he would 'be satisfied to belong to that class of less worthy workmen who procure from the quarry the marble' that later gifted sculptors may use to create 'masterpieces'.[16] Facts must be established first before useful theories can be constructed. In mechanics, Galileo maintained that 'anyone may invent an arbitrary type of motion and discuss its properties', but he wished to 'consider the phenomena of bodies falling with an acceleration such as actually occurs in nature and to make this definition of accelerated motion exhibit the essential features of observed accelerated motions'.[17]

Boyle often used Galileo's work as an example of how experience could correct the judgements of reason. In the *Christian Virtuoso*, for example, he discussed a number of Galileo's discoveries, including the moons of Jupiter, the mountains on the Earth's moon, and the acceleration of falling bodies as cases 'wherein we assent to experience, even when its information seems contrary to reason'.[18] He noted that previous philosophers had been misled in their formulation of laws governing falling bodies because they had relied upon their reason without taking account of the information of experience. Given the principle of gravity, for example, 'that determines falling bodies to move towards the center of the earth', it would seem 'very rational to believe', as the followers of Aristotle did, that 'in proportion as one body is more heavy than another, so it shall fall to the ground faster than the other'. But, 'notwithstanding this plausible ratiocination', Boyle argued that 'experience shows us' the falsity of such speculations.[19]

In opposition to the Aristotelian doctrine that different bodies fall at different rates because of an innate difference in their natures, Galileo showed that the resistance of the medium was responsible for the observed phenomena. Boyle's work in pneumatics and hydrostatics was in part designed as a way to extend Galileo's work by considering the nature of the mediums through which bodies move. Mathematical skills were required for determinations in such mechanical disciplines because, as Boyle noted, there are 'doubtless many properties and uses of natural things that are not like to be observed by those men, though otherwise never so learned, that are strangers to the mathematicks'.[20] Yet, Boyle was strongly opposed to the idea that mathematics would be the only tool required for the development of natural philosophy.

Mathematics was useful for describing the regularities that hold between the actions of bodies, but it could not provide natural philosophers with the reasons why bodies act as they do. The 'theorems and problems' of hydrostatics, for example, most of which are 'pure and handsome productions of reason, duly exercised on attentively considered subjects', had to be expanded to include the physical properties of bodies.[21] Boyle was critical of 'those mathematicians, that (like *Marinus Ghetaldus*, *Stevinius*, and *Galileo*) have added anything considerable to the Hydrostaticks,' because they 'have been wont to handle them rather as geometricians, than as philosophers, and without referring them to the explication of the phaenomena of nature'.[22] When Boyle designed his own hydrostatical experiments concerning the 'ingenious proposition (about floating bodies)' that was 'taught and proved, after the manner of mathematicians, by the most subtle *Archimedes*', he did so in order 'to manifest the physical reason, why it must be true'.[23] For Boyle, the importance of hydrostatical investigations extended beyond proving that certain propositions were true to the explanation of why they 'ought to be so'.[24] The first task was largely mathematical. The second was the true province of natural philosophy.

In addition to its lack of philosophical content, mathematics could sometimes be misleading. Boyle's negative reaction to those who relied exclusively upon mathematics can be seen in his assessment of Pascal's work in hydrostatics. In his *Physical Treatises*, Pascal seemed to follow Galileo's notion of 'experimental proof' when he maintained that 'in physical science, experience is far more convincing than argument'.[25] When Boyle examined his treatise 'On the aequilibrium of liquors', however, he found that, despite his experimental rhetoric, the trials reported by Pascal could not have been actually performed, since 'they require brass cylinders, or plugs, made with an exactness that, though easily supposed by a mathematician, will scarce be found obtainable by a tradesman'. According to Boyle, the 'experiments proposed by Monsieur *Pascal*' were thus 'more ingenious than practicable'.[26] Although Pascal's conclusions were 'consonant to the principles and laws of the Hydrostaticks', they lacked the experiential basis necessary for adequate proof.[27]

Thought experiments produced by mathematicians may be 'speculatively true', but they 'may oftentimes fail in practice'.[28] In order to overcome the prejudices of reason and 'discover deep and unobvious truths', Boyle insisted, as Bacon and Galileo had before him, that 'intricate and laborious experiments' must be actually performed.[29] In addition, as a corrective for the overly mathematical emphasis of the new mechanical philosophy, Boyle suggested that there were 'scarce any experiments that may better accommodate the Phaenician [corpuscular] principles, than those that may be borrowed from the laboratories of the chymists'.[30] Instead of speculating about the internal configuration of bodies from their manifest appearances, nature ought to be actually 'dissected' into its minute parts.

Chemical analysis would be ideally suited for this task because in such trials bodies are made 'more simple or uncompounded than nature alone is wont to present them to us'. Through such experiments, one could learn not only about the material ingredients of bodies but also about the efficacy of particular ingredients because the circumstances of the trials are carefully controlled. Most are performed in closed transparent vessels, for example, so that one may know 'what nature contributes' to the trials and therefore 'better know what concurs to the effects produced, because adventitious bodies . . . are kept from intruding upon those, whose operations we have a mind to consider'.[31] In a similar manner, the production of artificial compounds could provide important information about nature's processes, such as in the case of the production of 'factitious vitriol' where 'our knowing what ingredients we make use of, and how we put them together, enables us to judge very well how vitriol is produced'.[32]

As a 'science of signs', chemistry illustrated well the Baconian inverted order of demonstration. In chemical experiments one gains the ability to manipulate a substance and thus learn how it is disposed to act and react in combination with other substances, which in turn provides hints about the powers or qualities possessed by the substance by virtue of which it is able to effect changes in other bodies.[33] When we are consistently able to produce such effects artificially, then it is a good sign that we have understood the particular causal powers inherent in the substance and thus have gained an insight into its nature or essence. Such knowledge was, for Boyle, the promise of chemistry. But before it could fulfil its promise, the practice had to be put on a more rational basis.

Boyle's discussions concerning the rational grounding of chemistry have led to some confusion in the secondary literature. As Lawrence Principe effectively argues in Chapter 6 of this volume, Boyle never lost interest in, or rejected, the more mystical or spiritual aspects of traditional alchemical methods and goals. Indeed, in some works he actually wrote in the 'obscure, ambiguous, and almost aenigmatical way' for which he had criticised chemists in his other works.[34] When he spoke of the applications of chemistry to natural philosophical inquiry, however, he argued for the need for a more straightforward reporting style. In such works, Boyle sought to show 'the way experiments are made, so that they lose their mystery'.[35] He hoped that by bringing the chemists' experiments 'out of their dark and smoky laboratories', and by criticising the doctrines that they based upon these experiments, the chemists would be forced 'to speak plainer than hitherto'.[36]

It was also necessary to show the proper way in which to draw inferences from such experiments. Boyle noted that many chemists, such as apothecaries, were mere 'sooty empirics' who had 'been much more happy in finding experiments than the causes of them'. At the other extreme, however, were the chemists who moved too quickly to form general theories upon their 'substantial and noble experiments'. Paracelsus and his followers, for example, had 'fathered upon such excellent experiments' theories that were 'phantastic and unintelligible'.[37] Although Boyle believed that the Paracelsians were better than the Aristotelians in their analysis of the production of qualities in bodies, and their experiments had provided sufficient evidence for a definitive refutation of the Aristotelian theory of four elements, it was an 'ill-grounded supposition of the chemists' to think that their theory of the *tria prima* was confirmed because it could explain things not explained by Aristotle's followers.[38] As Boyle noted, to make this 'argumentation valid, it must be proved, (which I fear it never will be) that there are no other ways, by which those qualities may be explicated'.[39]

Of all the chemists, Boyle liked the Helmontians best, but they did not escape his criticism. Again, there was a problem with their advocacy of general theories that had been too quickly extrapolated from an insufficient number of experiments. Based upon his work on digestion, for example, Helmont had maintained that he could explain all natural processes in terms of the sympathy and antipathy of two material principles, acids and alkalis. Boyle thought that this theory was better than the Paracelsian *tria prima*, and that it could be of some use to physicians, yet he could 'not acquiesce in this hypothesis of alcali and acidum, in the latitude wherein I find it urged'.[40] These chemical 'duellists', just as the Paracelsians and Aristotelians before them, were guilty of having 'arbitrarily' assigned 'offices to each of their two principles', and they only found these principles in all bodies because their theory prejudiced their interpretation of the experimental results.[41]

Aside from their hasty theorising and obscure manner of writing, Boyle's major objection to the chemistry of his day was that its practitioners relied upon interpretive principles that were 'too few and too narrow'.[42] By focusing exclusively upon material principles, the chemists' theories did 'not . . . perform what may be justly expected from philosophical explication' because they only identified the agent and 'not the manner of operation' whereby the agent 'produces the effect proposed'.[43] In addition to material principles, a full explication of the qualities of bodies would have to take account of the 'coordination and contrivance' of the parts of bodies.[44] In the end, no matter how much they tried, the chemists' ingredients would never be able to provide the explanatory resources necessary for a comprehensive corpuscular philosophy.

> The chemists and other materialists, must (as indeed they are wont to do) leave the greatest part of the phaenomena of the universe unexplicated by the help of the ingredients (be they fewer or more than three) of bodies, without taking in the mechanical, and more comprehensive affections of matter, especially local motion.[45]

According to Boyle, matter and motion are both relevant to the explanation of physical processes, and so information from both areas must be collected and compared before a general theory would be deemed acceptable. Although he was critical of the chemists, he nonetheless sought to make chemistry a part of natural philosophy. His interest in chemistry extended beyond the contribution that he believed it could make to the corpuscular philosophy, however. As we saw above, he was also interested in the production of 'useful' knowledge. He was particularly concerned with the production of 'beneficial effects' in the area of medicinal remedies, and he maintained that chemical investigations had produced a number of concrete results in this area. As Boyle had urged the corpuscularians to pay serious attention to the chemists, so too did he attempt to show physicians the ways in which chemical knowledge could be useful for the improvement of their practice.

Although he was not a trained physician, Boyle had read the classical medical texts of Galen and Hippocrates, and he was acquainted with a number of leading medical men, including Harvey, Highmore, Lower, Petty, Stubbe and Sydenham.[46] In his essays on medical topics, Boyle noted that both Galen and Paracelsus had admitted that the complexities involved in medical practice made the acquisition of certain knowledge difficult, and thus he believed it to be 'no great presumption, if a man should attempt to innovate in any part' of medicine.[47] Advances in therapeutics could be made if physicians would more freely use those medicinal remedies prepared by chemists.[48] Diagnostics and prognostics could also be improved by a knowledge of chemistry. In his essay on the 'Semeiotical part of physick', Boyle discussed how a natural philosopher acquainted with chemical trials could 'assist the physician to make more certain conjectures from the signs he discovers of the constitution and distempers of his patient,' because 'he that better knows the nature of the parts and juices of the body, will be better able to conjecture at the events of diseases.'[49]

Although he urged physicians to take account of the new learning in chemistry, however, Boyle also urged chemists to pay more attention to the practical experience possessed by physicians. Before a drug could be effectively administered, for example, a physician had to make a correct diagnosis of the nature of a patient's illness based upon his familiarity with the symptoms of various diseases.[50] In addition, the proper dosage of a drug would have to be determined on the basis of all of the particular circumstances surrounding a case, including the patient's age, weight, gender and mental attitude.[51] Once again, a cooperative effort among different areas of learning would be required. Information from chemical trials had to be combined with knowledge gathered from anatomical dissections, physiological experiments and clinical experience before significant improvements would be made in medicine.

Instead of arguing from causes to their effects, medical professionals made use of an inverted order of demonstration whereby inferences are made from effects to their causes, and then predictions, based upon the inferred causes, are made concerning the course of the disease and any possible cures. Because medicine was a 'low science' that was not reducible to fundamental maxims, but consisted of probable conjectures based upon the interpretation of signs, some philosophers, such as Mersenne, viewed it as a defective science.[52] Boyle, on the other hand, saw the medical tradition as encapsulating the very mode of reasoning upon which the experimental philosophy was to be based. And it was in this tradition that he found an example of a theory, both descriptive and causal, that had been indisputably established upon the basis of experimental evidence. Boyle would often cite Harvey's theory of the circulation of the blood as one of the most important discoveries of his century.[53] The theory was important in its own right for the contribution that it made to knowledge about the human body. The manner by which it had been confirmed was equally important, however, because it provided a paradigm for how future discoveries could be made.

In his *Usefulness of Natural Philosophy*, Boyle noted that, although Harvey's description of the circulation of the blood had been generally accepted, there was still a 'controversy about the cause and manner of the heart's motion' among 'modern anatomists', some of whom preferred Harvey's theory of the contraction of the heart, and others who preferred Descartes' alternative explanation.[54] In his *Discourse on Method*, Descartes maintained that when venous blood entered the heart it was heated by a subtle flame and the rarefaction which resulted caused the heart to expand until it reached such a limit that the blood would be forcefully expelled by the built up pressure. In further support of his theory, Descartes argued that the more vivid colour of arterial blood indicated that it was warmer than venous blood, and thus had been heated in its passage through the heart.[55] Both theories were intelligible, and further observations would only reveal what was already known – that expansion and contraction are activities of the heart. To determine which was the efficacious activity, experiments had to be expressly designed to test the alternative theories.

Boyle described a number of anatomical vivisections that he and Robert Hooke had

performed where the flow of blood to the heart was artificially interrupted. On
Descartes' assumption, if the heart was emptied of blood, then its motion should cease,
yet when the experiments were performed the hearts continued to contract and expand
for a short while.[56] These results made Descartes' theory questionable, but they did not
produce a complete refutation. The final piece of evidence that would prove decisive
was an observation, originally made by Richard Lower, which showed that venous
blood would turn to the colour of arterial blood when exposed to the air, which in turn
led to new experiments designed to show that the colour change of the blood occurred
in the lungs and not the heart.[57] According to Boyle, these and similar experiments
served as 'collateral arguments' that when taken together provided a 'concurrence of
probabilities' in favour of the unconditional acceptance of Harvey's theory. As he
described it, Harvey's theory

> perhaps would never have prevailed, if this so strongly opposed truth had not been
> vigorously seconded by several collateral and subsequent experiments, that were
> made and inforced by ingenious and dexterous men who by the differing trials they
> made, supplied impartial men with so many mediums conspiring to prove the same
> conclusion, that the truth, though much opposed, could not be supprest, or
> hindered from being at length triumphant.[58]

Boyle developed his epistemological criterion for the acceptance of theoretical claims
in a manner consistent with the way in which he constructed his experimental
philosophy in general. A concurrence of probabilities is achieved when 'the most
information procurable that is pertinent to the things under consideration' supports a
particular conclusion and there is no evidence to the contrary that would militate
against it. Concurrence is a somewhat vague epistemological criterion, however, in
large part because it is based upon considerations of relevancy that in turn will depend
upon the current state of knowledge concerning a particular subject matter. Boyle
was aware of the fact that such a 'judgment of reason' would therefore be context-
dependent, and that the acceptability of knowledge claims could change as the context
of knowledge changed.[59] Experimental proof is a complex and dynamic process. For
this reason, he wrote that he would 'not debar' himself from revising his opinions in the
event that 'further progress in history shall suggest better hypotheses or explications'.[60]
Although Boyle spoke of an experiential grounding for the new philosophy, it should
be clear from the above discussion that he did not advocate a simple empiricist
epistemology.[61] From antiquity to the present day, empiricism has been characterised
as the belief in the reliability of sense perceptions and a sceptical probabilism
concerning any knowledge claims purporting to describe unobservable causal
processes.[62] Boyle saw experimentalism as an alternative to empiricism. Despite his
use of the term 'probability', he did not believe that knowledge must be limited to the
simple ordering of phenomenal appearances. Harvey's theory had been proven
conclusively. Although the concurrence of probabilities produced for the proof was not
logically compelling in the classical deductive sense, it was 'strong enough to deserve a
wise man's acquiescence'.[63]

The proof of Harvey's theory was successful in part because it was somewhat easy to identify the relevant information that would be required for proof, and in part because the technology was available by which the experiments could be performed in order to obtain the required information. Clearly the procurability of evidence is also context-dependent, and developments in experimental technique could significantly affect the amount and kind of information upon which theories are to be justified. Boyle's cautious attitude toward most of the theories of his day was a result not of a probabilistic epistemology, but of his appreciation of the problems involved in learning how to perform experiments and in judging which experimental results could be used as reliable indicators about processes operative in nature.

Experimental activity

According to Boyle, if the experimental method was 'to be performed as it ought to be', it would 'in many cases, besides some dexterity scarce to be gained but by practice, require sometimes more diligence, and oftentimes too more cost than most are willing, or than many are able to bestow'.[64] Although Boyle addressed the complexity involved in experimental practice on numerous occasions, little attention has been paid to these methodological discussions because historians and philosophers have been more interested in looking at the ways in which experiments were used to establish theoretical knowledge. Recently, however, Shapin and Schaffer made an attempt to rectify this situation by looking in detail at the manner by which experiments could be said to establish the matters of fact upon which subsequent theorising would be based.

In their study, Shapin and Schaffer sought to analyse the 'method-statements' of Boyle and his contemporaries in terms of the purposes for which they were made.[65] Their main concern, however, was to show the socio-political interests at work in the construction of an experimental community, and they thus dismissed the epistemic dimensions of Boyle's methodological works. They characterised two of his *Certain Physiological Essays*, for example, as 'a repertoire of excuses for the failure of particular experiments' that 'could be brought into play so that a given expectation or theory need not be invalidated by experimental failure'.[66] They then turned their attention to Boyle's notion of the spring of the air and used their previous analysis to illustrate how he was able to defend his conjecture in the face of anomalous experimental results.[67]

Shapin and Schaffer were right to point out that the problems identified by Boyle entailed that particular experimental results could not be used in isolation to refute a given theory, but their social analysis failed to provide the full story about why this would be the case. This is not to say that experimental science involves no more than the forms of reasoning I discussed above. Clearly, it is also an activity that has an important social dimension. Indeed, in contrast to Hobbesian or Cartesian individualism, Boyle's criterion of concurrence entailed that experimental proof could only be achieved through the collaboration of a number of investigators working in different areas of inquiry.

Social analysis is not the only appropriate means by which to understand the performances involved in such a practice, however. Actors' reasons for what they do are often drawn from the internal dynamics of an activity itself, distinct from any ulterior goals or purposes. A number of Boyle's essays contain extensive discussions concerning the dynamics of experimental activity. The following analysis will examine the epistemic and practical aspects of experimentation as identified by Boyle. The focus will be upon how Boyle learned about the intricacies of experimentation from his exposure to the practical activities of others and his own extensive laboratory experience, and how he used what he had learned to develop methodological strategies for the improvement of experimental practice.

Boyle's eclectic curiosity was not confined to the works of his learned predecessors. He also sought to acquaint himself with the practical procedures employed by tradesmen and artisans in their manipulations of nature. In the preamble to his second volume of *Usefulness* essays, Boyle stressed the importance of such experience when he noted that 'those, that are mere scholars, though never so learned and critical, are not wont to be acquainted enough with nature and trades'.[68] In order for natural philosophy to progress, the experiential basis upon which it was to be grounded would have to be broadened, which entailed that the pool of witnesses from whom observations are collected would have to be expanded.

According to Boyle, illiterate practitioners not only make excellent witnesses, but sometimes they are better than those who have learned from the schools, because they 'are usually more diligent about the particular things they handle' since 'their livelihood depends upon it'.[69] Boyle argued that the 'phaenomena afforded by trades' must be made a 'part of the history of nature', because they may 'both challenge the naturalist's curiosity, and add to his knowledge'.[70] He sought reports about successful practices from every sector of society and from all parts of the globe, and he advised his readers that such information could lead to significant 'additions and alterations' in the received opinions of natural philosophy.[71]

In a similar fashion, 'the knowledge of physicians might be not inconsiderably increased,' if they would be 'a little more curious' about

> the observations and experiments, suggested partly by the practice of midwives, barbers, old women, empiricks, and the rest of that illiterate crew, that presume to meddle with physick among our selves; and partly by the Indians and other barbarous nations, without excepting the people of such parts of *Europe* it self, where the generality of men are so illiterate and poor, as to live without physicians.[72]

Boyle noted, for example, that, although the practice of medicine in China was quite different from that in the West, it was apparently successful, and a further investigation into the practice could lead to a revision of common beliefs about what was healthful.[73] One could also learn about less radical forms of therapy from the Brazilian Indians, who used 'gums and balsams' to cure 'the limbs of soldiers wounded with gun-shot' that would otherwise have 'been cut off by the advice of our European surgeons, both Dutch and Portuguese'.[74] According to Boyle, European physicians were sometimes hampered in their practice by their 'too great reliance on the Galenical, or other ancient opinions' that led them to 'neglect useful remedies, because presented by persons, that ignore them, and perhaps too hold opinions contrary to them'.[75] He cautioned his readers 'to beware of relying so much upon the yet disputable opinions of physicians, as to despise all practices, though usually successful, that agree not with them'.[76]

If a practice is successful, it should be included in the history of nature regardless of the educational, social or ethnic class of the practitioner. In addition to gaining useful knowledge about the manner by which nature may be manipulated for human purposes, however, such practices could also yield reliable knowledge about the properties of natural bodies. Tradesmen, for example, 'often observe in things they deal about, divers circumstances unobserved by others, both relating to the nature of the things they manage and to the operations performable upon them'.[77] Boyle maintained that he had learned about the properties of different kinds of wood from 'carpenters, joiners and turners', and he had learned more about the 'kinds, distinctions, properties, and consequently of the nature of stones, by conversing with two or three masons' than he ever could have 'from *Pliny*, or *Aristotle* and his followers'.[78]

Boyle also learned that the artificial manipulations of tradesmen and artisans produce 'the most instructive condition, wherein we can behold' nature.[79] Experiments provide a way to avoid 'the superficial account given us of things by their obvious appearances and qualities'.[80] As the level of artificiality in an experiment increases, the level of superficiality decreases.[81] Boyle often used an analogy with the processes employed to test the genuineness of gold coins to illustrate the virtues of artificially contrived trials. He described how different the situations would be, for example, 'if a piece of coin, that men would have pass for true gold' were to 'be offered to an ordinary man, and to a skilful refiner'. The refiner

> will examine it more strictly, and not acquiesce in the stamp, the colour, the sound, and other obvious marks, that may satisfy a shop-keeper, or a merchant; yet when he has tried it by the severer ways of examining, such as the touchstone, the cupel, aqua fortis, etc., and finds it to hold good in those proofs, he will readily and frankly acknowledge, that it is true gold, and will be more thoroughly convinced of it, than the other person, whose want of skill will make him still apt to retain a distrust, and render him indeed more easy to be persuaded, but more difficult to be fully satisfied.[82]

The experimenter in natural philosophy has the same advantage as the refiner. By going beyond the superficial appearances of things, the experimenter is able to achieve a degree of certainty at the factual level that would be lost to a more empirical investigator. In an unpublished note, Boyle explained how the process of testing coins provided an instance of a concurrence of probabilities for the proof of a factual claim that was similar to the type of proof used above for the confirmation of Harvey's theory. When 'proving that a piece of yellow coin is a true guinea', it may

> be no concluding argument either that tis yellow, or that tis malleable, or that it touches well upon the stone or that tis heavier than brass, or that it sustains the

cupel or that tis not dissoluble in AF [aqua fortis] (though I say) to prove any one of these be not a sufficient argument that the proposed piece of coin is true gold, yet the evincement of each of these particulars, is of considerable use towards the compiling of a demonstration of the grand conclusion this being the result of all these proofs taken together.[83]

Although the artifice employed by practitioners provided Boyle with numerous examples of the virtues of experimental manipulation, he also learned that experiments are subject to a number of problems that only a deep and prolonged exposure to the practice would reveal. According to Boyle, experiments are 'seldom solitary'.[84] The complexity of the experimental method, which makes artificial manipulation especially suited for discovering the complexity of natural processes, also gives rise to a number of problems.

> He that knows not the nature or properties of all the other bodies [in the experiment], wherewith that on which the experiment proper is made, . . . can hardly discern what effects the experiment may possibly concur to produce, for many inventions and experiments consist as it were of several parts or require distinct actions, to some one of which, though it should happen to be the principal, a thing may not be useful, which by being requisite to another is of use to the experiment in general though not to each distinct part of it.[85]

Two of Boyle's *Certain Physiological Essays* of 1661 were devoted to a detailed account of the number of 'contingencies' to which experiments are liable. The first essay, entitled 'The unsuccessfulness of experiments' contained 'divers Admonitions and Observations (chiefly Chymical)' about problems that could arise from the state of the materials used in experimental trials.[86] All substances 'prepared by art', are subject to slight differences either from the fraudulent addition of impure ingredients or from a lack of skill in their preparation. Natural products could also be deceptive at the level of appearances. Metals and minerals, for example, that 'almost all men without hesitancy' look upon as being 'of the same nature as well as denomination', often differ between themselves as much as 'vegetables and animals of the same species' do.[87] The differences between animals of the same species are 'wont to be more obvious to the eye,' whereas minerals may appear to be 'perfectly similar,' yet it may often be the case that small quantities of other minerals 'lie concealed' in a sample and the presence of such adulterating minerals 'is hardly to be discerned, before experience have discovered it', either by 'exquisite separations' or by their 'unexpected operations' in experimental trials.[88]

The second essay, entitled 'Of unsucceeding experiments', contained a discussion of all the other 'contingencies to which experiments are obnoxious'.[89] In addition to its purity, the quantity of a material used in a trial should be considered, because 'divers experiments succeed, when tried in small quantities of matter, which hold not in the great'.[90] All of the circumstances surrounding an experimental trial, including the temperature, the time of year and the duration of time taken to produce a result, have to be carefully noted. The instruments employed by an experimenter also must be assessed for reliability because they are 'framed by the hands and tools of men' and thus may 'in divers cases be subject to some if not many imperfections upon their account'.[91]

At the most general level, there are numerous problems associated with the correct determination of what has actually been produced by an experimental trial. Sometimes an experimenter could be misled into believing that artificial manipulation had been responsible for an effect whose production had in fact been entirely natural.[92] On other occasions, the effect produced may be a result of the experimental manipulation of nature, yet other experimenters may not be able to replicate the phenomenon success-fully because of their failure to understand all of the circumstances necessary for its production.[93] In some experiments, the seeming contingency may proceed from 'easily discoverable causes', whereas in others 'the cause of the contingency is very abstruse and difficult to be discerned', but, even in the latter case, Boyle was confident that such causes would be discovered 'by men's future skill and diligence in observation'.[94]

Because of the complexity of experimental practice, Boyle noted that 'in divers cases such circumstances as are very difficult to be observed, or seem to be of no concern-ment to an experiment . . . may yet have a great influence on the event of it'.[95] From his frequent visits to the shops and workhouses of tradesmen, Boyle found that all such practices are liable to contingencies, and he sought to acquaint his readers with these problems so that they would appreciate the complexity of the new method and not 'impute a needless contingency' to nature when their experiments failed.[96] One could learn from smiths and farriers, for example, that something as seemingly inconse-quential as a few moments of time could have a very significant effect on the outcome of a trial. He noted that he had often 'observed, how artificers in the tempering of steel, by holding it but a minute or two longer or lesser in the flame' were able to 'give it very differing tempers as to brittleness or toughness, hardness or softness,' and 'none but an artist expert in tempering of iron would suspect that so small a difference of time of its stay in the flame could produce so great a difference in its tempers'.[97]

In another instance, Boyle discussed how glassworking is a 'trade that obliges the artificers to be assiduously conversant with the materials they employ' and yet, 'even to them, and in their most ordinary operations' there could occur 'little accidents' that keep them 'from doing sometimes what they have done a thousand times'. He recounted one of his visits to a 'glass-house', where he had gone to pick up some vessels that the 'eminently skilful workman, whom I had purposely engaged' had not been able to produce, in order to make the contingencies of practice vivid for his readers and advise them that

> it need to be no such wonder, if philosophers and chymists do sometimes miss of the expected event of an experiment but once, or at least but seldom tried, since we see tradesmen themselves cannot do *always* what, were they not able to do *ordinarily*, they could not earn their bread.[98]

As he ended his essays, Boyle noted that despite the vast catalogue of difficulties presented there, all that he had said should only be 'fit to persuade' his readers to a 'watchfulness in observing experiments, and wariness in relying on them; but not at all to such a despondency of mind, as may make you forbear the prosecution of them'. All practices are fallible, but the possibility of failure is not a sufficient reason for rejecting a practice. A physician, for example, does not 'renounce his profession, because divers of the patients he strives to cure are not freed from their disease by his medicines, but by death,' and neither does the 'husbandman forsake his cultivating of the ground, though sometimes an unseasonable storm or flood spoils his harvest, and deprives him of the expected fruit of his long toils'.[99] Although the problems surrounding the practice were a cause for concern, Boyle maintained that 'it is much more easy to under-value a frequent recourse to experiments, than truly to explicate the phaenomena of nature without them'.[100]

Experimental results are merely signs that provide hints about how nature works. Natural philosophers must exercise great caution when they attempt to interpret these signs, however, and, to 'avoid mistaking the causes' of experimental outcomes, 'inquirers must take notice of abundance of minute circumstances'.[101] Experiments are often performed to test theoretical knowledge, but, because of the numerous practical problems surrounding experimental performance, inquirers must be cautious when they attempt to draw theoretical conclusions from particular experimental results. Experiments must be considered 'in consort'.[102] Just as a single successful experimental result cannot be taken as conclusive proof for a theoretical claim, so also a failed experiment cannot be taken as an immediate refutation. Failed experiments could provide significant insights into the processes operative in nature, however, by suggesting new lines of research or improved experimental techniques that would perhaps be more decisive. As Boyle noted,

> in philosophical trials, those unexpected accidents, that defeat our endeavours, sometimes cast us upon new discoveries of much greater advantage, than the wonted and expected success of the attempted experiments would have proved to us.[103]

From his failed experiments, Boyle learned something about natural processes, but more importantly he also learned about how to improve the experimental method itself. From reflection upon his own work and that of his numerous and various predecessors, Boyle developed a set of practical strategies designed to aid naturalists in the acquisition of experimental expertise, the lack of which he had identified as 'the most obvious cause of error'.[104] In particular, Boyle told his readers that they 'must often make and vary experiments'.[105] The two practical strategies of variation and repetition would provide the best means by which one could insure the reliability of the evidence that would compose the informational basis of the new philosophy.

To increase the quantity of information required as a basis for rational inference, experimenters ought to perform a variety of trials, much as was done by refiners in their tests of the genuineness of gold coins. Variety may also serve as a check on the unheeded prejudice that results from an adherence to the interpretive principles of a particular area of learning. According to Boyle, the 'true philosophy' should include 'a comprehension of all the sciences, arts, disciplines, and other considerable parts of useful knowledge'.[106] Boyle favoured this eclectic approach to the study of nature because

> the more things one knows by experiment, the greater number he has of different touch-stones (if I may so speak) or ways of *discovering* and *judging*; or (which is the result of both) of *knowing*, the nature of anything proposed, that relates to it.[107]

Boyle gave many examples of the way in which a variation of the circumstances of a trial could produce radically different outcomes. In one instance, he discussed his own initial failure to replicate an experiment reported by Dr Brown, who had claimed 'that aqua fortis will quickly coagulate common oil'. At first Boyle suspected that his failure was due to the temperature at which the mixture was made, and so he attempted the trial again first by keeping the mixture 'in a cool place, and after in a digesting furnace'. When these variations failed to produce the desired result, he then suspected that his lack of success could be due to 'some peculiar though latent quality' in the materials employed, 'whereupon changing those liquors, and repeating the experiment, we found after some hours the oil coagulated'.[108]

Boyle's adherence to variation as a methodological strategy can be seen in his other works as well. In his *Usefulness* essays, for example, he discussed the 'ingenious attempts by *Sanctorius* in his *Medicina Statica*' to determine the wholesomeness of different types of foods by observing the amount of perspiration produced after their consumption. Boyle did not doubt the accuracy of Sanctorius's observations, but he maintained that he would withhold his assent from the latter's general conclusions until similar observations had been made in subjects 'of differing ages, sexes, complexions, and with variety of circumstances'.[109] Experimental subjects, whether human or inanimate, must be varied. It could easily be the case that a particular experimental result is not generalisable because of some unsuspected peculiarity within the subject chosen for trial.

Repetition, however, was the most important methodological strategy. As Boyle told his readers,

> the instances we have given you of the contingencies of experiments, may make you think yourself obliged to try those experiments very carefully and more than once, upon which you mean to build considerable superstructures either theoretical or practical, and to think it unsafe to rely too much upon single experiments.[110]

In a draft for a planned work on the 'Uses and bounds of experiments', Boyle discussed four ways in which the repetition of experiments could be useful to 'the heedful naturalist'.[111] A repeated experiment is performed at a different time, possibly with a different sample of material or a different piece of apparatus, and thus 'almost always presents some novelty', which could fulfil the above requirement of variation. But repetition is not done solely to increase the quantity of information admitted to the factual basis. It is also absolutely necessary as a way to check the quality of the information before it is admitted. One must make certain that the result is not a one-time chance occurrence, and thus experiments ought to be repeated in order to 'assure' the experimenter 'of the truth of the matter of fact, and that it may be relied on'.

Even when the experimenter is assured of the reliability of the main result, repetition could still be of use because experiments 'are made up of several partial experiments which as so many new phaenomena, are exhibited in the process or operation, though they were not from the beginning directly or particularly designed by the experimenter'. Repetition of a complex experiment would allow for a more detailed inspection of all of the subsidiary parts that contribute to the main experiment and thus could lead to the discovery of useful information about nature's processes and about the ways that might be used to learn more about these processes. Finally, repetition could also be of direct benefit to the acquisition of experimental expertise, because it 'confers on the maker of the trial a faculty of making it well, and oftentimes suggests to him expedients to compass his end more speedily or more easily, or more cheaply, or more certainly'.

Unlike the rhetoric employed by some of his contemporaries, Boyle wrote in order to make the difficulty of experimental practice vivid for his readers and to advise them that 'our way is neither short nor easy'.[112] The methodological strategies that he suggested did not consist of a set of hard and fast rules. Rather, they were merely general guidelines designed to ensure that natural philosophers would appreciate the fact that 'experiments are troublesome to make' and not view the experimental establishment of facts or theories in an overly simplistic manner.[113] He was learning his practice as he went along, and he shared his failures as well as his successes with his readers.

Boyle envisioned his experimental philosophy as a dynamic learning process. Progress would be made, but it would depend upon the accumulation of discoveries made by many workers over many years, and would require that these workers have a cautious, critical attitude. Flexibility is a prerequisite for knowledge acquisition. One must be willing to revise previous opinions about nature or about the experimental method itself, in the light of new factual or theoretical information.[114] Boyle was keenly aware of the limitations that faced him in his practice. Yet he was confident that if he set out all of the details of his practice, and included reports of his failures as well as his successes, then he would have laid the foundation upon which future generations would be able to build. He valued his method highly for the promise that it held. Yet he did not value his own discoveries very highly because he was sure that the future would much surpass him. As he wrote in his Postscript to *The Natural History of Human Blood* (1684):

> I presume, that our enlightened posterity will arrive at such attainments, that the discoveries and performances, upon which the present age most values itself, will appear so easy, or so inconsiderable to them, that they will be tempted to wonder, that things to them so obvious, should lye so long concealed to us, or be so much prized by us; whom they will perhaps, look upon with some kind of disdainful pity, unless they have either the equity to consider, as well the smallness of our helps, as that of our attainment; or the generous gratitude to remember the difficulties this

age surmounted, in breaking the ice, and smoothing the way for them, and thereby contributing to those advantages, that have enabled them so much to surpass us.[115]

Acknowledgements

Research for this paper was partially funded by the National Science Foundation, Grant 8821116.

NOTES

1 BP 38, fol. 80. This short manuscript is printed in M. B. Hall, *Robert Boyle on Natural Philosophy: An Essay with Selections from his Writings* (Bloomington: Indiana University Press, 1965), p. 177.

2 *Ibid.*

3 *Ibid.* Boyle reserved most of his criticisms for Aristotle's followers, not the philosopher himself. See his assessments of Aristotle in *Forms and Qualities*, *Works*, iii, 5, 8–9, 40–1, 75; and *Notion of Nature*, *Works*, v, 166. Boyle's attitude toward Aristotle was similar to that of other modern philosophers. See, for example, Desmond Clarke's discussion of Descartes' Aristotelianism in *Descartes' Philosophy of Science* (Pennsylvania State University Press, 1982). For more detailed accounts of Boyle's Aristotelianism see Peter Alexander, *Ideas, Qualities and Corpuscles: Locke and Boyle on the External World* (Cambridge University Press, 1985); and Norma Emerton, *The Scientific Reinterpretation of Form* (Ithaca: Cornell University Press, 1984), esp. pp. 143–50.

4 *Excellency of Theology*, *Works*, iv, 57. On p. 59 Boyle grouped the work by Harvey and Gilbert with that of Bacon, and likened Gassendi to Descartes. Margaret J. Osler, 'John Locke and the changing ideal of scientific knowledge', *JHI* 31 (1970): 3–16, has argued that Boyle fashioned his model of empirical inquiry after Gassendi. But clearly, although Boyle did devote considerable attention to Gassendi's atomistic theories, he did not categorise him as an empiricist. Throughout his works he always referred to him as a mathematician, or, what is almost the same, as an astronomer. For the non-empirical elements in Gassendi's philosophy, see Lynn Joy, *Gassendi the Atomist* (Cambridge University Press, 1987).

5 Francis Bacon, *The New Organon*, in *The Works of Francis Bacon*, eds. James Spedding *et al.* (Cambridge: Riverside Press, 1863), vol. 8, p. 32; and Book II, aphorism lii, pp. 347–8. In the Preface to *The New Organon*, p. 37, Bacon wrote that he 'rejected' all 'premature human reasoning which anticipates inquiry'.

6 *Usefulness, I, Works*, ii, 46. I have argued in more detail for Bacon's influence on Boyle in 'Robert Boyle's Baconian inheritance', *Studies in History and Philosophy of Science* 17 (1986): 469–86. See also Alexander, *Ideas, Qualities and Corpuscles* (n. 3); and G. A. J. Rogers, 'Descartes and the method of English science', *Ann. Sci.* 29 (1972): 237–55.

7 'Proemial essay', *CPE, Works*, i, 307.

8 *Defence, Works*, i, 121.

9 *Forms and Qualities, Works*, iii, 75. See Bacon, *New Organon* (n. 5), p. 33, and Book I, aphorism xcix.

10 Bacon, *New Organon* (n. 5), p. 53, and Book I, aphorism cxxiv.

11 *Usefulness, II, sect. 2, Works*, iii, 423. See also *Forms and Qualities, Works*, iii, 12; and *CPE, Works*, i, 311.

12 BP 28, p. 293.

13 *Works*, i, xxxiv. Boyle to Marcombes, 22 October 1646.

14 *Usefulness, II, sect. 2, Works*, iii, 467.

15 'Account of Philaretus during his Minority', *Works*, i, xxiv, and BP 8, fol. 135r. For some of his many references to Galileo in his published works see, *Spring, Works*, i, 81; *Usefulness, II, sect. 2, Works*, iii, 459–60; and *Excellency of Theology, Works*, iv, 56–60.

16 Galileo Galilei, *Dialogues Concerning Two New Sciences*, trans. Henry Crew and Alfonso de Salvio (NY: Dover Publications, 1954), p. 194.

17 *Ibid.* p. 160. On p. 84, Galileo discussed how the pendulum and inclined plane experiments could be used to make intervals of time 'not only observable but easily observable'.

18 *Christian Virtuoso, I, Works*, v, 527. A Rupert Hall, 'Galileo's thought and the history of science', in *Galileo, Man of Science*, ed. Ernan McMullin (NY: Basic Books, 1967), pp. 67–81, discusses the significance of Galileo's theoretical work for English science, but maintains, on p. 79, that Galileo did not have a 'decisive formative influence' on the development of experimental science. Boyle's frequent references to Galileo make this claim doubtful.

19 *Christian Virtuoso I, Works*, v, 527. Also, in *Excellency of Theology, Works*, iv, 61, Boyle discussed how the limits of pumps had to be found out by experience, much as Galileo had described in *Two New Sciences* (n. 16), p. 16.

20 *Usefulness, II, sect. 2, Works*, iii, 477.

21 *Hydrostatical Paradoxes, Works*, ii, 739.

22 *Ibid.* p. 740.

23 *Ibid.* p. 763.

24 *Ibid.* p. 746.

25 Blaise Pascal, *The Physical Treatises of Pascal*, trans. I. H. B. and A. G. H. Spiers (Columbia University Press, 1937), p. 31.

26 *Hydrostatical Paradoxes, Works*, ii, 746.

27 *Ibid.* p. 745.

28 *Ibid.* p. 759. On pp. 743–4, Boyle gave an example of an experiment described by mathematicians that failed in practice because of an 'unheeded physical circumstance'.

29 *Christian Virtuoso I, Works*, v, 523–4.

30 *CPE, Works*, i, 358. See Antonio Clericuzio, 'A redefinition of Boyle's chemistry and corpuscular philosophy', *Ann Sci.* 47 (1990): 561–89; and William Newman, 'Boyle's debt to corpuscular alchemy', ch. 5, this volume, for discussions about how Boyle's chemical interests influenced his version of corpuscularianism.

31 *CPE, Works*, i, 358. See the correspondence between Baruch Spinoza and Henry Oldenburg, concerning the issue of chemical experiments, in *Oldenburg*, i, 458–74.

32 *Forms and Qualities, Works*, iii, 51.

33 Douglas Lane Patey, *Probability and Literary Form* (Cambridge University Press, 1984), pp. 35–40, has discussed how the practice of reasoning from signs used by the ancient medical theorists was taken up by the chemists in the sixteenth century.

34 *Sceptical Chymist, Works*, i, 460. See also *Usefulness, II, sect. 1, Works*, ii, 232. See Lawrence Principe, 'Boyle's alchemical pursuits', ch. 6, this volume; and Michael Hunter, 'Alchemy, magic and moralism in the thought of Robert Boyle', *BJHS* 23 (1990): 387–410, for discussions concerning Boyle's lifelong interest in alchemy.

35 *Colours, Works*, i, 662. This work is often discussed as if it were a treatise on optics, but it is really more a work on the physico–chemical properties of coloured bodies.

36 *Sceptical Chymist, Works*, i, 461.

37 *Ibid.* p. 584.

38 'Imperfection of the chymists' doctrine of qualities', *Mechanical Origin of Qualities, Works*, iv, 274.

39 *Ibid.* p. 275.

40 'Reflections upon the hypothesis of alcali and acidum', *Mechanical Origin of Qualities, Works*, iv, 284. He stated his preference for Helmont in *Sceptical Chymist, Works*, i, 461.

41 'Reflections', *Mechanical Origin of Qualities, Works*, iv, 285. He made the same point in his *Sceptical Chymist, Works*, i, 510.

42 'Imperfection of the chymists' doctrine of qualities', *Mechanical Origin of Qualities, Works*, iv, 277.

43 'Reflections', *Mechanical Origin of Qualities, Works*, iv, 289. See *Usefulness, II, sect. 1, Works*, ii, 101, where Boyle noted that Paracelsus was neither a great reasoner nor a logician.

44 'Imperfection of the chymists' doctrine of qualities', *Mechanical Origin of Qualities, Works*, iv, 279. He noted that the alchemists' 'way of judging by material principles, hinders . . . the assignation of causes from being satisfied.' See also 'Excellency and grounds of the mechanical hypothesis', *Works*, iv, 74.

45 'Excellency and grounds of the mechanical hypothesis', *Works*, iv, 73. In 'Reflections', *Mechanical*

Origins of Qualities, *Works*, iv, 288–9, he noted that material elements could not explain qualities such as 'springiness', light and electricity, and that this objection alone was 'enough to shew' that the chemists' hypotheses were 'insufficient for the explication of qualities'.

46 References to Galen, Hippocrates and Harvey are scattered throughout Boyle's works, esp. *Usefulness, I* and *II, sect 1*, *Works*, ii. See *Works*, i, xlvii, for Highmore's 1651 dedication to Boyle; and p. lxxxvi, for Sydenham's 1666 dedication. Boyle's motivation for the study of medicine was in part a product of his own ill health. See the account of his various illnesses in *Medicinal Experiments*, *Works*, v, 315–16,

47 *Usefulness, II, sect. 1*, *Works*, ii, 153.

48 *Ibid.* p. 167.

49 *Ibid.* p. 89. The 'Semeiotical part of physick', was Essay III in the collection.

50 *Ibid.* p. 84. At Boyle's suggestion, his neighbour in Pall Mall, Thomas Sydenham, published detailed clinical descriptions of various diseases beginning in 1666 with his *Methodus curandi febres*. See G. G. Meynell, *A Bibliography of Dr. Thomas Sydenham (1624–1689)* (Folkestone: Winterdown Books, 1990).

51 *CPE*, *Works*, i, 340. In his manuscript collections of particular remedies, Boyle was very careful to note the desired dosage and the circumstances that could alter the dosage given. See BP 25 and 27.

52 See Patey, *Probability and Literary Form* (n. 33), p. 43; and Peter Dear, *Mersenne and the Learning of the Schools* (Ithaca: Cornell University Press, 1988), p. 110.

53 See Richard A. Hunter and Ida Macalpine, 'William Harvey and Robert Boyle', *NRRS* 13 (1958): 115–27, for Boyle's frequent references to Harvey.

54 *Usefulness, II, sect. 1*, *Works*, ii, 69. This work was published in 1663, but according to Boyle parts of it had been written ten to twelve years earlier. Gassendi had been one of the last to accept the circulation of the blood. He finally acquiesced in the circulation but preferred Descartes' theory that the function of the circulation was for the distribution of heat from the heart. See G. S. Brett, *The Philosophy of Gassendi* (London: Macmillan and Co., 1980), p. 111; and Robert Willis, 'Introduction', in *The Works of William Harvey*, trans. Robert Willis (London, 1846; reprinted Annapolis, Md: St John's College Press, 1949), p. li.

55 René Descartes, *Discourse on Method*, in *The Philosophical Works of Descartes*, trans. Elizabeth S. Haldane and G. R. T. Ross, 2 vols (Cambridge University Press, 1931), i, 109–13 (part V).

56 *Usefulness, II, sect. 1*, *Works*, ii, 69–71. The experiments were performed on fish and frogs.

57 Lower's observation was communicated to Boyle in a letter dated 24 June 1664, printed in *Works*, vi, 472–3. Lower discussed the further experiments that he designed and performed in Richard Lower, *Tractatus de Corde*, trans. K. J. Franklin, in *Early Science in Oxford*, ed. R. W. T. Gunther, vol. 9 (1932; reprinted London: Dawson's, 1968).

58 BP 17, fol. 40. Immediately following this passage, Boyle noted that the Torricellian and air-pump experiments were at the moment producing the same type of concurrence of probabilities in support of his theory about the weight and spring of the air. See also *Defence*, *Works*, i, 123.

59 *Reason and Religion*, *Works*, iv, 181.

60 *Mechanical Origin of Qualities*, *Works*, iv, 235.

61 I have discussed Boyle's notions of experience and concurrence in more detail in 'Scientific experiment and legal expertise', *Studies in History and Philosophy of Science* 20 (1989): 19–45.

62 Boyle has frequently been described as an empiricist. See, for example, J. E. McGuire, 'Boyle's conception of nature', *JHI* 33 (1972): 523–42; Osler, 'John Locke' (n. 4); and Shapin and Schaffer, *Leviathan and the Air-Pump*. See Richard H. Popkin, *The History of Skepticism from Erasmus to Spinoza* (University of California Press, 1979), for an historical account of empiricism; and Bas van Fraassen, *The Scientific Image* (Oxford: Clarendon Press, 1980), for a recent defence of empiricism.

63 *Things above Reason*, *Works*, iv, 450.

64 *Defence*, *Works*, i, 122; see also *Usefulness, I*, *Works*, ii, 5.

65 Shapin and Schaffer, *Leviathan and the Air-Pump*, p. 14.

66 *Ibid.* p. 185.

67 *Ibid.* pp. 186–7.

68 *Usefulness, II, sect. 2*, *Works*, iii, 395.

69 *Ibid.* p. 443.

70 *Ibid.* p. 442. He added that the excuse made by those who neglected this part of the history of nature

because it must be learned from 'illiterate mechanicks', was 'indeed childish, and too unworthy of a philosopher, to be worthy of a solemn answer'.

71 *Exp. Obs. Physicae*, *Works*, v, 569. He collected a vast number of such reports, especially those which he found particularly 'odd', in an 'Outlandish Book' preserved in BP 27, pp. 1–274. Boyle's business contacts gave him access to a large number of these reports. He was a member of the Council for Foreign Plantations; the Corporation for Propagating the Gospel in New England; the Company of the Royal Mines; the East India Company; and the Hudson's Bay Company. Boyle discussed general guidelines that travellers should follow when making and recording observations of this nature in his 'General heads for the natural history of a country', printed as an appendix to *General History of Air*, *Works*, v, 733–43. This work was originally printed in *Phil. Trans.*, no. 11. A draft of it can be found in BP 26, fols. 47–9.

72 *Usefulness, II, sect. 1*, *Works*, ii, 162.

73 *Ibid.* pp. 161–2.

74 *Ibid.* p. 117. Boyle took this account from Gulielmus Piso, *De Medicina Brasiliensis Libri 4* (1648).

75 *Ibid.* p. 163.

76 *Ibid.* p. 164.

77 *Usefulness, II, sect. 2*, *Works*, iii, 444.

78 *Ibid.* pp. 444–5. The information that Boyle obtained from tradesmen often played an important role in his theoretical writings. See, for example, his essay on 'The intestine motions of particles of quiescent solids', in *CPE*, *Works*, i, 454–5.

79 *Usefulness, II, sect. 2*, *Works*, iii, 443.

80 *Usefulness, I*, *Works*, ii, 9. See also BP 9, fol. 25r.

81 Boyle often found it necessary to apologise for the lack of a sufficient degree of artificiality in the experiments that he reported. See, for example, *Cold*, *Works*, ii, 474, 512; and *Spring, 1st Continuation*, *Works*, iii, 177.

82 *Christian Virtuoso I*, *Works*, v, 538. For other places where he used the analogy see, *CPE*, *Works*, i, 308; *Sceptical Chymist*, *Works*, i, 591; 'A chymical paradox', in *Icy Noctiluca*, *Works*, iv, 498; and *Notion of Nature*, *Works*, v, 159.

83 BP 5, vol. 18v.

84 BP 9, fol. 9.

85 BP 10, fol. 7. This passage is from a page headed 'Use of Experiments'.

86 *CPE*, *Works*, i, 318–33.

87 *Ibid.* p. 321.

88 *Ibid.* p. 322.

89 *Ibid.* pp. 334–53.

90 *Ibid.* p. 335.

91 *Ibid.* p. 347. See his discussion of the problems that he encountered with his engine in *Spring*, *Works*, i.

92 *CPE*, *Works*, i, 335–6. As an example, Boyle discussed how Bacon had been mistaken in his belief that his pruning of rose bushes had been responsible for their flowering in the autumn.

93 *Ibid.* p. 341. He discussed the practice of grafting as an example of this problem.

94 *Ibid.* p. 342.

95 *Ibid.* p. 340.

96 *Ibid.* p. 342.

97 *Ibid.* p. 341.

98 *Ibid.* p. 339.

99 *Ibid.* p. 352.

100 'Animadversions upon Mr. Hobbes's problemata de vacuo', *Hidden Qualities of Air*, *Works*, iv, 105.

101 *Forms and Qualities*, *Works*, iii, 75.

102 *Exp. Obs. Physicae*, *Works*, v, 567. In *Colours*, *Works*, i, 708–24, Boyle described fifteen experiments in consort designed to test his conjectures concerning the production of black and white.

103 *CPE*, *Works*, i, 353.

104 *Ibid.* pp. 318–19.

105 *Forms and Qualities*, *Works*, iii, 75.

106 *Christian Virtuoso, I, Appendix*, *Works*, vi, 700.

107 BP 9, fol. 111.

108 *CPE, Works*, i, 349–50.
109 *Usefulness, II, sect. 1, Works*, ii, 112.
110 *CPE, Works*, i, 348–9.
111 BP 9, fol. 12.
112 *Sceptical Chymist, Works*, i, 499.
113 *CPE, Works*, i, 358.
114 As Boyle wrote in his *High Veneration, Works*, v, 150, 'our knowledge is a progressive or discursive thing'.
115 *Human Blood, Works*, iv, 758–9.

Carneades and the chemists: a study of *The Sceptical Chymist* and its impact on seventeenth-century chemistry

ANTONIO CLERICUZIO

Like other works of Robert Boyle, *The Sceptical Chymist* had a long and difficult gestation. In February 1654 Hartlib referred to Boyle's writing a treatise 'de Chemia et de Chymicis'. This was very likely the early version of *The Sceptical Chymist* to which Boyle referred in his 'Author's preface' of 1661:

> It is now long since, that to gratify an ingenious gentleman, I set down some of the reasons, that kept me from fully acquiescing either in the peripatetical, or in the chymical doctrine of the material principles of mixed bodies.

The preface informs us that Boyle circulated the manuscript among 'some learned men', who received it favourably and prompted the author to 'make it public '.[1] A letter from Clodius answering Boyle's doubts about the nature of metalline mercuries leads us to infer that Boyle circulated the manuscript among members of the Hartlib Circle.[2]

What is evidently a copy of this treatise, entitled 'Reflexions on the experiments vulgarly alledged to evince the 4 peripatetique elements, or ye 3 chymicall principles of mixt bodies', survives in Henry Oldenburg's commonplace book. It was published in 1954 by Marie Boas, who maintained that the treatise was written between 1651 and 1657.[3] This dating has been questioned by Charles Webster, who suggested that analogies between Boyle's water culture experiments and those performed by Robert Sharrock in 1659 appeared to indicate that Boyle wrote his 'Reflexions' between 1659 and 1660.[4] A comparison between the account of the experiment given in the 'Reflexions' and that in the published work shows that the experiment – and *a fortiori* the treatise itself – dates back to several years before the composition of the final version of *The Sceptical Chymist*. In the 'Reflexions' Boyle recorded that he had endeavoured to repeat van Helmont's experiment 'last summer', whereas in *The Sceptical Chymist* he refers to the same experiment as having been made 'divers years since'.[5] It is very likely that Boyle wrote the 'Reflexions' a few years after writing the first part of *The Usefulnesse of Experimental Philosophy* in the early 1650s, a work which is thoroughly imbued with Helmontian iatrochemistry.[6] If this is the case, then in the mid-1650s Boyle criticised the chemists' theory of principles with arguments drawn mainly from van Helmont's *Ortus Medicinae*. His objections to the traditional interpretation of fire analysis, his interest in the operations of the universal solvent (the *Alkahest*), as well as the long section devoted to the Helmontian water theory, make it apparent that the 'Reflexions' were indebted to – indeed represented a development of – van Helmont's chemistry. In addition, the passage devoted to seminal principles – presented here as a kind of comment on the water culture experiment, but which disappears in *The Sceptical Chymist* – is evidence of Boyle's acceptance of one of the key notions of Helmontian natural philosophy.

The Sceptical Chymist took definitive shape in Oxford. The dry and straightforward form of the text of the 'Reflexions' – which were written for a friend, possibly somebody who could repeat the chemical experiments which it described – disappears in the published text, where Boyle's attack on the Aristotelians' doctrine of the elements and the Spagyrists' theory of principles took the form of a dialogue. Boyle sought the opinion of friends and colleagues – not, as he related, the same as those who had read the 'Reflexions' – about the new work. Robert Sharrock was probably one of the readers of Boyle's manuscript. His competence in chemistry, and the fact that during 1659 and 1660 he had access to Boyle's manuscripts (he was engaged in editing and translating certain of Boyle's works), makes him a likely member of the group of readers who advised Boyle to give *The Sceptical Chymist* to the press as soon as possible.[7] This would explain the analogies between Boyle's and Sharrock's water culture experiments.

It is apparent that, when Boyle consigned it to the press, *The Sceptical Chymist* was far from being complete. Possibly only the first part – that is, the 'physiological considerations' – was in a definitive form. The remaining sections (by far the greater part of the work) were published, as the preface tells us, 'maimed and imperfect', a consequence of the pressure exerted on Boyle by his friends, who urged him to publish the text as it was. As a result, the dramatic structure of the dialogue is virtually lost after the first part. Once the occasion of the meeting has been described and the four interlocutors have first put forward their arguments, the dialogue takes a different turn: Themistius and Philoponus no longer intervene, and Carneades does not argue simply as a sceptic, but proposes certain positive arguments concerning the composition of mixed bodies. The largest part of the work is in fact a monologue by Carneades, interspersed with occasional observations of Eleutherius.

I believe that Boyle's decision to send *The Sceptical Chymist* to the press was partly influenced by the wide diffusion of the chemical theory of principles among the Oxford physiologists as attested by Willis's *De Fermentatione*, published in 1659, but already circulating in manuscript form in 1658.[8] There can be little doubt that the following statement in Boyle's preface alludes to those physiologists, like Willis, Bathurst and Glisson, who had embraced the chemical theory of matter:

For I observe, that of late chymistry begins, as indeed it deserves, to be cultivated by learned men, who before despised it . . . whence it is come to pass, that divers chymical notions about matters philosophical are taken for granted and employed, and so adopted by very eminent writers both naturalists and physicians. Now this, I fear, may prove somewhat prejudicial to the advancement of solid philosophy.[9]

As we have seen, the hasty publication of *The Sceptical Chymist* was responsible for the imperfect form of the dialogue. It is, however, of some interest to consider the reasons which prompted Boyle to give his work the form of a dialogue. Clearly, the dialogue form was well suited to a work which, according to the author's original project, was primarily intended to suggest doubts on the Paracelsian view of principles, not to assess new theories. I believe that there was an additional reason which led Boyle to give his work the shape of a dialogue, namely his effort to clear chemistry of the often-repeated charge that it was a purely practical discipline. The philosophical dialogue was widely adopted in the Renaissance. It was a respectable literary tradition to which Boyle wanted his work to be linked.[10] Boyle's decision to write a dialogue shaped after Cicero's model – whose *De Natura Deorum* is referred to in the preface – was part of his attempt to make chemistry palatable to those Oxford professors who were suspicious of those, like John Webster, who championed chemistry in opposition to the traditional University curricula.[11]

It is apparent that the style of the work was an integral part of Boyle's critical assessment of the chemical literature. In his reflections on style in *Certain Physiological Essays* (1661), Boyle censured the 'dull and insipid way of writing, which is practised by many chymists'.[12] The tone of the conversation, the emphasis on the style of philosophical discussion among gentlemen, the statement that the book was written by a gentleman and that only gentlemen were introduced as speakers, were part of Boyle's attempt to reform chemistry. These features of the dialogue were all meant to make chemistry a socially and intellectually respectable discipline. He presented his dialogue as a model of 'civility' in philosophical disputation, as opposed to the method of argument and controversy often adopted by chemists, who, as he put it, 'rail instead of arguing'.[13] Boyle described the conversation among the gentlemen gathered in Carneades's garden as an example of how 'a man can be a champion of truth, without being an enemy of civility'.[14] It is apparent that Boyle's opposition to contemporary chemistry was not purely conceptual. It also involved the status of the discipline and the social standing of its students. The ideal audience of Boyle's work was the community of 'learned gentlemen', well-read in philosophy and conversant with chemical experiments. The purpose of Carneades's arguments was to make those who had been 'bewitched' by the chemists' doctrines aware of their mistakes.[15]

Furthermore, Boyle's intention was not merely to warn natural philosophers against any reliance on the experiments of 'vulgar chemists'. He also wished to establish a line of demarcation between such unphilosophical 'vulgar chemists' and learned chemists. The former were mere practitioners, who confined themselves to performing experiments, whereas the latter were those who, like van Helmont, 'give philosophical accounts of them'.[16] This distinction Boyle applied to alchemists as well: he often extolled the true adepti, learned men by whom, he wrote, 'could I enjoy their conversation, I would both willingly and thankfully be instructed'.[17] I do not think that Boyle's purpose in *The Sceptical Chymist* was 'to provide the tools to construct experimental essays in chemistry'.[18] His purpose – as is unambiguously expressed by Carneades at the beginning of the work – was 'to question the very way of probation employed by Peripatetics and Chymists, to evince the being and number of the elements'.[19] This issue, which underlies all the sections of *The Sceptical Chymist*, is particularly evident in Carneades's criticism of the chemists' uncritical interpretation of the results of their fire analysis. Following van Helmont, he questioned the standard view of analysis by fire and suggested more sophisticated methods of chemical analysis.

The contrast between *The Sceptical Chymist* and the chemical textbooks which proliferated in the first half of the seventeenth century is manifest. It is not, however, the one Golinski has pointed out, i.e. the opposition between a model of experimental essay structured upon the language of an ideal laboratory and the systematic, methodical style of chemical textbooks. Boyle did not dismiss textbooks like Beguin's *Tyrocinium* on the grounds that – in Golinski's words – they were not 'founded on narrative descriptions of experimental "facts"'.[20] For Boyle, their fault was of a different nature: they failed to provide the philosophical interpretation of experiments. They were the product of – and in turn they reproduced – the traditional status of chemistry as a purely practical discipline. By arguing for the philosophical status of chemistry, Boyle was attempting to set the educated chemists above the unlearned ones in public estimation. He aimed to raise chemistry from the rank of an unphilosophical, operative discipline and to show that it was an important component of natural philosophy, since it could yield great insight into the constitution of bodies.

Closely related to his attack on the Paracelsian doctrine of principles was Boyle's criticism of the received chemical terminology. This criticism did not occur in the 'Reflexions', but was brought out in *The Sceptical Chymist* and further developed in the *Appendix* added to the second edition of 1679/80, *The Producibleness of Chemical Principles*. Boyle complained of the 'intollerable ambiguity [chemists] allow themselves in their writings and expressions . . . and the unreasonable liberty they give themselves of playing with names at pleasure . . . '. He went on to state how 'even Eminent Writers, (such as Raymund Lully, Paracelsus and other) do so abuse the termes they employ, that as they will now and then give divers things, one name; so they will oftentimes give one thing, Many Names . . . even in Technical Words or Termes of Art they refrain not from this Confounding Liberty, but will, as I have Observ'd, Call the same Substance, sometimes the Sulphur, and Sometimes the Mercury of a Body.'[21] It is noteworthy that in *The Sceptical Chymist* Boyle made a firm distinction between the communication techniques of chemistry and those of alchemy. Boyle found it legitimate that alchemists should 'write darkly, and aenigmatically, about the preparation of their *Elixir*, and some few other grand *Arcana*, the divulging of which they may upon grounds plausible enough esteem unfit'. On several occasions, Boyle himself adopted the alchemists' practice of secrecy.[22] However, such an obscure and allusive style was not appropriate in natural philosophy.

Not only did Boyle criticise the darkness of current chemical language; he also questioned the generally accepted chemical terminology, since it did not account for the variety of properties found in the substances which chemists usually denoted under the same name.[23] The great 'disparity and unlikeness' among salts, as well as among sulphurs and mercuries – an obvious consequence of the fact that they are not simple and elementary, but compound, bodies – was, according to Boyle, generally overlooked by chemists. The ambiguity of current chemical terminology followed as a consequence of this. Boyle, whose interest in natural languages is well known, aimed to reform chemical terminology by giving different chemical substances specific names, in order to account for their differing properties.[24] In *The Sceptical Chymist*, and more thoroughly in *The Producibleness of Chemical Principles*, he investigated various kinds of salts and spirits and censured both the terminology and the system of classification employed by chemical textbooks and lexicons.[25] This critique was an integral part of Boyle's attempt to provide a new chemical classification, a task which he pursued above all in *The Producibleness of Chemical Principles*.

In *The Sceptical Chymist* Boyle laid down the theoretical foundations of such classification, in terms of the hierarchy of corpuscles. Historians have, in general, paid little attention to this aspect of Boyle's theory of matter – which in fact contains helpful hints for understanding the relationship of chemistry to corpuscular philosophy.[26] I shall consider Boyle's arguments as they are brought out in the section of *The Sceptical Chymist* devoted to the *mixtio*. Here he states that particles of the universal matter stick together to form clusters endowed with distinct qualities. These are the chemical (not purely mechanical) corpuscles which take part in chemical reactions. Boyle calls them *prima mixta*. He distinguishes two kinds of 'clusters' of particles. One is formed of particles which 'are so minute', and adhere so tightly, that in the compounded body the cluster retains its own nature. The other one is made of particles whose contexture is not so close as in the former kind of cluster. This means that particles from a different body can insinuate themselves between them, disjoin them and form a different cluster.[27] Therefore, it may happen that the corpuscles which concurred to form a given mixed body cannot be recovered by 'any known way of analysis'. This undermines the chemical doctrine of principles, which postulated that the *tria prima*, being the ingredients of all mixed bodies, could always be extracted by fire analysis.

Boyle's view of principles is well elucidated by the following statement:

> If you allow of the discourse I lately made you, touching the primary associations of the small particles of matter, you will scarce think it improbable, that of such elementary corpuscles there may be more sorts than either three, or four, or five. And if you will grant, what will scarce be denied, that corpuscles of a compounded nature may, in all the wonted examples of chymists, pass for elementary, I see not why you should think it impossible, that aqua fortis, or aqua regis, will make a separation of colliquated silver and gold, though the fire cannot: so there may be

some agent found out so subtle and so powerful, at least in respect of those particular compounded corpuscles, as to be able to resolve them into those more simple ones, whereof they consist, and consequently increase the number of distinct substances, whereunto the mixed body has been hitherto resoluble.[28]

From this passage, it becomes apparent that Boyle did not rule out the existence of elementary and homogeneous chemical substances. What he firmly denied was the chemists' procedure for establishing their number, and their methods of drawing them from mixed bodies.

Boyle's classification of corpuscles has an additional consequence for the theory of chemical composition, namely that a mixture can be made of corpuscles of different levels of complexity, a view which was to be developed by Joachim Becher. It is possible, Boyle states, that 'some bodies are made up of mixt bodies, not all of the same order, but of several; as (for instance) a concrete may consist of ingredients, whereof the one may have been a primary, the other a secondary mixt body . . . '.[29]

Historians have maintained that Boyle's chemical ideas (and in particular those contained in *The Sceptical Chymist*) had little impact on late seventeenth-century chemists who – it is commonly asserted – ignored his censure of the Paracelsian doctrine of principles and continued to uphold the chemical theory of matter. This is true, but it is only a part of the truth. I contend that the impact of Boyle's work was not negligible, albeit difficult to recognise. *The Sceptical Chymist* was widely known, often quoted, though very seldom properly understood. Some chemists adopted specific aspects of the work, though very few thoroughly developed the ideas therein contained.

The impact of Carneades's arguments on chemical courses was somewhat slight. Lefebvre's *Traité de la Chimie* (2nd edn, 1662) and Glaser's *Traité* (1668) kept the traditional distinction between the theoretical and the practical part. The few pages in these books devoted to chemical theory reproduced the received view of principles as well as the traditional chemical terminology. The case was different with Lémery's *Cours de Chimie*, first published in 1675.[30] Echoing Boyle's arguments, Lémery complained of the obscurity of current chemical language and provided a detailed classification of different kinds of spirits and salts. Unlike other chemical courses published in the same period, Lémery's attempted to legitimate chemistry among French natural philosophers by vindicating its dignity against persistent opposition and scepticism. Like Boyle, he declared that chemistry was a discipline which explained 'la manière dont la Nature se sert dans ses opérations'. The theoretical part of the *Cours* shows that Lémery had assimilated some of Boyle's arguments: he denied that the five principles remain in their former condition in the mixed body, and stated that they could not be obtained again by fire analysis.

The fifth edition of Lémery's *Cours*, published in 1683 (three years after the publication of *The Producibleness of Chemical Principles*), contains a long discussion of Boyle's objections against the theory of principles. Although he did not follow the Boylean view that the chemical principles were the product of fire, Lémery – who in 1683 was in England – explained the status of principles in a way which reveals his adoption of some of Boyle's criticisms of traditional chemical theory. He warned readers that in his *Cours* he employed the term 'Chemical Principle' as a 'working tool', not in a strict sense. On the chemical principles, Lémery adopted an 'instrumentalist' point of view. He asserted that they could be called principles only insofar as chemists were not able to progress in the analysis of bodies. In theory, they could be further decomposed into particles which would properly deserve the name of principles. Lémery's ideas and his interpretation of Boyle's doctrines played an important part in the development of chemistry in France. This was due both to the success of his *Cours* and to the position Lémery occupied. As a fellow of the Paris Académie des Sciences, he dominated the chemical studies in the Academy in the late seventeenth and early eighteenth centuries. His 'moderate' view of the chemical principles was adopted by William Homberg in papers written for the *Memoires* of the *Académie des Sciences* in 1699. Similarly, Du Hamel, secretary of the Académie des Sciences, echoed Boyle's critique of the chemical theory of principles in his *Consensus Veteris et Novae Philosophiae*, a popular work which was reprinted several times in the seventeenth century.[31]

The diffusion of *The Sceptical Chymist* in Germany took place in a context in which Aristotelianism was still very much alive. The result was a compromise of Aristotelian doctrines, corpuscular philosophy and chemistry, a compromise originally advocated by Daniel Sennert (1572–1637) in the early seventeenth century.[32] This is well exemplified by Joachim Becher. In his *Physica subterranea*, Becher maintained that the three Paracelsian principles were not simple substances, but mixed bodies: for him, the only universal principles were water and earth.[33] He supported this statement with arguments taken from Boyle's *Sceptical Chymist*. Becher also introduced a classification of compound bodies which was in part a development of Boyle's, although the corpuscular theory of matter played only a marginal role in his work. Becher distinguished 'simplicia' (homogeneous substances), 'composita', 'decomposita' and 'superdecomposita' in increasing order of complexity.[34] Stressing the importance of classification and definitions – in a manner which ultimately stems from the Aristotelian view of science – Becher endorsed Boyle's appeal for a reform of chemical terminology.

In England, *The Sceptical Chymist* brought about three different reactions. Whereas the Oxford physiologists failed to espouse Boyle's arguments against the chemical doctrine of principles, the Helmontians interpreted Boyle's work as a mere development of van Helmont's ideas, producing a synthesis of Helmontian and Boylean chemistry which had wide diffusion. Only one English chemist fully adopted and developed Boyle's chemical views as contained in *The Sceptical Chymist*, Daniel Coxe, a Fellow of the Royal Society and correspondent of Boyle. Taking the Oxford physiologists first, their reception of *The Sceptical Chymist* was somewhat lukewarm. In the third edition of *De Fermentatione* (1662), Thomas Willis inserted a discussion of the status of the five principles, which was meant to meet the objections contained in *The Sceptical Chymist*. Although he allowed the possibility that they could not be the ultimate components of mixed bodies, he did not modify his view that they were their constituents. Richard Lower, in his *Vindicatio* (1665), defended Willis's doctrine of the five principles. John Mayow, in his *Tractatus Quinque* (1674), did not question the chemical doctrine of principles.

The *Sceptical Chymist* played an important role in the Helmontian–Galenist controversy which took place in the 1660s. The Helmontians (in particular Marchamont Nedham, George Starkey and William Simpson) used Boyle's work to reinforce and validate their arguments against the Galenists. However contrived this might appear, such an interpretation was not entirely illegitimate. In *The Sceptical Chymist* the Helmontians could find a number of eulogies of van Helmont, which they often quoted. They also referred to the water culture experiment to support the theory of water as the material principle, whilst ignoring Carneades's objections to the Helmontian doctrine.[35] Though Boyle never supported the Helmontians in their attack on the College of Physicians, he did not disavow their use of *The Sceptical Chymist*. As a result, in the 1670s and in the early 1680s such Helmontians as John Webster, George Thomson, Thomas Shirley, William Simpson, George Acton and William Bacon combined Helmontian and Boylean chemistry, interpreting the notions of *archeus*, seeds, *blas* and *gas* in corpuscular terms.[36]

Among the Fellows of the Royal Society, Daniel Coxe is the only one who thoroughly understood Boyle's outlook on chemical analysis. In his letter to Boyle of 19 January 1666 [n.s.], Coxe unambiguously expressed his dissatisfaction with the chemists' methods of analysis. Following Carneades's arguments, Coxe argued:

> No Considerable progresse can bee made in disquisitions Concerning any Concrete in nature, especially Mineralls unlesse we are masters of some Excellent Menstrua, from whose assistance we may derive many Advantages. Either analyzing ye bodies wee operate on & thereby Informe our selves what theire Constituent principles are.

For Coxe, the methods of analysis adopted by chemists failed to yield the elementary components of mixed bodies, since 'they divide into Integrall only'. In the same letter Coxe asserted that the 'Texture of the first Principles being more close, they are not separable but by some Extreeme Subtle Analyzer which by reason of the minutenesse brisk motion, & Convenient Figuration of itt's parts may Disunite them'.[37] According to Coxe, these were the properties of van Helmont's Alkahest. Coxe stated that, after carefully perusing the passages of van Helmont's *Ortus Medicinae* dealing with the universal solvent, he came to the conclusion that: 'Although they may not instruct us how to make itt yet from them all put together I apprehend we may derive some information concerning its nature'.[38] As we can see, Coxe shared Boyle's principal argument against the chemists' doctrine of principles, namely that the traditional methods of analysis failed to reveal the actual components of mixed bodies.

The somewhat meagre impact – particularly in England – of Carneades's arguments against the Paracelsians' doctrine of principles, together with the encouragement of friends and the solicitation of the printer, persuaded Boyle to prepare a new, enlarged version of *The Sceptical Chymist*.[39] According to Boyle, he abandoned this project after the death of the printer of the original edition in 1669. Instead, in January 1679/80 he republished *The Sceptical Chymist* with virtually no alterations, adding an appendix.[40] This was originally meant to include four sections: 'The producibleness of chemical principles', 'The uncertainty of analysis made by distillation', 'The various effects of fire according to the differing ways of employing it', and 'Doubts whether there be any element or material principle of mixt bodies, one or more, in the sense vulgarly received'.[41] The planned appendix was so bulky that he decided to publish only one section, namely *The Producibleness of Chymical Principles*.

In the published appendix Boyle did not confine himself to reasserting his objections to the doctrine of chemical principles. He also thoroughly criticised the theory of water as the material principle, evidently in order to distance himself from the Helmontians. In addition, Boyle provided a new classification of chemical substances – developing the views contained in *The Sceptical Chymist*. In the section devoted to salts, Boyle stated that there were three distinct 'families' of salts: acid salts (vinegar and spirit of salt); volatile salts (salt of hartshorn, salt of urine and of blood); and alkalies or lixiviate salts (salt of tartar and potash). Such a classification of salts corresponded to the differing chemical properties of each class.[42]

One of the most ambiguous terms in seventeenth-century chemistry ·was 'spirit'. Paracelsians like Severinus and Croll put much emphasis upon the role of the spirit in nature. They regarded spirit as the active substance *par excellence*.[43] Lefebvre's *Course* contained a long section on spirit, which was described as a homogeneous substance and the 'source' of all terrestrial bodies – including the *tria prima*. The notion of spirit played a central role in Willis's *De Fermentatione*. Willis, along with the Paracelsians, developed the concept of spirit as an explanatory category for a wide range of phenomena, both chemical and physiological. He wrote that spirits 'determinate the Form and figure of every thing, prefixed, as they were, by Divine designation: they conserve the bonds of the mixture by their presence; and open them, by their departure at their pleasure . . . '. His definition of spirits was vague and ambiguous. As he put it:

> Spirits are Substances highly subtil, and Aetherial Particles of a more Divine Breathing, which our Parent Nature hath hid in this Sublunary World, as it were the instruments of Life and Soul, of Motion and Sense, of every thing.[44]

It is not surprising then, that Boyle felt it necessary to begin the section on the producibleness of spirits by putting his finger on the weakness of the chemists' notion of spirit:

> As for what the chymists call spirits, they apply the name to so many differing things, that this various and ambiguous use of the word seems to me no mean proof, that they have no clear and settled notion of the thing. Most of them are indeed wont, in the general to give the name of spirit to any distilled volatile liquor, that is not insipid, as is phlegm, or inflammable, as oil.

Boyle pointed out that under this general term chemists included liquors of quite different and contrary natures: both acid substances (as spirit of nitre, spirit of salt, and vinegar) and volatile alkalies (as urinous spirits). In addition, Boyle distinguished a third kind of spirits – very subtle and penetrant – which he called vinous or inflam-mable. In his views they were neither acid nor alkaline.[45]

The most interesting (and intriguing) chapter of *The Producibleness* is the one devoted to mercury. The conclusions Boyle reached after a long discussion – largely based on the writings of Paracelsus and those attributed to Raymond Lull – differs from that of the chapters on the other principles. Boyle did not peremptorily deny that mercury was a simple and homogeneous substance. Moreover, dealing with the metalline mercuries, Boyle suggested that they existed, yet they had not been extracted from metals. In his view they were 'magisteries', namely, substances produced in an intermediate stage during the process of transmutation. His definition of *magisterium* is explicitly borrowed from Paracelsus:

> It is a preparation, whereby there is not an analysis made of the body assigned, nor an extraction of this or that principle, but the whole, or very near the whole body, by the help of some additament greater or less, is turned into a body of another kind.[46]

After considering the arguments against the preexistence of metalline mercuries, Boyle took into account those supporting the chemists' view, finding two of them particularly conclusive. One was the consideration that mercury easily amalgamates with other metals, evidently due to the 'cognation' between quicksilver and the mercurial part of metals. The second and stronger argument – which Boyle's 'philosophical candour' forbade him to conceal – was that

> the gravity of a metal cannot reasonably be supposed to proceed from the whole body of the metal, but only from some one ingredient heavier in specie than the rest, and, than the metal itself. And this ingredient or principle can be no other, than the most ponderous body, mercury.[47]

In the same section Boyle unambiguously declared that metals were not perfectly homogeneous bodies.

What is apparent from these somewhat puzzling notions is that there was a strong link between Boyle's investigations of the compositions of mixed bodies and his alchemical interests, the evidence for which is surveyed by Lawrence Principe in the next chapter. Although Boyle himself insisted on keeping the language of chemistry separate from that of alchemy, it would be misleading to demarcate clearly between his alchemical interests and his chemical theories. This is attested as much by the above mentioned observations on the mercuries contained in metals as by Boyle's well-known experiments on the incalescence of mercury. Indeed, *The Producibleness of Chemical Principles* contains one of Boyle's few public vindications of alchemy. In it Boyle differentiated the 'adepts', able to transmute base metals into gold, from 'those that are wont to write courses of chymistry or other books of that nature'.[48]

The inclusion of sections like these in his work makes it less surprising that Boyle's chemical legacy was not a simple one. This underlines the fact that a straightforward view of him as 'sceptical chemist' who attacked Paracelsian principles on the basis of the mechanical philosophy, depends on a highly selective reading of his writings. It also again demonstrates the fallaciousness of a characterisation of Boyle purely as advocate of experimental chemistry. It is apparent that Boyle's target was not alchemy, but practical chemistry, as exemplified by the chemical courses of his day. They were lacking in the proper combination of practice and theory which – paradoxically – some alchemists achieved.

NOTES

1 *Sceptical Chymist, Works*, i, 458–9.

2 BL 2, fol. 88. The date is uncertain; Maddison has suggested 1652–4.

3 Marie Boas, 'An early version of Boyle's *Sceptical Chymist*', *Isis* 45 (1954): 153–68 (hereinafter referred to as 'Reflexions').

4 C. Webster, 'Water as the ultimate principle of nature: the background to Boyle's *Sceptical Chemist*', *Ambix* 13 (1966), 96–107.

5 *Works*, i, 494.

6 See A. Clericuzio, 'From van Helmont to Boyle: a study of the transmission of Helmontian chemical and medical theories in seventeenth-century England', *BJHS* 26 (1993): 303–34.

7 R. Sharrock, *The History of the Propagation and Improvement of Vegetables* (Oxford, 1660), p. 77, which certainly alludes to Boyle's manuscripts. Letters from Sharrock to Boyle contain evidence of Sharrock's work of translation: BL 5, fols. 92–102.

8 See Samuel Hartlib to Boyle, 16 December 1658: *Works*, vi, 115.

9 *Works*, i, 459.

10 Renaissance views of the typology of dialogue are summarised in Carlo Sigonio, *De Dialogo* (Venice, 1562), and in Torquato Tasso, *Discorso dell'arte del dialogo*, in *Rime e Prose* (Venice, 1586). On the dialogue form of *The Sceptical Chymist*, see R. Multhauf, 'Some nonexistent chemists of the seventeenth century', in A. G. Debus and R. Multhauf, *Alchemy and Chemistry in the Seventeenth Century* (Los Angeles: William Andrews Clark Memorial Library, 1966), pp. 31–50.

11 *Works*, i, 462. See A. G. Debus, *Science and Education in the Seventeenth Century. The Webster–Ward Debate* (London: Macdonald, 1970).

12 *CPE, Works*, i, 304.

13 *Works*, i, 461.

14 *Ibid.* p. 462.

15 *Ibid.* p. 510.

16 *Ibid.* p. 522. Boyle stated: 'There is great difference betwixt the being able to make experiments, and the being able to give a philosophical account of them'.

17 *Ibid.* p. 463.

18 J. V. Golinski, 'Robert Boyle: scepticism and authority in seventeenth–century chemical discourse', in *The Figural and the Literal: Problems of Language in the History of Science and Philosophy, 1630–1800*, eds. Andrew E. Benjamin, Geoffrey N. Cantor and John R. R. Christie (Manchester University Press, 1987), p. 68.

19 *Works*, i, 473.

20 Golinski, 'Robert Boyle' (n. 18), p. 73.

21 *Works*, i, 520. See J. V. Golinski, 'Chemistry in the Scientific Revolution: problems of language and communication', in *Reappraisals of the Scientific Revolution*, eds. D. C. Lindberg and R. S. Westman (Cambridge University Press, 1990), pp. 382–90.

22 *Works*, i, 521. See L. Principe, 'Robert Boyle's alchemical secrecy: codes, ciphers, and alchemical concealments', *Ambix* 39 (1992): 63–74. For secrecy and natural philosophy, see William Eamon, 'From the secrets of nature to public knowledge', in *Reappraisals of the Scientific Revolution* (n. 21), pp. 333–66. One of the reasons why Boyle was often reluctant to make his alchemical knowledge public was his lifelong obsession with plagiarism.

23 *Works*, i, 523, 530.

24 For Boyle's interest in natural languages, see, e.g., Boyle to Hartlib, 19 March 1647, *Works*, i, xxxvii–iii.

25 *Works*, i. 532–6.

26 The only exception is T. S. Kuhn, 'Robert Boyle and structural chemistry in the seventeenth century', *Isis* 43 (1952): 12–36. On Boyle's theory of matter, see A. Clericuzio, 'A Redefinition of Boyle's chemistry and corpuscular philosophy', *Ann. Sci.* 47 (1990): 561–89.

27 *Works*, i, 506.

28 *Ibid.* p. 515.

29 *Ibid.* pp. 524–5.

30 For Lefebvre, Glaser and Lémery, see H. Metzger, *Les Doctrines Chimiques en France du début du XVIIe siècle à la fin du XVIIIe* (Paris, 1923); and A. G. Debus, *The French Paracelsians. The Chemical Challenge to Medical and Scientific Tradition in Early Modern France* (Cambridge University Press, 1991), pp. 125–30, 147–8.

31 On William Homberg, see J. R. Partington, *A History of Chemistry*, 4 vols (London: Macmillan, 1961–70), iii, 42–7. For Jean Baptiste Du Hamel (1624–1706), see G. Santinello, *Storia delle Storie Generali della Filosofia* (Brescia: La Scuola, 1979), ii, 22–31.

32 Daniel Sennert, *De Chymicorum cum Aristotelicis et Galenicis Consensu ac Dissensu* (Wittenberg, 1619) and *Hypomnemata Physica* (Frankfurt, 1636). On Sennert, see Partington, *History* (n. 31), ii, 271–6.

33 Johann Joachim Becher, *Actorum Laboratorii Chymici Monacensis, seu Physicae Subterraneae Libri duo* (Frankfurt, 1667), pp. 123, 138.

34 *Ibid.* p. 525.

35 See Clericuzio, 'From van Helmont to Boyle' (n. 6).

36 The fusion of Helmontian and Boylean doctrines was not confined to England. It was adopted by Michael Ettmuller (1644–83), in *Medicina Hyppocratis Chymica* (Leipzig, 1670) and by David van der Beck, *Experimenta et Meditationes circa Naturalium Rerum Principia* (Hamburg, 1674). In a review of this book, published in *Phil. Trans.* 9–10 (1674): 60–4, the reviewer disclaimed the author's statement that Boyle shared the Helmontian doctrine of water and seminal principles as the beginnings of all mixed bodies.

37 BL 2, fol. 54.

38 *Ibid.*

39 *Works*, i, 588.

40 Four Latin editions of *The Sceptical Chymist* had been issued before that date: two in 1662, one in 1668 and one in 1677. Some manuscript papers possibly related to Boyle's planned revised edition are in BP 28, p. 406.

41 *Works*, i, 588. Manuscript notes related to the unpublished sections survive in the Boyle Papers: I hope to publish an account of them at a later date.

42 *Ibid.* p. 596.

43 See A. Clericuzio, 'The internal laboratory', in *Alchemy and Chemistry in Renaissance and Early Modern Times*, eds P. M. Rattansi and A. Clericuzio, forthcoming.

44 Thomas Willis, *The Remaining Medical Works* (London, 1681), p. 3.

45 *Works*, i, 609.

46 *Ibid.* p. 637. According to Martin Ruland, magisterium is 'a Chemical state in which matter is developed and exalted by the separation of its external impurities.' M. Ruland, *Lexicon Alchemiae* (Frankfurt, 1612).

47 *Works*, i, 639.

48 *Ibid.* p. 590.

Boyle's alchemical pursuits

LAWRENCE M. PRINCIPE

Although Newton's alchemy has received an enormous amount of attention during the last twenty years, these studies have engendered few similar ones for other important figures of his age. The contemporary of Newton most suited to such study is surely Robert Boyle, for his long-term interest, activity and belief in traditional alchemy has remained largely unpublicised. Though there have always been a few visible signs of Boyle's alchemical endeavours, 'one is apt to overlook what an obsession [alchemy] was' for him, as L. T. More remarked almost fifty years ago.[1] Indeed, the marks of Boyle's alchemy have been largely glossed over, save by a few authors, and in spite of the recent resurgence of interest in Boyle, his alchemical pursuits continue to stand largely unintegrated into our portrayals of him.[2] Much of the standard image of Robert Boyle persists from an earlier historiographic tradition which sought to identify heroic figures in the development of modern science. This tradition's dismissal of alchemy facilitated an extension of Boyle's occasional criticisms of alchemical methodology and epistemology to a condemnation of alchemy in general, and Boyle thus became a major point of transition from alchemy to chemistry.[3] But neither the emergence of chemistry nor the demise of alchemy is so tidy, nor is Boyle's role so simplistic. Boyle's works and papers teem with alchemical references, theories, practices and processes. Until Boyle's alchemical pursuits are incorporated into his historical image, that image will remain distorted by a magnification of his work on 'modern, reputable' topics of atomism, pneumatics and such like, at the expense of 'archaic, disreputable' topics like alchemy.

In order to assess the role and importance of alchemy in understanding Robert Boyle, three interrelated questions need to be addressed: (1) What was the depth and duration of Boyle's involvement with alchemy – was it a passing interest or a long-term programme, and how much of its traditional claims did he believe? (2) Did Boyle's interest in alchemy extend only to its practical or experimental portion, or did he also employ or accept some of its theories, conventions or spiritual/mystical dimensions? and (3) What role does alchemy play with respect to Boyle's overall career? I shall argue that alchemy occupied an enormous portion of Boyle's time and experimental activity, that he believed its traditional claims, adopted its conventional secrecy and employed parts of its theory as explanatory principles. Further, in the latter part of his career, alchemy, by virtue of its spiritual facets, functioned as a link between Boyle's two main spheres of activity, the advancement of experimental natural philosophy and the defence of the Christian faith.

Cataloguing Boyle's alchemical pursuits requires first that we be able to identify alchemical activities accurately, and especially to distinguish them satisfactorily and consistently from more properly *chemical* activities. Admittedly, in some seventeenth-century situations, the distinction is neither warrantable nor valuable, but if we wish to use the label 'alchemy' to describe a set of activities and beliefs (and we should be able to do so) we need to be clear and consistent about its referents. Texts can be differentiated based on their motives and theoretical assumptions, but practical processes by themselves present a greater problem. The very same process (for example, the calcination of lead) may be properly alchemical or chemical (i.e. non-alchemical) depending upon the intent, goals and interpretation of the operator. Occasionally it has been all too easy to classify experimental activities as 'alchemical' or 'chemical' based upon preconceived notions of the experimenter, usually revolving around his received 'credibility' or 'modernity'. For negotiating this difficult terrain, I believe it is safe and legitimate to call endeavours aiming at traditional alchemical desiderata – the Philosophers' Stone, the extraction of mercuries and sulphurs, the alkahest, etc. – as alchemy, although it does not follow that alchemy should be restricted to such topics.

Boyle himself draws a distinction in the middle of a troublesome area – trans-mutation. Boyle's belief in metallic transmutation has been recognised for many years, but only in a *non-alchemical* sense.[4] Boyle's transmutation has been seen as a logical consequence of his corpuscularian hypothesis, as revealed in an oft-quoted passage from the *Origine of Formes and Qualities*:

> . . . since bodies, having but one common Matter, can be differenc'd but by Accidents, which seem all of them to be the Effects and Consequents of Local Motion, I see not, why it should be absurd to think, that . . . almost of any thing, may at length be made Any thing: as, though out of a *wedge* of Gold one cannot immediately make a *ring*, yet by either Wyre-drawing that Wedge by degrees, or by melting it, and casting a little of it into a Mould, That thing may easily be effected.[5]

As a result, it has been possible to frame *apologia* for apparently alchemical senti-ments around this corpuscularianism-derived transmutation. But what has not been hitherto recognised is Boyle's belief in *two separate types of transmutation* – one mechanical, the other properly alchemical. Boyle clearly differentiates mechanical transmutation from alchemical projection later on in the *Origine of Formes*:

> . . . there may be a real Transmutation of one Metal into another, even among the perfectest and the noblest Metals, and that effected by Factitious Agents in a short time, and, if I may so speak, after a Mechanical manner. I speak not here of Projection . . . because, though Projection includes Transmutation, Transmutation is not all one with Projection, but far easier than it.[6]

The 'projection' to which Boyle refers is the traditional alchemical method of transmuting metals by *projecting* (i.e. casting on, from *proiecere*) a minute quantity of the Philosophers' Stone upon a crucible full or molten lead or hot mercury, thereby converting it in a few minutes into pure gold.

Boyle's belief in the reality of alchemical transmutation (by projection of the Philosophers' Stone) is made abundantly clear in his unpublished *Dialogue on the Transmutation of Metals*. This text recounts a debate set at a 'Most Noble Society', quite possibly the Royal Society, concerning the possibility of transmutation by projection.[7] The 'Anti-Lapidist' party, headed by a character named alternately as Erastus or Simplicius, argues that alchemical theories are unclear and undemonstrated, that the accounts of transmutation are unreliable and that the entire idea of the Philosophers' Stone and transmutation by projection is unbelievable. On the other side, Zosimus, heading the 'Lapidist' party, counters that alchemical theories may well be deficient, but alchemical claims do not transgress the bounds of possibility, and that there are sufficient accounts of transmutations and related phenomena to credit them. Arguments and counter-arguments are presented, until Pyrophilus enters and recounts his first-hand experience with an 'Anti-Elixir' which when projected upon molten gold successfully transmuted it into a base metal. (Boyle published this conclusion to the unpublished *Dialogue* anonymously in 1678 as *An Historical Account of a Degradation of Gold*.)[8] Pyrophilus's account systematically confounds all of Erastus's objections, and the dialogue closes with victory for the Lapidists. Indeed, the dialogue takes on an epideictic tone unusual for Boyle's writings; the rejection of Erastus's views is much stronger than the rather lukewarm refutations found in *The Sceptical Chymist*. The Society rebukes Erastus soundly, and Boyle underscores this chief anti-Lapidist's erroneous position by naming him Simplicius in a reference to that hapless character so embarrassingly refuted in the Galilean dialogues. Clearly, Boyle was convinced of the reality of the Philosophers' Stone and the validity of alchemical claims for its powers.

Boyle's belief in the Philosophers' Stone led him to try to make it for himself. Early in his career in March 1646/7, upon receiving in pieces the chemical furnace he had ordered, Boyle lamented to his sister Katherine, that 'I see I am not designed to finding out the Philosophers' Stone, I have been so unlucky in my first attempts at Chymistry'.[9] Of course, this mention might be dismissed as youthful enthusiasm or a joking complaint. After all, in the apologetic preface to his 'Essay on nitre', written in the mid-1650s and published in 1661, Boyle protests that he laboured at 'Chymical Tryalls . . . without seeking after the Elixir, that Alchymists generally hope and toyl for, but which they that know me know to be not at all in my aime'.[10] In spite of this protestation, many less obvious instances leave no doubt about Boyle's continuing attempts to confect the Stone. One of these appears in brief intimations in *On the Unsuccessfulness of Experiments*, *The Sceptical Chymist* and more fully in *Origine of Formes*.[11] In the last, Boyle describes a powerful solvent he calls his *menstruum peracutum* which can dissolve gold and transmute a portion of it into silver. He uses this process to show the enormous changes which can be effected on stable bodies, and deploys it, at least overtly, to expound his corpuscularian hypothesis. But his careful and deliberate choice of expressions and vocabulary creates an allusive sub-text, accessible to readers conversant in alchemical literary styles, which reveals the original goal of the process as the preparation of the Philosophers' Stone. For example, when remarking upon the dissolution of gold in his solvent, he writes that it is 'wont to melt as it were naturally . . . without Ebullition (almost like Ice in luke-warm water)'.[12] This seemingly unremarkable simile is actually a canonical phrase of the alchemical literature reserved exclusively for the dissolution of gold in Philosophical Mercury, the first crucial step in confecting the Stone. Alchemical writers view the dissolution of gold 'like ice in warm water' as a sign distinguishing the 'true' menstruum. Examples of this stock phrase occur in Ripley, Arnold of Villanova, Sendivogius, Philalethes, Siebmacher and others.[13] Even the use of the adverb 'naturally' reinforces the alchemical connection, as the 'true, philosophical' dissolvent of gold had to act naturally, returning gold to its first principles. Boyle also uses the phrase 'to destroy gold' recalling the alchemical axiom of Roger Bacon (which Boyle elsewhere quotes) *facilius est aurum facere quam destruere*.[14]

The connection of this gold process to the preparation of the Philosophers' Stone is underscored by its origin in the enigmatic alchemical text commonly known as the *Twelve Keys* of Basilius Valentinus.[15] This work promises to reveal the process for making the Philosophers' Stone by means of twelve emblematic figures. The first four keys, deciphered out of their cryptic style, provide a process very nearly identical to that described by Boyle. Boyle's 'gold process' was initially an attempt to make the Philosophers' Stone by following Basil's obscure directions.[16]

Of further interest is Boyle's use of alchemical explanatory principles. To explain the conversion of yellow gold into white silver, Boyle suggests that 'our Menstruum may have a particular Operation upon some Noble, and (if I may so call them) some Tinging parts of the Gold'.[17] Boyle here borrows from the alchemical theory of the *tria prima* (mercury, sulphur, and salt) which attributed qualities such as colour, smell, and flammability to the essential sulphur. Alchemists endeavoured to extract this sulphur of gold in order to implant it in other metals, thus 'tinging' them, according to the alchemical jargon, into gold. This notion of 'transplanting' the golden sulphur is ridiculed by Erastus/Simplicius in the *Dialogue on Transmutation*, but an example of it is described matter of factly by Boyle in *Unsucceeding Experiments*.[18] Boyle suggests 'in a sober sense' the real existence and extractability of 'some such thing' as the '*Tinctura Auri*, or *Anima Auri*', names synonymous with '*Sulphur Auri*'.[19] This alchemical explanation stands at odds with Boyle's better-known attribution of colour to the mechanical effects of texture. Yet the two theories – mechanical changes in texture and extraction of the alchemical sulphur – exist side-by-side when Boyle offers an explanation of the menstruum's effects:

> . . . some more noble and subtle Corpuscules, being duely conjoyn'd with the rest
> of the Matter, whereof Gold consists, may qualifie that Matter to look Yellow, . . .
> yet these Noble parts may either have their Texture destroy'd by a very piercing

Menstruum, or by a greater congruity with its Corpuscles, then with those of the remaining part of the Gold, may stick more closer [*sic*] to the former, and by their means be extricated and drawn away from the latter.[20]

These 'noble and subtle Corpuscules', whether extracted or altered, constitute the equivalent of the alchemical sulphur of gold by virtue of their colour-producing role in heterogeneous gold, and by the fact that their removal transmutes gold into silver, and their introduction transmutes silver into gold.

Even the casual observer of alchemy knows that when the preparation of the Philosophers' Stone is mentioned, the topic of secrecy cannot be far distant. Secrecy is a crucial element of alchemy – all alchemists command their students to keep silence, and they themselves shroud their knowledge in an obscure metaphorical double-talk which continues to deter and confound readers. In a fragment, possibly from the *Dialogue on Transmutation*, Boyle complains that

Divers of Hermetic Books have such involv'd Obscuritys that they may justly be compared to Riddles written in Cyphers. For after a Man has surmounted the difficulty of deciphering the words & terms, he finds a new & greater difficulty to discover ye meaning of the seemingly plain Expression.[21]

Many historians have been at pains to distance Boyle from the secrecy of earlier traditions, and to portray him as a major figure in the promotion of open communication and the public validation of knowledge.[22] Although these analyses do have some merit, they are not applicable to all of Boyle's work, and in particular they are not representative of his alchemical pursuits. There is, in fact, no shortage of evidence attesting to Boyle's commitment to *secrecy* in alchemical matters, for when dealing with alchemy Boyle adopted much of the secretive style of traditional alchemical writers. I have found that Boyle employs no fewer than ten different codes, ranging from simple word-, letter-, or number-replacements, to more complex nomenclators and ciphers, throughout his papers and correspondence.[23] Significantly, these concealment techniques occur only in the context of traditional alchemical pursuits. As an example, consider the following extract from Boyle's laboratory notes:

There is a way but not without sopsis to make a kind of Balisma of ye distilled Liquor of our Lithoi . . . those same Ebanim may be so order'd as to perform the same things with the regian Tirsus of Nogirus & with the Bombast too.[24]

Obscure passages such as this one occur frequently throughout the Boyle Papers, and strike the reader as rather more like Paracelsus than Robert Boyle. Curiously, they are often interspersed with clearly expressed notes and memoranda regarding addenda for his books, travel accounts, and non-alchemical processes.

Boyle is often purposefully obscure even in published writings. I have already mentioned the corpuscularian text and alchemical sub-text of the gold process. In addition, several works, including *Sceptical Chymist* and the *Usefulness of Experimental Philosophy*, use the principle of dispersion – a traditional alchemical method of concealment whereby dissevered parts of a single process are scattered disconnectedly through large stretches of unrelated text.[25] Boyle's reticence to publish clear and complete accounts of odd 'chemical tryalls' can provoke frustration. Leibniz complained to Henry Oldenburg that Boyle 'is to be implored to express everything candidly, and not to conceal [*supprimere*] so many things as he is accustomed to do in his other writings'.[26] For example, at the conclusion of the *Dialogue on Transmutation*, after the discussion has been concluded and the President has risen from his seat, Pyrophilus ends the text with the off-handed teaser 'yet I have not (because I must not do it) as yet acquainted you with the Strangest effect of our Admirable Powder'.[27]

Boyle's remarkable paper 'Of the incalescence of quicksilver with gold' should be mentioned in this regard.[28] This paper which appeared in the 21 February 1675/6 issue of *Philosophical Transactions* describes a specially prepared mercury which grew hot when mixed with gold, contrary to the normal behaviour of quicksilver. But Boyle is so evasive about describing this mercury that the publication is far more frustrating than informative. He never explains how the mercury is prepared, but only recounts its effects. Contrary to the norms of Royal Society discourse, and certainly contrary to the prevailing depictions of Boyle's commitment to open communication, Boyle flatly refuses to answer any questions regarding the mercury.

Boyle believed that this incalescent mercury held great power in reference to the preparation of the Philosophers' Stone. Boyle's interest in this one substance alone spanned nearly forty years, from its first preparation in 1652 to its inclusion in a memorandum (*ca.* 1691) listing the most important items to be discharged before his death.[29] He explicitly connects this mercury with the alchemical sources, writing that he found in alchemical texts 'some dark passages, whence I then ghess'd their knowledge of it, and in one of them I found, though not all in the very same place, an Allegorical description of it'.[30] In a typically alchemical manner, he excuses his secrecy by citing the 'political inconveniences that may ensue if it prove to be one of the best kind and fall into ill hands'.[31]

The incalescent mercury paper, like the gold process, reveals Boyle's continuing attempts to confect the Philosophers' Stone, and, like the codes, ciphers and dispersions, attests to the secrecy with which he covered his alchemical pursuits. But it also indicates another important aspect of Boyle's alchemy – his attempts to contact alchemists. We can speak, rather loosely, of an 'alchemical community' – a loose confederation of laborants and *adepti* working largely independently, yet joined by common beliefs, practices, goals and linguistic conventions. These bonds of commonality were strengthened by the hand-to-hand circulation of letters and manuscripts, which, not intended for public view, recounted alchemical secrets in a far more straightforward manner, largely free from the obscuring metaphor and allegory of published texts. Boyle refers to this hidden community as 'a sett of *Spagyrists* of a much higher order . . . being able to transmute baser Metalls into perfect ones', and confesses belief in their existence and the powers of their arcana.[32] Boyle had, in fact, obtained the receipt for the incalescent mercury in 1651 from just such a private alchemical communication. A letter to Boyle from George Starkey describes a process for 'animating' mercury; this process (employing the star regulus of antimony), when carried out in a modern laboratory, provided a mercury which did in fact grow hot when mixed with gold, confirming this letter as the source of Boyle's incalescent mercury.[33]

The incalescent mercury paper should be read not as a source of information, but as a request for it from the alchemical community. Boyle was apparently unsure of how to prepare the Philosophers' Stone from the mercury.[34] The solution was to obtain direct contact with knowledgeable alchemists rather than wandering in the uncertainty of obscure alchemical texts. He had already questioned 'several prying Alchymists' who 'of late years travelled into many parts of *Europe* to pry into the Secrets of Seekers of Metalline Transmutations' regarding this substance.[35] Boyle addresses the paper to as large an audience as possible by publishing it in English and Latin in parallel columns – a publishing practice both unprecedented and unrepeated in *Philosophical Transactions*. He unambiguously indicates the alchemical nature and importance of the paper by connecting his results to alchemical work published previously under allegorical disguise. Boyle then demonstrates his acceptance of alchemical rules by committing himself to secrecy, and culls out responses to the paper from non-adepts by refusing to accept questions. Finally, having targeted his audience and proven his communion with them, he asks the elite adepti of the alchemical community for further information – hoping that he

> may safely learn . . . what those that are skilful and Judicious enough to deserve to
> be much considered in such an affair, will think of our Mercury . . . The knowledge
> of the opinions of the wise and skilful about this case, will be requisite to assist me
> to take right measures in an affair of this nature. And till I receive this information,
> I am obliged to silence.[36]

How many responses Boyle received is unclear – we know of one from Isaac Newton in a letter of 26 April 1676 to Henry Oldenburg.[37] Nonetheless, in another publication, Boyle reveals his conclusions when he states that he used up the last 'remnant of a certain Quicksilver which I intend never to make againe, (and of which for the sake of Mankind, I resolve never to teach the preparation)'.[38]

The foregoing examples demonstrate several points about Boyle's alchemical work: (1) Boyle believed in the existence and powers of the Philosophers' Stone, and endeavoured to prepare it; (2) Boyle adopted parts of alchemical theory for use as explanatory principles – sometimes in conflict with his now more-celebrated mechanical principles; (3) Boyle used codes and blinds to shroud his alchemical pursuits in secrecy; and (4) Boyle attempted to contact the alchemical community.

Many other points germane to Boyle's alchemy warrant further investigation. For example, his rarely read tract *On Celestial Magnets* contains alchemical notions of 'magnets' which can attract and fix celestial influences and effluvia in striking similarity to those which B. J. T. Dobbs found important to Newton, and which are ultimately derived from Sendivogius, Philalethes and others.[39] In the essay 'Observations on the growth of metals' Boyle mentions a 'metalline Seed or Ferment' whereby metals might be increased.[40] Many works contain comments upon, or references to the extraction of the sulphurs and mercuries of metals, yet quite without the type of condemnation which, according to the prevalent perception, Boyle heaped upon such notions in the *Sceptical Chymist*.[41] Indeed, the appendix to the second edition of the *Sceptical Chymist* contains so much matter which is thoroughly alchemical (especially regarding the mercuries of metals) that it calls for a reevaluation of the message and impact of Boyle's best-known work.[42]

Besides the Philosophers' Stone, Boyle sought out lesser alchemical desiderata, for example the extraction of a colourless, water-like liquid from common quicksilver, and mentions that he has, in fact, accomplished this feat.[43] To this category may also be added the reduction of glass into a malleable body, which Boyle claims allusively.[44] There are many examples of Boyle's experimental attempts to follow alchemical texts. His fascination with and frequent processes on vitriol are suggestive of a debt to those of Basilius Valentinus, whom he calls 'that great and candid Chymist', and to whom had been attributed an acrostic naming vitriol as the source of the Philosophers' Stone.[45] Finally, Boyle's famous 'redintegration' experiments on nitre, amber, turpentine, vitriol and other materials may stem from spagyric notions and Boyle's attempt to pursue such principles.[46] (According to the spagyric theory, a material was to be divided into its essentials, those essentials purified, and then recombined to reform the original material in an exalted form.)

Having catalogued these representative examples of Boyle's alchemical pursuits, they can now be condensed into a brief chronology demonstrating Boyle's long-term alchemical commitment. In the early 1650s, Boyle was intensely active in alchemy. During these years he made his first incalescent mercury (1652), prepared the *menstruum peracutum* (*ca.* 1654–5), followed Basilius Valentinus's cryptic text in an attempt to make the Stone, and read the works of other alchemists and iatrochemists including van Helmont and Paracelsus.[47] He carried on alchemical correspondence with the American alchemist George Starkey and Hartlib's son-in-law Frederick Clodius. At this time he was more than 'the amateur repeating elementary operations' as one biographer suggested; he was already tackling the *arcana maiora*.[48] Boyle's alchemical activity seems to have slackened around the end of the decade, and remained less intense through the 1660s. There are, nevertheless, still a fair number of alchemical laboratory operations in the Papers and a few pieces of alchemical corre-spondence.[49] By the 1670s, however, Boyle's alchemy was again vigorous, as marked by his composition of the *Dialogue on Transmutation* (early 1670s), the publication of its conclusion as the anti-elixir tract (1678), the revelation of the incalescent mercury (1675/6), the alchemical sentiments of *Producibleness* (1677), and the commencement of extensive correspondence with continental alchemists, which continued well into the 1680s. Voluminous laboratory records from the 1680s attest to Boyle's continued alchemical interest throughout that decade, at the end of which he successfully moved to have the Act against Multipliers repealed (1689).[50]

During these forty years, Boyle's motives for the study of alchemy changed and developed. During the active period of the 1650s, Boyle was in contact with the Hartlib Circle, within which alchemical projects were carried out by some of Boyle's correspondents, notably Clodius and Starkey, alias Eirenaeus Philalethes.[51] These associates would have fostered Boyle's alchemical activities. Writing at this time, Boyle compared 'Chymical Furnaces' to anatomical knives in regard to their utility in studying 'the Book of Nature'.[52] One achievement for which Boyle has been duly credited is his application of chemistry to the investigation of natural phenomena; alchemy, with its emphasis on separation, analysis and change, could have been interwoven into the same purpose.

But fully to explain Boyle's continued interest in alchemy requires an exploration of why any scholar, past or present, devotes his life to a chosen field. The choice of a field is contingent upon both internal and external forces. The latter of these can be addressed by a consideration of the historical influences – intellectual, social and political – of the thinker's environment; for example, what prevailing needs made a certain topic important, relevant, advantageous, or prestigious to pursue, under whose influence or tutelage the person develops and so forth. It is these factors which have traditionally been accorded the lion's share of attention simply because they are traceable and treatable by rational historical argument. The former influences, however, consist of some internal disposition towards a particular field, an intangible correspondence between searcher and search, which is not readily amenable to, or explicable by, the techniques of historical investigation. The simple fact that human beings fail to respond similarly to similar environments argues for the role of such internal forces. Clearly, to say that a figure pursued some particular subject because he found it interesting or engaging is no historical argument, yet it must be conceded that such considerations are still part (and perhaps no small part) of the choice of a field of study.

Boyle quickly favoured alchemical and chemical studies, finding great interest and satisfaction in them. Soon after his first sallies into the field, his letters reveal an almost palpable excitement. In a letter (23 July 1649) presumably to Samuel Hartlib and intended to be the introduction to the 'Invitation to Communicativeness', Boyle regrets having time 'snatcht from my new-erected furnaces'.[53] A month later (31 August 1649) Boyle enthusiastically informs his sister Katherine that

> Vulcan has so transported and bewitch'd mee that as the Delights I tast in it, make me fancy my Laboratory a kind of Elizium; so as if the Threshold of it possest the Quality the Poets ascrib'd to that Lethe their Fictions made men taste of before their entrance into those seates of Blisse, I there forget my Standish and my Bookes and almost all things.[54]

At the other end of his career, Boyle held quite a different view of alchemy. Drawing upon Bishop Gilbert Burnet's notes from an interview conducted during the last few years of Boyle's life, Michael Hunter has convincingly shown that by that time Boyle had linked alchemy closely with the spirit realm.[55] When Boyle began to adopt this further dimension of alchemy is still unclear; nonetheless, with this link in place, motives for the study of alchemy become more compelling for Boyle, and alchemy expands to special significance in his thought.

Traditional alchemy always held close links with the metaphysical and the supernatural; thus, as Boyle became more involved in traditional alchemy, and increasingly
convinced of its claims, he would have found such connections in alchemical texts. He
probably began his alchemy–spirit realm link with the unanimous testimony of the
adepti that alchemical knowledge came by God's revelation. Alchemists considered the
Philosophers' Stone as a *donum Dei*; it could not be prepared solely by the sweat of the
laborant, but required cooperative revelation by God. Biblically, God's messages often
came through angels or other spiritual means, thus these beings could likewise be the
imparters of alchemical truths. Indeed, Boyle's early alchemical collaborator George
Starkey claimed in a letter to Boyle (26 January 1651/2) that his knowledge of the
secrets of alchemy was revealed to him in a dream by a good spirit [*Eugenius*] sent by
God.[56] It was partially this sacred nature of alchemical knowledge which demanded
secrecy from the alchemists, as illicit communication of such secrets would (to use an
alchemical phrase) cast the pearls of divine revelation before the swine and, in the
words of Elias Ashmole, 'render one *Criminall* before *God*, and a *presumptuous violator*
of the *Caelestiall Seales*'.[57] This same consideration may likewise have helped provoke
Boyle's secrecy. The indiscretion of open alchemical communication might be
accounted the sin against the Holy Spirit (through Whom the knowledge had been
selectively imparted) which Boyle so feared.[58]

A fragmentary dialogue reveals one motivation for the study of alchemy at this
period to be a desire to contact rational spirits. There Boyle states that 'the acquisition
of the Philosophers-Stone may be an inlett into another sort of knowledge and a step
to the attainment of some intercourse with good spirits'.[59] This claim exceeds those
of nearly all alchemists, who, although they stressed the holy status of alchemical
knowledge and its metaphysical overtones, did not endow the *products* of their art with
supernatural/magical powers but rather emphasised alchemy's natural operation. Yet
Boyle's conceptions are not wholly isolated or novel, for a branch of alchemy which
sought supernatural alchemical products developed in England during the seventeenth
century. In this context appears the notion of 'supernatural Stones', more precious than
the alchemists' usual *summum bonum*, the natural, transmutatory Philosophers' Stone.
For example, Elias Ashmole describes a 'Magicall Stone' which allows the user to see
any person he wishes anywhere in the world and to understand the language of animals,
as well as an 'Angelicall Stone', which 'can neither be *seene*, *felt*, or *weighed*; but *Tasted*
only' and 'affords the *Apparition* of *Angells*, and gives a power of conversing with
them'.[60] An anonymous manuscript in the British Library (dated 1660) adds to these
the 'yet more stupendous, more immediately Divine' *Lapis Evangelicus* which
transmutes the whole soul into one of perfect charity and godliness, 'giveing a heart
larger than the Sand of the Sea'.[61] Boyle presumably had these rather recent developments in alchemy in mind when he wrote (or planned to write) his 'Of the supernatural
arcana pretended by some chymists', which he lists among his hermetic tracts.[62]
Exactly how much of this supernatural alchemy Boyle accepted or rejected is yet
unclear. Boyle may even have entertained the belief that spirits were responsible for
alchemical change itself. Hunter has found that Boyle brought Bishop Burnet along

with him to witness a projection; the presence of Burnet would be difficult to understand unless Boyle expected (or feared) some spirit activity at the time of the projection
itself.

It has often been noted that accounts of witchcraft and magic were valuable because,
if proven, they would demonstrate the existence of spirits against the assaults of
atheists. This was Glanvill's mission, and Boyle supported it.[63] Alchemy, if its objects
had the power to summon rational spirits, could provide a superior weapon. Whereas
witchcraft was notoriously difficult to demonstrate either in its operation or its effects,
the red powder of projection and its products could be demonstrated (though not
prepared) by anyone. Further, if spirits were involved in alchemical change, the
success of a spirit-assisted transmutation would provide an instance of immaterial,
incorporeal substances inducing changes in material, corporeal substances. This would
silence those who doubted that an incorporeal God could affect matter and that an
immaterial human soul could actuate the human body. Both points were on Boyle's
mind, as many fragments among the Boyle Papers present arguments refuting those
who deny the action of the incorporeal on the corporeal.[64] In reverse, if, as Boyle
suggested, angels were in fact attracted by the Philosophers' Stone by 'congruities
or magnatisms capable of involving them', this fact would be an instance of the
incorporeal being affected by the corporeal.[65]

On a wider scale, a spirit-related alchemy would tie together both spheres of Boyle's
thought by overlapping experimental philosophy on the one hand and theology on the
other. This sort of alchemy allowed for both the existence and operation of spirits and
the analysis of physical evidence. Ordinarily, Boyle employed mechanical arguments to
explain physical properties and changes without resorting to spiritual agencies. He
rejected, for example, Henry More's *Principium Hylarchicum*, or Spirit of the World, as
an explanation of natural phenomena (including Boyle's own pneumatic observations)
in favour of a mechanical explanation. Yet this was not a rejection of spiritual involvement *per se*. Rather, it was based upon the fact that More's incorporeal agent was
unnecessary.[66] Boyle 'heartily wish[ed] him [More] much success' in his 'grand and
laudible design . . . of proving the existence of an Incorporeal substance', even though
'our Hydrostaticks do not need it'.[67] Furthermore, the existence of More's irrational
operative spirit was questionable, unlike Boyle's rational spirits, whose existence was
proven by the testimony of Scripture. Alchemy was a field where spiritual involvement
might not be superfluous.

Boyle was concerned, especially in his later years, about the possible misapplication
of the New Philosophy, which, by succeeding in banishing occult explanations, might
be used to deny the existence of the spirit realm as well. Boyle was acutely aware that,
though the 'tumultuous Justlings of Atomical Portions of senseless Matter' could
explain the origin of forms and qualities, they could not explain the origin of the world
– although they could describe the composition of a vinous spirit, they could not
describe the composition of a rational, incorporeal spirit.[68] The concern over the
New Philosophy's possible tendency to atheism is well-known, and underlies much
of Boyle's *Christian Virtuoso*. Indeed, the preface of that work addresses the

inapplicability of mechanical views to 'Incorporeal and Rational Beings' while asserting the importance of studying them.[69] In sum, Boyle found himself in a potential dilemma: his conspicuous success in applying mechanical explanations to the world resulted in a tension with his theological commitments, as these new explanatory models threatened to distance spiritual agents (including God) from all contact with the natural world, possibly rendering them superfluous. Boyle's late-career view of alchemy would have helped bridge the gap between experimental philosophy and theology. Alchemy complemented and corrected the system of natural philosophy that Boyle supported. Alchemy was the interface between the rational, mechanical functioning of the corporeal world and the supra-rational, miraculous workings of the spiritual world.

As the foregoing examples have documented, Robert Boyle spent an enormous amount of time in the study and pursuit of traditional alchemy. Boyle's works abound with alchemical references and processes. He did not 'rapidly eschew alchemy and mysticism' as one writer claims, but rather continued his alchemical studies over a span of forty years, and what that writer labels as 'mysticism' became, in fact, increasingly important to Boyle in the latter part of his career.[70] Boyle's interests ranged from the preparation of the sulphurs and mercuries of minerals and metals to the confection of the Philosophers' Stone, of whose existence and transmutatory powers he was certain. In the earlier part of Boyle's career, alchemy functioned as a means of exploration and analysis, providing him with new experimental phenomena and occasionally with theoretical principles for the explanation of those phenomena. During the later part of his career, alchemy shifted its importance by virtue of the link he had made between it and the spirit realm. In that context, alchemy constituted a middle term operating between his two potentially adversarial philosophical commitments: his largely mechanical, corpuscularian world and his spiritual, theological world. The inclusion of Robert Boyle's alchemical pursuits and practices in our overall portrayal of him restores an important facet of his complex (but remarkably coherent) career and thought. Boyle needs to be viewed in this totality if the transformations in seventeenth-century science and Boyle's role in directing them are to be rightly appreciated and understood.

NOTES

1   L. T. More, *The Life and Works of the Honourable Robert Boyle* (New York: Oxford University Press, 1944), p. 214.

2   The following deal directly with Boyle's alchemy: L. T. More, 'Boyle as alchemist', *JHI* 2 (1941): 61–76, slightly altered as ch. xi of his *Life and Works* (n. 1), pp. 214–30; Muriel West, 'Notes on the importance of alchemy to modern science in the writings of Francis Bacon and Robert Boyle', *Ambix* 9 (1961): 102–14; L. M. Principe, 'The gold process: directions in the study of Robert Boyle's alchemy', in *Alchemy Revisited*, ed. Z. R. W. M. von Martels (Leiden: E. J. Brill, 1990), pp. 200–5; Michael Hunter, 'Alchemy, magic and moralism in the thought of Robert Boyle', *BJHS* 23 (1990): 387–410. See also A. J. Ihde, 'Alchemy in reverse: Robert Boyle on the degradation of gold', *Chymia* 9 (1964): 47–57; H. Fisch, 'The scientist as priest: a note on Robert Boyle's natural theology', *Isis* 44 (1953): 252–65.

3   See the closely related issue of Boyle's posthumous transformation into a rigid mechanist as described in A. Clericuzio, 'A redefinition of Boyle's chemistry and corpuscular philosophy', *Ann. Sci.* 47 (1990): 561–89, on pp. 561–3.

4   Marie Boas [Hall], *Robert Boyle and Seventeenth-Century Chemistry* (Cambridge University Press, 1958), pp. 102–6.

5   Boyle, *Origine of Formes and Qualities* (London, 1666), pp. 95–6 [*Works*, iii, 35]. (In the notes to this chapter, Boyle's writings have been cited from their original seventeenth-century editions; for the convenience of the reader, cross-references to the 1772 *Works* have been added in square brackets.)

6   *Ibid.* pp. 364–5 [*Works*, iii, 47].

7   The reconstructed and annotated *Dialogue on the Transmutation and Melioration of Metals* (which survives as scattered fragments among the Royal Society Boyle Papers) is currently being prepared for publication by the present author.

8   Boyle, *An Historical Account of a Degradation of Gold, made by an Anti-Elixir; A Strange Chymical Narative* (London, 1678) [*Works*, iv, 371–9]; see also Ihde, 'Alchemy in reverse' (n. 2). R.S. MS 198, fol. 143v, demonstrates the connection of *Degradation* with the dialogue.

9   Transcribed in *Works*, i, xxxvi. The original, as with many Boyle letters, although it was transcribed in the *Works*, was not among the papers given by Miles to the Royal Society or by Birch to the British Museum.

10 Boyle, *A Physico-Chemical Essay . . . touching . . . Salt-petre*, in *Certain Physiological Essays* (London, 1661), preface, [ii] [*Works*, i, 354].

11 Boyle, *On the Unsuccessfulness of Experiments* in *CPE* (n. 10), pp. 62–3 [*Works*, i, 354]; *The Sceptical Chymist* (London, 1661), pp. 39–40, 222, and 407 [*Works*, i, 475, 526, 578]; *Forms and Qualities* (n. 5), pp. 349–73 [*Works*, iii, 93–100]; partially repeated in *Of the Mechanical Origine and Production of Volatility*, in *Mechanical Origine of Qualities* (London, 1675), pp. 45–7 [*Works*, iv, 303].

12 Boyle, *On the Unsuccessfulness of Experiments*, in *CPE* (n. 10), p. 63 [*Works*, i, 332].

13 G. Ripley, *The Compound of Alchymie*, in *Theatrum chemicum Britannicum*, ed. Elias Ashmole (London, 1652; reprint edn, ed. Allen G. Debus, New York: Johnson Reprint Co., 1967), pp. 114, 136; M. Sendivogius, *Novum lumen chemicum*, in *Musaeum hermeticum* (Frankfurt, 1678), pp. 584, 586–7; E. Philalethes [G. Starkey], *Introitus apertus ad occlusum regis palatium*, in *Musaeum hermeticum*, pp. 682, 678; J. A. Siebmacher, *Wasserstein der Weisen* (Frankfurt and Leipzig, 1760), p. 44.

14 Boyle, *Sceptical Chymist* (n. 11), pp. 177, 407 [*Works*, i, 513, 578]; *Forms and Qualities* (n. 5), p. 363 [*Works*, iii, 96].

15 Basilius Valentinus, *Von dem grossen Stein der Uhralten* (Eisleben, 1599); the work became better known through the Latin edition: *Practica, cum duodecim clavibus et appendice de magno lapide antiquorum sapientum*, in Michael Maier, *Tripus aureus* (Frankfurt, 1618). An English edition first appeared in London in 1657 bound with Valentine's *Last Will and Testament*.

16 For a fuller account of this 'gold process', see Principe, 'The gold process' (n. 2).

17 Boyle, *Forms and Qualities* (n. 5), p. 360 [*Works*, iii, 96].

18 BP 25, pp. 203–5; *Of the Un-succeeding Experiments*, in *CPE* (n. 10), p. 68 [*Works*, i, 334].

19 Boyle, *Forms and Qualities* (n. 5), p. 358 [*Works*, iii, 95]; see also *Of the Un-succeeding Experiments*, in *CPE* (n. 10), p. 68 [*Works*, i, 334].

20 *Ibid.* p. 359 [*Works*, iii, 95–6].

21 BP 25, p. 295.

22 Boas [Hall], *Boyle and Chemistry* (n. 4), *passim*; *Robert Boyle on Natural Philosophy* (Bloomington: Indiana University Press, 1965), p. 276, see also p. 86; Shapin and Schaffer, *Leviathan and the Air-Pump*, esp. pp. 71, 78; Steven Shapin, 'The house of experiment in seventeenth-century England', *Isis* 79 (1988): 373–404; 'Pump and circumstance: Robert Boyle's literary technology', *Social Studies of Science* 14 (1984): 481–520; Jan Golinski, 'A noble spectacle: phosphorus and the public culture of science in the early Royal Society', *Isis* 80 (1989): 11–39.

23 L. M. Principe, 'Robert Boyle's alchemical secrecy: codes, ciphers, and concealments', *Ambix* 39 (1992): 63–74.

24 BP 17, fol. 60r. The alchemical nature of this obscure passage is clear from the mention of 'our Lithoi' and 'Tirsus of Nogirus', which translate to 'our silver' and 'spirit of mercury'. A complete decipherment list will be published in due course.

25 See Principe, 'Boyle's alchemical secrecy' (n. 23).

26 *The Correspondence of Isaac Newton*, H. W. Turnbull *et al.* (eds), 7 vols (Cambridge University Press, 1959–77), ii, 64. The letter is dated 27 August 1676 (N.S.).

27 Boyle, *Degradation of Gold* (n. 8), p. 17 [*Works*, iv, 379].

28 Boyle, 'Of the Incalescence of Quicksilver with Gold', *Phil. Trans.* 10 (1676): 515–33 [*Works*, iv, 219–30].

29 *Ibid.* p. 521 [*Works*, iv, 223]; BP 36, fol. 87.

30 Boyle, 'Incalescence' (n. 28), p. 521 [*Works*, iv, 223]. This reference is probably to E. Philalethes [G. Starkey], *Introitus apertus* (n. 13), pp. 658–9.

31 Boyle, 'Incalescence' (n. 28), p. 529 [*Works*, iv, 228].

32 Boyle, *Experiments and Notes about the Producibleness of Chymicall Principles*, published with the second edition of *The Sceptical Chymist* (Oxford, 1680), preface [*Works*, i, 590].

33 The letter, BL 6, fol. 99, was published by W. Newman, 'Newton's *Clavis* as Starkey's *Key*', *Isis* 78 (1987): 564–74; compare the plain text of this letter to the fanciful 'Allegorical description' published under Starkey's pseudonym Eirenaeus Philalethes, and cited by Boyle (n. 30). Laboratory replication of Starkey's process was carried out by the present author. A paper treating the incalescent mercury and its transmission from Starkey to Boyle and on to Locke, Oughtred and Newton is in preparation.

34 Boyle put the incalescent mercury through the manipulation cited by alchemists as appropriate for making the Stone, namely sealing 'the matter' hermetically in a 'glass egg' and digesting it for a long period. Boyle digested his mercury in this way for ten or twelve years! Boyle, *Experimenta et observationes physicae* (London, 1691), pp. 142–3 [*Works*, v, 599–600].

35 Boyle, 'Incalescence' (n. 28), p. 518 [*Works*, iv, 221].

36 *Ibid.* pp. 528–9 [*Works*, iv, 227–8].

37 Newton, *Correspondence* (n. 26), ii, 1–2; B. J. T. Dobbs, *The Foundations of Newton's Alchemy* (Cambridge University Press, 1975), pp. 194–6.

38 Boyle, *Producibleness* (n. 32), p. 214 [*Works*, i, 647]. Although the date of publication is 1680, the book was given the *imprimatur* on 30 May 1677.

39 Boyle, *On Celestial Magnets*, in *Tracts: Containing Suspicions about Some Hidden Qualities of the Air* (London, 1674) [*Works*, iv, 96–100]; Dobbs, *Foundations* (n. 37), p. 190; see also pp. 38–9, 153–5, 188.

40 Boyle, *Observations about the Growth of Metals in their Ore*, in *Tracts* (n. 39), esp. p. 23 [*Works*, iv, 79–84, esp. p. 84].

41 For example, Boyle, *Some Considerations touching the Usefulnesse of Experimental Naturall Philosophy* (Oxford, 1663), part I, p. 14 [*Works*, ii, 11]; *Producibleness* (n. 32), esp. pp. 117–242 [*Works*, i, 623–54].

42 This reevaluation has been begun by A. Clericuzio; see 'Carneades and the chemists: a study of *The Sceptical Chymist* and its impact on seventeenth-century chemistry', ch. 5, this volume.

43 Boyle, *Producibleness* (n. 32), pp. 234–7 [*Works*, i, 653]; *Possibility of the Resurrection*, in *Reason and Religion* (London, 1675), p. 19 [*Works*, iv, 198]; *Sceptical Chymist* (n. 11), pp. 134–5, 232 [*Works*, i, 501–2, see also p. 529].

44 See West, 'Importance of alchemy' (n. 2); see also Boyle, *Strange Reports*, in *Exp. Obs. Physicae* (n. 34), p. 9 [*Works*, v, 605].

45 Boyle, *On the Unsuccessfulness of Experiments*, in *CPE* (n. 10), p. 48 [*Works*, i, 324]; Valentine's acrostic was *Visita Interiora Terrae Rectificando et Invenies Occultum Lapidem.*

46 See Boyle, *A Physico-Chemical Essay . . . touching . . . Salt-petre*, in *CPE* (n. 10), [*Works*, i, 359–76]; *Forms and Qualities* (n. 5), pp. 250–69 [*Works*, iii, 61–5].

47 The first work published under Boyle's name bears reference to both van Helmont and Paracelsus as 'learn'd Expositors' of the 'Book of Nature'; Boyle, *Seraphic Love* (London, 1659), p. 56 [*Works*, i, 262]; see also A. Clericuzio, 'Robert Boyle and the English Helmontians', in *Alchemy Revisited* (n. 2), pp. 192–9.

48 More, *Life* (n. 1), pp. 67–8.

49 I am hesitant to ascribe degrees of activity to various periods based solely on the quantity of surviving papers. It is clear that the preservation of Boyle's *nachlass* was spotty at best, for it has been separated, recombined and purged. The loss of most of the Hartlib correspondence in the hands of Thomas Birch and Henry Miles is especially troubling. See Michael Hunter, *Letters and Papers of Robert Boyle: A Guide to the Manuscripts and Microfilm* (Bethesda, Md.: University Publications of America, 1991), pp. xviii–xix.

50 Maddison, *Life*, p. 176; Newton to Locke, 2 August 1691, Newton *Correspondence* (n. 26), ii, 217.

51 On Hartlibian alchemy, see Dobbs, *Foundations* (n. 37), pp. 62–80; on Eirenaeus Philalethes, see Newman, 'Newton's *Clavis*' (n. 33); 'Prophecy and alchemy: the origin of Eirenaeus Philalethes', *Ambix* 37 (1990): 97–115.

52 *Seraphic Love* (n. 47), p. 56 [*Works*, i, 262]. See also Clericuzio, 'Redefinition of Boyle's chemistry' (n. 3), pp. 564–6.

53 BL 6, fol. 3, printed in R. E. W. Maddison, 'The earliest published writing of Robert Boyle', *Ann. Sci.* 17 (1961): 165–73, on pp. 165–6.

54 BL 1, fol. 120 [*Works*, vi, 50].

55 Hunter, 'Alchemy, magic and moralism' (n. 2).

56 BL 5, fol. 133r; other contemporaneous alchemists comment on the necessity of help from good angels. See, for example, [Eusèbe] Renaudot, 'A conference concerning the Philosophers' Stone', in Samuel Hartlib, *Chymical, Medicinal and Chyrurgical Addresses* (London, 1655), pp. 101–12, on p. 112.

57 Ashmole, 'Annotations and discourses', in *Theatrum chemicum Britannicum* (n. 13), p. 439. The sentiment probably originates in R. Bacon, *Theatrum chemicum*, 6 vols, (Argentoratus [Strasburg], 1659; reprint edn, Torino: Bottega d'Erasmo, 1981), v, 856.

58 Hunter, 'Alchemy, magic and moralism' (n. 2), p. 399.

59 BP 7, fols 134–50, on fol. 138. This dialogue was first cited by Hunter, 'Alchemy, magic and moralism' (n. 2), pp. 397–8.

60 Ashmole, *Prologomena*, in *Theatrum chemicum Britannicum* (n. 13), sig. Bv.

61 British Library MS Sloane 648, fols 99–100, on fol. 99v.

62 R.S. MS 198, fol. 143v. No tract of such a tract remains, unless perhaps the fragmentary dialogue of n. 59 is part of it.

63 M. E. Prior, 'Joseph Glanvill, witchcraft and seventeenth-century science', *Modern Philology* 30 (1932): 167–93; *Works*, i, 57–9; a letter to Glanvill dated 18 September 1677.

64 BP 2, fols 100–3.

65 In the dialogue on spirits and alchemy, BP 7, fols 134–50 (see n. 59), fol. 135. Eleutherius later questions how 'a little powder that is as truely corporeall as pouder of post or of brick should be able to attract or invite incorporeall and intellegent beings' (fol. 142). Although Arnobius and Cornelius assert that Parisinus has 'sayd so much to weaken the arguments we are speaking of [that the Philosophers' Stone cannot so operate]' (fol. 144), the content of those promised responses is not extant.

66 R. A. Greene, 'Henry More and Robert Boyle on the spirit of nature', *JHI* 23 (1962): 451–74.

67 Boyle, *An Hydrostatical Discourse occasion'd by some Objections of Dr. Henry More* (London, 1672), p. 142 [*Works*, iv, 627–8].

68 Boyle, *Christian Virtuoso I* (London, 1690), pp. 9, 29–30 [*Works*, v, 514, 519].

69 *Ibid.* preface, sigs A4v–r [*Works*, v, 510].

70 Boas [Hall], *Boyle and Chemistry* (n. 4), p. 26.

Boyle's debt to corpuscular alchemy

WILLIAM R. NEWMAN

In the preface to his *Origin of Forms and Qualities*, Robert Boyle spells out his attitude toward contemporary chemists in the following terms: ' . . . though I have a very good opinion of chymistry itself, as it is a practical art; yet, as it is by chymists pretended to contain a system of theorical principles of philosophy, I fear it will afford but a very little satisfaction to a severe inquirer . . . '.[1] In this fashion Boyle distances himself from the theoretical principles underlying Paracelsian iatrochemistry and alchemy, while at the same time allowing for his appropriation of the alchemists' techniques in the laboratory. Despite Allen Debus's fundamental article of 1967, in which he showed that Boyle's *Sceptical Chymist* owed a major conceptual debt to Jan Baptist van Helmont, Boyle's dismissal of alchemical theory as an influence on his own work still remains largely unchallenged.[2]

In this chapter I will argue for a radically different position. It is my view that Boyle knew and absorbed the material theory of alchemy to such a degree that it forms a seamless part of his own work. I do not refer here primarily to Boyle's attitude toward the transmutation of metals, which is obviously derived from alchemy. Rather I have in mind his corpuscular theory, which lies at the very heart of Boyle's interpretation of the mechanical philosophy.[3] In order to substantiate this claim, I will first have to present the alchemical matter theory to which I refer. After that, I will show in some detail how Boyle appropriated not only alchemical processes, but the very corpuscular explanations of those processes employed by alchemists.

The corpuscular theory to which I refer came into being in the late thirteenth century, with the *Summa perfectionis* attributed falsely to the Ismāʿīlī alchemist Jābir ibn Ḥayyān, called in Latin 'Geber'. The *Summa* of 'Geber' has been called the 'Bible' of the medieval alchemists, and indeed its doctrines underlie the huge alchemical *corpora* attributed to Albertus Magnus, Arnold of Villanova and Ramon Lull, as I have shown in my book, *The Summa perfectionis of pseudo-Geber*.[4] Among the most influential doctrines of the *Summa* one finds: the so-called 'mercury alone' theory, according to which mineral sulphur is only a corrupter of metals, whereas the noble metals are really composed of almost pure mercury; the theory of three ascending degrees or orders of alchemical medicines, of which only the third effects real transmutation into a noble metal; and finally the theory that the metals, minerals and indeed elements are composed of insensible particles. All of these doctrines in their alchemical incarnation can be traced back through the early modern period and late Middle Ages, to their origin in the *Summa perfectionis*.

Now let me describe the corpuscular theory of matter delineated by the *Summa*. 'Geber's' theory of matter relies on a technical use of the Latin terms *pars*, *subtilis* and *grossus*, which may at first seem strange vehicles for the conveyance of a corpuscular philosophy. A *subtilis pars*, or 'subtle part', however, means simply a small particle or corpuscle, whereas a *grossa pars*, that is a 'gross part', is a large particle. The proper understanding of this terminology is absolutely crucial to Geberian alchemy. It is these tiny *partes* or particles, which are insensibly small whether they be gross or subtle, that make up the structure of the material world. Thus in describing the composition of the metallic principles mercury and sulphur, 'Geber' states the following:

> . . . we will say that each of these [principles] *in genere* is of very strong composition and uniform substance. This is so because the parts of the earth are united *per minima* – through the smallest – to the aerial, watery, and fiery parts in such a way that none of them can separate from the other during their resolution. But each is resolved with the other on account of the union that they mutually have received *per minima*.[5]

In this passage 'Geber' tries to explain why sulphur and mercury can be sublimed without leaving a residue, or, to put it another way, why these two metallic principles cannot be analysed by means of sublimation. The answer is that the smallest particles of the four elements making them up have bonded so tightly that mere fire cannot separate them. This is what the author means by a union *per minima* – through the smallest parts. Moreover, mercury and sulphur are composed of a 'uniform substance', meaning that the second order particles made up of all four elements are homoeomerous. I shall now show that this homoeomerity does not refer merely to the relative proportion of fire, air, water and earth in each corpuscle of mercury, for example, but also to the size of the mercurial particles.

In his long chapter on the alchemical technique of sublimation, 'Geber' relates that 'When fire elevates something, it always elevates the subtler parts; therefore it leaves the grosser ones behind.'[6] A heterogeneous mixture may therefore be separated by means of sublimation, because the fire will be incapable of raising the 'grosser', that is larger, and hence heavier particles. It is precisely because mercury is composed solely of very subtle particles – that is very small particles of uniform size – that they all pass over during sublimation without the deposition of much, if any, residue.

Such 'subtlety' of particles is also invoked by 'Geber' to explain the passage of a substance from wet to dry. When describing the formation of metals within the earth, the *Summa* states that these start out as volatile fumes, but that continual cooking separates the smaller particles until a liquid metallic substance, and finally a solid, remains behind. Interestingly, 'Geber' puts this into the form of a universal precept –

... we [here] pass on a general rule to you, dearest son – that the thickening of any moisture does not come about unless first there be an exhalation from the humid of its most subtle parts, and a preservation in the humid of its grosser parts.[7]

In part this is an extension of 'Geber's' foregoing remarks on sublimation, inasmuch as here too he assumes that smaller particles sublime off easier than larger ones. But here such a separation is explicitly linked to the passage from moisture to solidity, and indeed liquidity itself is here made a function of small particle size.

'Geber' finds a further use for 'subtlety' in his explanation of the great specific gravity of gold. Let us quote him here *in extenso*:

> ... the very subtle substance of quicksilver led forth to fixation, and the purity of the same, along with the very subtle, fixed, unburning matter of sulphur, is the whole essential matter of gold ... Because [the quicksilver in gold] had subtle, fixed parts, its parts could therefore be much compressed; and this was the cause of its great weight.[8]

Here the *Summa* tells us that gold is made up of tiny, uniform particles of mercury and sulphur, whose very small size allows them to be tightly packed together, thus avoiding the creation of large interstices. It is this absence of big gaps between its particles that makes gold so heavy. The same idea is generalised in a slightly later passage of the *Summa*:

> The cause of great weight, therefore, is the subtlety of the substance of the bodies, and its uniformity in essence. For through this, their parts can be compacted, since nothing may separate them; and the compaction of the parts is the induction of weight, and its perfection.[9]

The *Summa* could not be less equivocal here. 'Subtle parts', that is, small, uniform particles, can be tightly compacted without leaving the large pores that would exist between large ones. Thus, metals composed of such fine corpuscles will obviously be heavier than those made up of larger ones. Let us pause for a moment and see where 'Geber's' matter theory has taken us. First he has told us that matter is of a particulate nature, made up of particles varying in size. Secondly, he informed us that such particles, when mixed *per minima*, that is through the smallest of them, combine to form a so-called 'strong union', as exists in mercury and sulphur. Thirdly, we learned that smaller particles are sublimed more easily than larger ones, and fourthly that liquidity is a function of small particle size. Fifthly, 'Geber' asserted that great weight is the product of many small uniform particles being closely packed into a constricted union. Let us now observe how 'Geber' uses these and related ideas to explain in detail the alchemical procedures of sublimation, calcination and transmutation.

Sublimation, as stated before, is the process of separating smaller particles from larger ones by means of heat. But it is more complicated than that. As Geber says when discussing the sublimation of sulphur and arsenic sulphide,

If . . . arsenic or sulphur be sublimed, it is necessary that these be sublimed by a
remiss fire. For since they have subtle parts conjoined uniformly to the gross, their
whole substance would ascend without any purification, but rather burnt and
blackened.[10]

The idea here is simply that native sulphur or arsenic sulphide is a mixture of
heterogeneous particles. If a large fire is used in sublimation, all the particles will
sublime, but a partial separation of them can be effected by employing a small heat,
since only the small particles will then rise up. But 'Geber' then introduces another
technique of separation, which he calls the application of *faeces* or 'dregs'. By adding a
given substance to the *sublimandum*, the alchemist can further prevent the grosser
heterogeneities from sublimating along with the subtle. As he says himself,

. . . in order for anyone to separate the dirty, earthy substance, it is necessary to find
ingenious techniques . . . [such as] cleansing with an admixture of dregs. For the
admixture of dregs combines with the gross parts and holds them down at the base
of the aludel, not permitting them to ascend.[11]

The addition of dregs is necessary because these substances, meaning primarily
calces of the metals and so-called 'prepared salts' combine with the gross particles in
the *sublimandum*, making them bigger still. Then the principle that fire more easily
raises smaller particles can be utilised to greater effect in the retention of the gross
heterogeneities at the bottom of the sublimatory. One can see then how 'Geber' has
given a corpuscular explanation not only to sublimation *per se*, but to the artisanal
techniques employed by alchemists in refining sublimation to its maximum
sophistication.

Let us now pass to the *Summa*'s description of calcination. In modern terms,
calcination is simply the oxidation of a metal by exposure of it to heat in an oxygen
rich environment. To 'Geber', however, calcination occurs when a porous metal is
penetrated by fire particles, which then force out its corpuscles of sulphur. What is left
after a metal has been calcined is therefore primarily mercury which has been deprived
of its own humidity and 'fixed'. In order to make this more vivid, let us once again
quote the *Summa*:

. . . on account of their great quantity of earthiness, and large measure of burning,
fleeing sulphureity, [iron and copper] are easily brought into a calx. This occurs
because the continuity of the quicksilver is broken, due to the abundant earthiness
mixed into the substance of the said quicksilver; therefore a state of porosity is
created in them, through which the sulphureity, passing, can escape. Fire, on
account of this reaching up to it, can burn and elevate it. Through this it is also
given that the parts become rarer, and are converted into cinder because of the
discontinuity due to this rarity.[12]

'Geber's' explanation of calcination is entirely corpuscular. Such easily oxidised
metals as iron and copper are composed of mercury and sulphur, along with an earthy
impurity. The gross particles of earth 'break the continuity' of the mercury, to use

'Geber's' expression, hence creating large pores. The fire of calcination can communicate with the metals' sulphur via these pores; upon doing so, the fire drives out the sulphur, leaving discontinuous particles of dried out mercury in a cinereal, powdery state.

Let us pass now to a description of 'Geber's' theory of alchemical transmutation, in order to show that it, too, has a corpuscular basis. The final chapters of the *Summa* describe the three orders of alchemical medicines in some detail. To simplify, a first order medicine is one that effects only superficial, illusory change, such as the production of brass from calamine and copper. A second order medicine, on the contrary, works fundamental change, but only on one of the five specific differences constituting the species of 'perfection', namely lack of superfluities such as earthiness or excess sulphur, white or yellow colour, the melting point of gold or silver, resistance of the medicine to heating and the specific weight of gold or silver. An example of a second degree change employed by 'Geber' himself is that of calcination, which burns out excess sulphur or earth, but leaves all the other imperfections of a metal intact. A third order medicine, finally, works real change in all of these specific differences at once, thus producing real gold or silver from a base metal.[13]

How is it that these different medicines vary in their effectiveness? To simplify somewhat, 'Geber' states that nothing distinguishes a second order medicine from that of gold except 'the greater subtilisation of the particles' in the latter.[14] This suggests that there is no essential difference between the medicines of different rank, but that they are distinguished by particle size alone. This is confirmed by the *Summa*'s penultimate chapter, in which the repeated sublimation of mercury is used to attain a state of continually increasing 'subtlety'. The leading idea is that the particles of mercury must be cleansed of all their earthy *grossities*: this will result in a product of uniformly tiny corpuscles. In order for the perfecting mercury to enter into the *profundum* or deep structure of a base metal, so that it can engage in a true *mixtio per minima*, the mercury particles must be exceedingly small. Only then will they be able to infiltrate the pores of the metal and unite with the metallic microstructure at its most fundamental level. Such perfected mercury will not only penetrate and combine with the mercurial part of a base metal; it will also cover that mercurial component and protect it from burning. The *Summa* has a technical term for this covering: it is called a *palliatio* of the metal's pre-existent material.[15] The mercurial medicine added to the base metal provides a *defensio* of the metal's so-called 'pure substance' by means of this 'cloaking'. Hence the transmuted part of the metal upon further heating will divest itself of its excessive sulphur and earth, with the result that only pure gold or silver will remain.

Although I have given a simplified version of 'Geber's' transmutational theory here, it should be clear enough that he views transmutation itself, just as sublimation and calcination, in corpuscular terms. The *Summa* does not consider the essential differences of the metals to consist in unique substantial forms imposed by the Creator. The specific differences of the metals, such as weight and colour, are simply products of their differing components, mercury, sulphur and impure earth, and the interplay of these principles in turn can be explained in terms of their action on the metallic microstructure. The hylomorphic pair of matter and form are never invoked by the *Summa* to explain the behaviour or transmutation of metals: at every opportunity, 'Geber' introduces explanations of these phenomena in strictly corpuscular terms.

I have dwelt so long on the medieval *Summa* because of the unfamiliarity of its doctrines to most scholars. Although I do not wish to assert that Boyle derived corpuscular ideas *directly* from the *Summa*, there is no need to make that claim in order to assert that Boyle's chemistry has a 'Geberian' component. Due to its massive influence on the fundamental texts of Western alchemy one could hardly escape the *Summa*'s doctrines, even in the seventeenth century. Let us now therefore show some of the points at which Boyle betrays a 'Geberian' influence.

Boyle's writings on matter theory often display a surprising tendency to employ a language unrelated to that of atomism proper. Like 'Geber', Boyle often speaks of 'subtle parts' when he means 'small corpuscles', and of 'gross parts' when he means 'large corpuscles'. In the *History of Fluidity*, for example, Boyle argues that the ambient air contains various heterogeneous 'small parts' in a state of constant suspension.[16] Similarly, wine contains 'subtile and spirituous parts' that remain in solution among the 'more aqueous parts'.[17] In *Of the Producibleness of Chemical Principles*, similarly, Boyle speaks of mercury impregnated with the 'subtile parts of another body . . . '.[18] In the first section of the second part of *The Usefulness of Experimental Natural Philosophy*, on the other hand, Boyle refers to the detoxification of a drug 'by the avolation of some subtile parts driven away by the heat of the boiling water . . . '.[19] These subtle parts reappear in the same text when Boyle remarks that 'an unperceived recess of a few subtile parts' may convert wine into vinegar.[20] Finally, again in *The Usefulness*, Boyle refers to the greater activity of 'the subtile parts of a mixed body' when compared to the bigger corpuscles.[21]

The same sort of usages occur when Boyle relates the characteristics of 'gross parts', as in *The History of Fluidity*, where Boyle, like 'Geber', argues that fluidity is a function of small particles and solidity of large ones. To Boyle, the corpuscles making up a solid or 'firm' body are 'commonly somewhat gross'.[22] Such 'gross parts', being heavier than their 'subtle' counterparts, tend to sink in a fluid medium. Thus, in describing an alchemical recipe given to him by an acquaintance, Boyle states in *The Usefulness* that a substance left sealed up on a dunghill for forty days will in part liquefy, while the 'grosser parts' will subside.[23]

It is highly likely that Boyle's 'subtle' and 'gross parts' are nothing more than the *subtiles partes* and *grossae partes* of the *Summa*. A further example of such shared vocabulary is the expression 'mixture *per minima*', the characteristic phrase employed by the *Summa* to describe an intimate mixture of particles at the microlevel. Boyle uses this term at least five times in *Sceptical Chymist* alone, to describe the mixture of his *minima* or *prima naturalia*. Thus he says that 'impure silver and lead' fused together 'will thereby be colliquated into one mass, and mingle *per minima*, as they speak . . . '.[24] Likewise, the silver component of a gold–silver alloy dissolved in aqua fortis will dissolve, whereas the gold will precipitate, revealing that the gold 'retained its nature, notwithstanding its being mixt *per minima* with the other'.[25] Similarly, when ashes and sand unite to form glass, or antimony and iron yield *regulus martis*, this is a 'union *per minima* of . . . two . . . bodies of differing denominations.'[26] On the other hand, since the Philosophers' Stone converts lead to gold, 'it seems not always necessary, that the bodies, that are put together *per minima*, should each retain its own nature'.[27] Finally, in describing the 'exquisite blending' of mercury and sulphur to form cinnabar, he speaks of them as being 'mingled *per minima*'.[28] In all these passages the same idea is expressed – namely that of a mixture through the smallest particles of a substance or substances.

Although such terms as *subtilis* and *grossa pars* may have existed both within and outside of the alchemical context, further evidence of Boyle's debt to alchemical corpuscularianism may be found if we examine his explanations for particular chemical phenomena, such as the inability of fire to separate the components of sulphur and mercury, the workings of sublimation, the cause of higher specific weight in some metals than in others, and the transmutation of a base metal into a noble one. These are among the processes that we examined from the *Summa perfectionis*, and the reader will be surprised, I believe, at the similarity of Boyle's corpuscular explanations to those of 'Geber'.

First let us consider Boyle's explanation for the fact that sulphur is sublimed by the application of heat rather than broken down into more basic components. In *The Sceptical Chymist* Boyle states the following:

> . . . what can a naked fire do to analyze a mixt body, if its component principles be so minute, and so strictly united, that the corpuscles of it need less heat to carry them up, than is requisite to divide them into their principles?[29]

Here Boyle states that sulphur is sublimed rather than decomposed because its principles are 'strictly united' and so small that the fire sublimes them before they can be broken down. The reader will recall that 'Geber' explained the phenomenon of sulphur's volatility on precisely these same assumptions, namely that the elements making it up were united through the smallest in a *fortissima compositio*, that the particles of sulphur proper were very minute, and that smaller corpuscles sublime more easily than larger ones.

Let us then pass to Boyle's further comments on sublimation. In his *Physico-chemical Essay . . . touching . . . Salt-petre*, Boyle describes the decomposition and redintegration of nitre, in the following terms:

> . . . the most fugitive parts of concretes may in spight of their natural mobility be detained in bodies by their union and texture with the more sluggish parts of them, among which those lighter and more active ingredients may be so entangled, as to be restrained from avolation.[30]

In other words, by recombining the volatile component of nitre with its fixed residue, one can in fact render the volatile part fixed. What is happening on the microlevel is that the smaller volatile components get trapped among the bigger non-volatile ones and hence lose their escape path.

Boyle reiterates this explanation in a different context in his *Experiments and Notes About the Mechanical Origin and Production of Volatility*. In describing a number of methods for lessening the volatility of substances, Boyle says the following:

> The fourth way of producing fixity in a body, is by putting to it such an appro-priated additament, whether fixed or volatile, that the corpuscles of the body may be put among themselves, or with those of the additament, into a complicated state, or entangled contexture.[31]

Let us now think back to 'Geber's' explanation of the use of 'dregs' in sublimations. The addition of 'prepared salts' and other compounds holds back the volatile component by combining with its smaller particles, or, to use a term more closely related to the Latin, by 'comprehending' them. In both authors we see the same explanation for a fixed agent 'holding back' a volatile one: the smaller particles are restrained by coming into combination or contact with the larger ones.

So far we have seen a rather striking agreement between Boyle's and 'Geber's' explanations of sublimation in general and the action of 'dregs' in particular. A similar pattern emerges when we inspect Boyle's explanation for high specific weight in metals. In *Producibleness*, Boyle states the following:

> I considered too, that it might be justly demanded, whence mercury itself, as well as whence metals, derive their great ponderosity; and I see not, why it may not be said, that both the one and the other owe it to the solidity, and close order, of the corpuscles they consist of.[32]

Here Boyle explains the great weight of metals and especially mercury, in terms of the close-packing of their corpuscles. The reader cannot help but be reminded of the *Summa*'s parallel passage, where 'Geber' said that the cause of 'great weight in general' was the *densatio* – the close-packing – of many minute particles. One could argue that no other explanation was available to the two authors, but this is hardly true: a standard scholastic analysis explained such phenomena in terms of relative proportions of fire, air, water and earth.

Having examined Boyle's corpuscular terminology, his treatment of sublimation and his description of metallic specific weight, let us now consider his views on trans-mutation. Here I shall restrict myself to two textual *loci*, the first of them provided by Boyle's work *Producibleness*. In his usual fashion, Boyle describes the alchemical production of a philosophical mercury there while dismissing the alchemists' own theoretical explanation of how the process works. As he says,

> I do not wonder to find, that divers philosophical Spagyrists themselves, before they proceed to more intimate preparations of mercury, order it to be several times previously incorporated and sublimed with acid salts or sulphurs, and then revived with alkalies. Since without examining their grounds it may be said, according to mechanical principles, that by diligent commixtures the

mercury is divided into exceedingly minute, if not invisible, globules, or such like parts.[33]

Here Boyle declines to comment on the alchemical theory behind the practice of repeatedly subliming mercury with various salts or sulphur as a first step to the preparation of a philosophical mercury. And yet the attentive reader will remember that this very process was described already in the *Summa perfectionis*, and with the same theoretical interpretation. The *Summa* recommended the multiple sublimation of mercury with 'prepared salts' in order to render the mercury particles ever more subtle, so that it could penetrate into the *profundum* of a base metal. Only when the mercury had been broken down into very tiny corpuscles could a true *mixtio per minima* of it and the base metal occur.

We should therefore be wary when Boyle usurps an alchemical process while dismissing its alchemical interpretation. The same process can be seen in his famous description of transmutation in *Forms and Qualities*. In describing the existence of a 'tincture' or 'soul of gold', Boyle concedes that 'some such thing may be admitted'.[34] He then distances himself from the 'wild fancies' of the 'chymists' which are as unintelligible as the Peripatetic substantial forms. Nonetheless, Boyle continues,

> I would not readily deny but that there may be some more noble and subtile corpuscles, that being duly conjoined with the rest of the matter whereof gold consists, may qualify that matter to look yellow, to resist aqua fortis, and to exhibit those other peculiar phaenomena that discriminate gold from silver.[35]

Once again, we find Boyle accepting an alchemical phenomenon, while ostensibly rejecting its alchemical interpretation. In its stead Boyle offers a corpuscular explanation, according to which certain specially 'subtle' corpuscles enter into the metallic structure of the metal and add the properties of gold. Despite Boyle's disclaimer, the reader of the *Summa* will at once note that Boyle has given an excellent summary of 'Geber's' corpuscular theory of *palliatio*. According to the *Summa*, mercury particles should be sublimed repeatedly so that they could be reduced in size and made capable of ingress into a base metal. Upon entering the *profundum* of the metal, its most intimate structural level, these particles united *per minima* with the metal's intrinsic mercury, and provided the *palliatio* necessary to make the metal resistant to flame and corrosives. The same mercury added the requisite specific weight and colour of gold, being a medicine of the third order.

This brief look at Boyle's transmutational theory finishes our comparison of Boyle and 'Geber'. We have compared the two authors' corpuscular terminologies, as well as their explanations of such fundamental phenomena as sublimation, varying specific weight and transmutation. It is therefore time to draw some conclusions. First, despite the obvious parallels, let me repeat that I am not arguing that Boyle derived these elements of his corpuscularianism directly from the *Summa*. 'Geberian' alchemy had already become so widespread by the seventeenth century that this supposition would be unnecessary in order to establish the *Summa*'s influence.

This said, Boyle's contemporaries were nonetheless still reading the *Summa* and understanding its corpuscular matter theory. One such author was Kenelm Digby, who lists his corpuscular forbears, whom he calls atomists, in his *Two Treatises* published in 1644. After listing some of these authorities, Digby states 'The same doe Hippocrates and Galen: the same, their Master Democritus; and with them the best sort of Physitians: the same doe Alchymistes, with their Master Geber; whose maxime to this purpose, we cited above . . . '.[36] The maxim to which Digby refers relates to 'Geber's' theory of *mixtio per minima*, showing that Digby was indeed cognisant of the *Summa*'s corpuscularism.

Although Boyle regarded Digby highly as a corpuscular philosopher, I do not wish to infer from this that Boyle, like Digby, had made a careful reading of the *Summa perfectionis*. It is not necessary to assume Boyle's familiarity with the *Summa* in order to maintain a knowledge of 'Geberian' alchemy. Boyle's early friend and co-worker George Starkey, in his Philalethan text *De metallorum metamorphosi*, presents a transmutational theory that is in all its essentials an elaboration of the one given by the *Summa*.[37] But Starkey got most of this theory not from the *Summa*, but rather from the late fourteenth-century *Epistola ad Thomam de Bononia* by Bernardus Trevirensis. Bernard's letter contains a full exposition of 'Geberian' alchemy that adopts the 'mercury alone' theory and most importantly the corpuscularism of the *Summa*, while ostentatiously rejecting 'Geber' himself as an obscurantist.[38]

Yet another avenue by which 'Geberian' alchemy may have reached Boyle is through the intermediary of Frederick Clodius, Samuel Hartlib's son-in-law. Clodius and Benjamin Worsley engaged in an extended polemic on the proper origin of the Philosophers' Stone in which the latter 'denied . . . that Quicksilver is the matter'.[39] Clodius, in his response, notes that he agrees with 'Geber, [in] saying that the matter of the arcanum is not mercury of the vulgar, or its whole substance, but rather its essence and the subtle power extracted therefrom'.[40] In another letter, perhaps to Boyle himself, Clodius further identifies himself with 'Geber', saying,

> . . . with Geber I believe that the mercury of the wise is quicksilver, not the vulgar sort, but extracted from that, for this is not mercury in its own nature, nor in its whole substance, but the medial and pure essence of that, which draws its origin from it [i.e. from quicksilver], and is created therefrom.[41]

A second, equally important, point to make is that the 'Geberian' matter theory cannot explain many of the salient elements in Boyle's own corpuscular philosophy. Nowhere does 'Geber' say anything about the 'figure' or shape of his corpuscles. Size is the all-important quality of the 'Geberian' corpuscle, and so we can expect an influence from the *Summa* only when Boyle is preoccupied with that particular characteristic. In the same vein, 'Geberian' alchemy has nothing to say about motion at the microlevel. Hence the second of Boyle's two 'great catholic principles, matter and motion' is not imported from 'Geberian' alchemy.

A third and final point is that the alchemy of 'Geber' had been absorbed not only by Digby, Philalethes and Clodius, but by more central figures in the Aristotelian tradition of *minima naturalia* as well. Thus Daniel Sennert, for example, speaks of mixture *per minima* in reference to the *minima naturalia*, or smallest bodies, existing in nature.[42] Like Digby, Sennert was incorporating alchemy with more mainstream Aristotelian traditions. Hence some of Boyle's alchemically oriented corpuscularism may derive from sources such as Sennert rather than works more centrally placed in the tradition of alchemy. This does not vitiate my thesis, however, since Sennert himself openly relied on the *Summa perfectionis*,[43] as indeed did other scholastic commentators such as Bartholomaeus Keckermann.[44] If Sennert served to transmit the ideas of 'Geber' to Boyle, the influence of 'Geber's' corpuscular alchemy was diminished by that not a whit.

When these points have all been conceded, the fact remains that many of Boyle's corpuscular interpretations of phenomena and processes related to alchemy derive from the alchemical sources themselves. Boyle scholars, having been blinded by Boyle's many aspersions cast upon the spagyrists, have failed to recognise the distinction to be made between Paracelsianism and the corpuscular alchemy of 'Geber'. The reigning inability to distinguish among alchemical traditions has made it possible to ignore the alchemical – but not Paracelsian – substratum that underlies many of Boyle's corpuscular ideas. Perhaps now that alchemy is being integrated into the legitimate concerns of the history of science, this misperception will be allowed to change.

NOTES

1 *Works*, iii. 13.
2 Allen G. Debus, 'Fire analysis and the elements in the sixteenth and the seventeenth centuries', *Ann. Sci.* 23 (1967): 127–47. See, however, the writings cited by Lawrence Principe's 'Boyle's alchemical pursuits', ch. 6, this volume.
3 Boyle himself was at times quite clear about the relationship between his corpuscularianism and 'the mechanical philosophy'. In the preface to his *Physico-chemical Essay . . . touching . . . Salt-petre*, *Works*, i, 356, he says the following: ' . . . the corpuscular [philosophy is] . . . an account of the phaenomena of nature by the motion and other affections of the minute particles of matter. Which because they are obvious and very powerful in mechanical engines, I sometimes also term it the mechanical hypothesis or philosophy.'
4 William R. Newman, *The Summa perfectionis of Pseudo-Geber* (Leiden: E. J. Brill, 1991), pp. 193–210.
5 *Ibid.* p. 322.
6 *Ibid.* p. 359.
7 *Ibid.* p. 279.
8 *Ibid.* pp. 473–4.
9 *Ibid.* p. 545.
10 *Ibid.* pp. 361–2.
11 *Ibid.* p. 362.
12 *Ibid.* pp. 427–8.
13 *Ibid.* pp. 546–8.
14 *Ibid.* p. 573.
15 *Ibid.* pp. 566–70.
16 *Works*, i, 382.
17 *Ibid.*
18 *Works*, i, 646.
19 *Works*, ii, 123.
20 *Ibid.* p. 170.
21 *Ibid.* p. 234.

22 *Works*, i, 401.
23 *Works*, ii, 147–8.
24 *Works*, i, 479.
25 *Ibid.* p. 504.
26 *Ibid.* p. 506.
27 *Ibid.* p. 508.
28 *Ibid.* p. 567.
29 *Ibid.* p. 483.
30 *Ibid.* p. 368.
31 *Works*, iv, 310.
32 *Works*, i, 639.
33 *Ibid.* p. 645.
34 *Works*, iii, 95–6.
35 *Ibid.*
36 Kenelm Digby, *Two Treatises: In the One of Which, the Nature of Bodies; in the Other, the Nature of Mans Soule, is looked into* (London, 1645), p. 425.
37 Eirenaeus Philalethes, *Philalethae tractatus de metallorum metamorphosi*, in *Bibliotheca chemica curiosa*, 2 vols, ed. J. J. Manget (Geneva, 1702), ii, 676–85.
38 Bernardus Trevirensis, *Bernardi Trevirensis ad Thomam de Bononia medicum Caroli Octavi Francorum regis responsio*, in Manget, *Bibliotheca* (n. 37), ii, 399–408. For Bernard, see Lynn Thorndike, *History of Magic and Experimental Science*, 8 vols (New York: Columbia University Press, 1923–58), iii, 611–27.
39 Hartlib Papers, University of Sheffield, 42/1/3, dated 'Dublin. 31. Octob. 1654'. See also Worsley's letter, evidently to Clodius, in Hartlib Papers 42/1/27.
40 Hartlib Papers, Sheffield, 42/1/37: '9. Quid sibi velit Geber, dicens, Materiam Arcani non esse mercurium vulgi, seu ejus totam substantiam, sed essentiam & subtilem vim ex eo extractam.'
41 Hartlib Papers, Sheffield, 16/1/7, dated 'Londini 4 Julii Anno 1654' – ' . . . cum gebro credidi, quot Sophorum mercurius est Argentum vivum, non tamen vulgare, sed ab eo extractum, scilicet mercurius ille [non] est argentum vivum in sua natura, Nec in tota sua substantia, sed media et pura ejus essentia quae ab illo originem sumsit, et ab illo creata est.'
42 Daniel Sennert, *De chymicorum cum Aristotelicis et Galenicis consensu et dissensu liber* (Wittenberg, 1629), pp. 212, 392, 394, *et sparsim*.
43 *Ibid.* Sennert refers to 'Geber' as an alchemical authority on pp. 12 and 396.
44 Bartholomaeus Keckermann, *Systema physicum, septem libris adornatum* (Hanover, 1623), p. 606.

Boyle and cosmical qualities

JOHN HENRY

Boyle published his *Tracts about the Cosmical Qualities of Things* in 1671 (although its imprimatur is dated 3 November 1669), as a kind of sequel to the *Origin of Forms and Qualities* of 1666, but unlike the latter work these tracts have attracted very little scholarly attention. If they have been noticed at all they have been given a seemingly unproblematic mechanistic interpretation. Boyle defines 'systematical or cosmical' qualities as those qualities of a body which do not derive from the sizes, shapes and motions of its constituent particles, but

> depend upon some unheeded relations and impressions which those bodies owe to
> the determinate fabrick of the grand system or world they are part of.

'A system [of the world] so constituted as ours is', Boyle wrote, 'whose fabrick is such that there may be divers unheeded agents, which, by unperceived means, may have great operations upon the body we consider', can give rise to special attributes of that body.[1] It is by no means immediately clear what Boyle has in mind here, but, relying upon standard assumptions about the nature of the mechanical philosophy and about Boyle's status as a leading mechanist, it might seem reasonable to suppose that Boyle is merely extending the discussion in the *Origin of Forms and Qualities* about the power of a key to open a lock.[2]

Boyle's point in his discussion of locks and their keys is that the power of a key to open a lock becomes the 'main part of the notion and description' of that key. The result is that the ability of the key to open the lock is 'looked upon as a peculiar faculty and power in the key', even though there is no real or physical entity corresponding to this power. The key is simply a specially shaped piece of metal designed to be able to turn through the wards of a lock and draw back its bolt. The power of the key, in other words, derives from the arrangement of the system of the lock, and this power, therefore, could be said to be 'a systematic quality' of the key (though Boyle does not use this phraseology in the earlier work). A stark physical description of the shape and size of the key is, of course, perfectly true, and in a purely physical sense is all there is to be said about the qualities of the key. Nevertheless, Boyle points out, such a description misses the crucial point about it – its ability to open a particular lock. The systematic quality of the key, therefore, should be acknowledged as an important element in its make–up.[3]

According to this interpretation, then, systematic or cosmical qualities are to be seen as complex extensions of this idea, based on the lock–key analogy, which was first put forward, after all, in Boyle's preceding book. The analogy in the *Origin of Forms and Qualities* was applied to comparatively simple natural examples, the solubility of gold in *aqua regia*, the poisonous properties of ground glass and the way bodies appear to be coloured. It would seem that the aim of the tract on *Systematical or Cosmical Qualities* was merely to indicate how the mysterious qualities of some bodies may similarly be seen as epiphenomena of the way they interact with their surroundings. In spite of seemingly occult appearances and experiences, Boyle wished to explain, the real physical qualities of these bodies derive ultimately from the mechanical affections of size, shape and motion.

If Boyle's *Cosmical Qualities* has failed to attract any special attention, it is almost certainly due to the fact that he expounds his views on these qualities in such a way that there seems to be no significant advance on what was explained in the *Origin of Forms and Qualities*. The examples he gives in the course of the essay are easily explained in mechanistic terms. A wedge will not cleave unless it is hit by a hammer, a knife will not attract a needle unless excited by a magnet. A smoothly polished slab of marble can be used to pick up another piece of polished marble, even one heavier than itself, because the pieces of marble adhere 'by virtue of the fabrick of the world, which gives the ambient air fluidity and weight'. Other examples tell of the magnetisation of a piece of iron by heating it and letting it cool while the iron is orientated north–south, or of the change of a piece of paper from opacity to translucence merely by soaking it with oil or 'even a fit kind of grease'. The speed of a ship under sail can be increased by wetting the sail, we are told, and common tartar which stays dry in the air and is only dissolved in water with difficulty is changed by moderate calcining, so that it spontaneously turns into that liquour which chemists call tartar per deliquium. Glass which is heated and then cooled too quickly cracks because 'the subtle bodies that are in it' cannot find a passage of escape, but if it is cooled slowly the glass is able to settle into the appropriate texture.[4]

So, if there is nothing here that could not have been included in the earlier work on the *Origin of Forms and Qualities*, why does Boyle want to give them a special designation as cosmical qualities? A closer look makes clear that, whatever our own superficial responses to *Cosmical Qualities* might be, Boyle himself believes that there is more to be said than anything which was covered in the *Origin of Forms*:

> I have in the *Origin of Forms* touched upon this subject already, but otherwise than I am now about to do: for whereas that which I principally (and yet but transiently) take notice of is, that one body being surrounded with other bodies, is manifestly wrought on by means of those among whom it is placed; that which I chiefly in this discourse consider, is the impressions that a body may receive, or the power it may acquire from those vulgarly unknown or at least unheeded agents by which it is

affected, not only upon the account of its own peculiar texture or disposition, but by virtue of the general fabrick of the world.[5]

This is still puzzling, however. With the exception of the example of magnetising an iron bar by cooling it in alignment with the Earth's magnetic flux, it is difficult to see why the general fabric of the world has to be invoked to explain Boyle's specific examples. Boyle himself describes the action of oil on paper in terms of its action on the pores in the paper, straightening and widening them to allow light particles through; similarly water on a sail causes the fibres of the cloth to swell and make the sail less permeable by the wind. He has already given an account of the adherence of two polished marble plates in his *Continuation of New Experiments . . . Touching the Spring and Weight of the Air* (1669) without feeling the need to explain the fluidity and weight of the air in terms of 'the fabrick of the world', and the behaviour of tartar, though inexplicable to Boyle and his contemporaries in any precise way, surely did not demand an explanation on a cosmic scale.

Certainly, all these examples cannot simply be explained in terms of what Frederick J. O'Toole has referred to as 'non-relational inherent properties', that is to say, in terms of the sizes, shapes and motions of the primary corpuscles of nature and the 'texture' or precise way in which these primary corpuscles are combined to constitute the particular bodies in question. It seems clear enough that these examples can only be explained by recourse to what O'Toole calls 'non-inherent relational properties'.[6] But these relational properties are, as O'Toole's analysis makes clear, simply those properties which Boyle did describe in the *Origin of Forms*, namely those which depend upon, or are functions of, the 'peculiar texture or disposition' of the relevant bodies, whose situations make it clear that they are 'manifestly wrought on' by one another. It seems that we have to go beyond O'Toole's account if we wish to understand what Boyle meant when he suggested that there are powers and capacities of bodies which can only be understood

> not barely upon the score of these qualities that are presumed to be evidently inherent in it, nor of the respects it has to those other particular bodies to which it seems to be manifestly related, but upon the account of a system so constituted as our world is.[7]

It is my contention that Boyle intended in *Cosmical Qualities* to provide hints towards what was effectively a new concept of qualities but failed to fully explicate what he meant by them. Furthermore, it is quite possible that Boyle's 'failure' in this regard was not accident but design. In what follows, therefore, I hope to show that there are sufficient indications in *Cosmical Qualities* that Boyle did have in mind something which is rather less compatible not only with O'Toole's analysis of the matter theory of the mechanical philosophy, but also with all standard accounts of the mechanical philosophy.[8] In addition I will try to reconstruct just what it was that Boyle only hinted at and to offer some tentative suggestions as to why he chose to obscure his own conception.

In the very opening paragraph of *Cosmical Qualities* Boyle seems to undermine any suggestion that such qualities are merely to be seen in terms of the key–lock analogy. The qualities of most bodies, he declared, 'do for the most part consist in relations, upon whose account one body is fitted to act upon others or disposed to be acted upon by them'. He then gives clear examples of the key–lock format, such as the ability of quicksilver to dissolve gold and silver. As we have already seen, however, he immediately tells us that his intention in his new book is to go beyond this to consider qualities which derive from the fact that bodies belong to 'a system so constituted as our world is'. The fabric of our world, Boyle goes on,

> is such that there may be divers unheeded agents, which, by unperceived means, may have great operations upon the body we consider, and work such changes in it, and enable it to work such changes on other bodies, as are rather to be ascribed to some unheeded agents than to those other bodies with which the body proposed is taken notice of.[9]

The real subjects of the tract, then, not brought out by the examples, are these 'unheeded agents' which work 'by unperceived means'. This is confirmed in the opening paragraph of *Cosmical Suspicions*, described as an 'Appendix' to the discussion of cosmical qualities, which suggests that 'there may be . . . peculiar sorts of corpuscles that have yet no distinct name, which may discover peculiar faculties and ways of working'.[10] It is perfectly clear from this statement that what Boyle is proposing cannot be fitted into our standard conceptions of the mechanical philosophy. The corpuscles themselves are said to have their own unique faculties and ways of working. These are obviously not inert particles which work only by impact and other contact actions to give the appearance of having powers and faculties, like the power to turn a lock, or the faculty of pulsification of the heart and arteries.

The inescapable conclusion is that these qualities are occult, and this is confirmed by Boyle's efforts to forestall criticism of his new conception:

> least you should think that under the name of Cosmicall Qualities I should introduce Chimaeras into Naturall Philosophy I must betimes advertise you, that you will meet with divers Particulars in the following Discourse, fit to show that these Qualities are not meerly fictitious qualities: but such, whose Existence I can manifest, not only by considerations not absurd, but also by real Experiments and Physical Phaenomena.[11]

The vocabulary here suggests that Boyle thought of cosmical qualities as occult qualities. In the rhetoric of the new philosophy, the occult qualities all too frequently invoked by scholastic philosophers were dismissed as 'fictitious' or as mere 'Chimaeras'. The experimental philosopher could escape these charges, however, by insisting that the qualities he discussed could be made manifest by their effects. When Leibniz described Newton's gravitational attraction as 'a chimerical thing, a scholastic occult quality', Newton rebutted the charge by insisting that gravity was a *manifest* quality whose cause only was occult.[12] It would seem from Boyle's apology that cosmical qualities were not, therefore, to be seen as mere epiphenomena of the mechanical arrangement of the world system (analogous to the power of the key to turn a lock). If they were, Boyle would scarcely have felt the need to enter into the rhetoric of occult versus manifest qualities. The fact that he did discuss cosmical qualities in these terms is a clear indication that he considered them as occult qualities whose reality could be demonstrated, experimentally, by the manifest nature of their effects.

Boyle's tract on *Cosmical Qualities* and its appendix cannot consistently be regarded as a work which adheres to strictly mechanistic precepts. Although the specific examples which Boyle gives in *Cosmical Qualities* do seem compatible with mechanistic readings, Boyle continually intimates that what he really intends to 'excite [our] curiosity and attention about', as he puts it,[13] are 'unheeded agents' which work by 'unperceived means', and which are hitherto unknown particles acting not merely by impact but by various occult means. The putative existence of such unknown particles is one of Boyle's main considerations in developing his notion of cosmical qualities. There are 'certain subtle bodies in the world', he suggests, 'that are ready to insinuate themselves into the pores of any body disposed to admit their action, or by some other way affect it', and it is the action of these subtle bodies which gives rise to the cosmical qualities of the bodies into which they have entered.[14] These agents are, Boyle proposes, 'unobserved sorts of effluvia in the air', which are as unknown now as the 'swarm of steams moving in a determinate course betwixt the north and the south' were before Gilbert demonstrated that the Earth was a giant magnet.[15] Although this last image might seem to suggest that Boyle is returning to standard mechanistic explanations in terms of streams (or steams) of effluvia, it is more likely, in view of his talk of 'unperceived means' of action and 'peculiar faculties and ways of working', that Boyle was thinking of magnetism as an exemplar of something which was not, *pace* Descartes, entirely amenable to strictly mechanistic explanations. As Boyle wrote for his *Disquisition about the Final Causes of Natural Things*,

> There are a great many things which . . . cannot with any convenience be immediately deduced from the first and simplest principles; namely, matter and motion; but must be derived from subordinate principles; such as gravity, fermentation, springiness, magnetism etc.[16]

By the same token, as he wrote in an intended 'Essay of various degrees or kinds of the knowledge of natural things':

> [it would] be backward to reject or despise all explications that are not immediately deduced from the shape, bigness and motion of atoms or other insensible particles of matter . . . [for those who] pretend to explicate every phenomenon by deducing it from the mechanical affections of atoms undertake a harder task than they imagine.[17]

Further support for this interpretation can be seen in the otherwise inexplicable fact that Boyle also speculates, in *Cosmical Suspicions*, that

> there may be a greater number even of the more general laws than have yet been distinctly enumerated, so I think that when we speak of the established laws of nature . . . they may be justly and commodiously enough distinguished; some of

them being general rules that have a very great reach, and are of greater affinity to laws more properly so-called, and others seeming not so much to be general rules or laws, as the customs of nature in this or that peculiar part of the world; of which there may be a greater number, and those may have a greater influence on many phaenomena of nature than we are wont to imagine.

A little later Boyle reiterates this extraordinary idea, saying,

I have some time suspected that there may be in the terrestrial globe itself, and the ambient atmosphere, divers, whether laws or customs of nature, that belong to this orb, and may be denominated from it, and seemed to have been either unknown to, or overseen by both scholastical and mathematical writers.[18]

Leaving aside the interesting proposed distinction between laws and customs of nature, it seems undeniable that Boyle is expressing dissatisfaction here with the extremely restricted, 'distinctly enumerated' laws of nature of Descartes' *Principia philosophiae*. Certainly, Descartes is the principal 'mathematical writer' that Boyle has in mind, and it seems clear that Boyle has been led by his own 'notions and observations' to doubt whether all physical phenomena can be explained by Descartes' three mechanistic laws of nature and his seven rules of impact.[19]

Whether it is true or not that Boyle's chosen examples of cosmical qualities do not adequately represent what he really had in mind, there is little we can do to explicate his meaning. What we can do, however, albeit to a limited extent, is to try to uncover what Boyle was thinking of when he wrote of those 'unheeded agents', those 'peculiar sorts of particles' with 'peculiar faculties and ways of working' which were held to produce cosmical qualities in other bodies. Boyle himself provides a clue to the nature of these agents when he tells us in chapter 2 of *Cosmical Qualities* that there are three main bodies in the fabric of the world 'whose more unobserved operations' have the most influence on cosmical qualities. These three are the subterraneal parts of the globe, the stars and aether about them and the atmosphere or air we live in.[20] He then tells us that experimental observations on each of these are to be dealt with in other tracts. A moment's inspection reveals that he does not mean the other tracts published with *Cosmical Qualities*. These deal only with the temperature of subterraneal and submarine regions, together with some hearsay reports about the bottom of the sea. If we wish to discover what these bodies are, therefore, we will have to search elsewhere in Boyle's output.

For guidance let us take those three bodies or groups of bodies in turn: the subterraneal parts of the globe, the atmosphere around us and the stars and aether. With regard to the first of these, Boyle has little to say beyond a few scattered speculations. In *Cosmical Suspicions* itself he points to the evidence for 'slow internal change' in 'the mass of the earth', such as the change of magnetic variation, and concludes that

there may be agents that we know not of, that have a power to give the internal parts of the terrestrial globe itself a motion; of which we cannot yet certainly tell according to what laws it is regulated, or so much as whether it be constantly regulated by certain laws or no.[21]

Similar speculations are reiterated in the appendix to *Hidden Qualities of the Air*, 'Of celestial and aerial magnets':

> there may be in those vast internal parts of the earth, whose thin crust only has been here and there dug into by men, considerable masses of matter, that may have periodical revolutions, or accensions, or eustations, or fermentations, or, in short, some other notable commotions, whose effluvia and effects may have operations yet unobserved, on the atmosphere, and on some particular bodies exposed to it; though these periods may be perhaps either altogether irregular, or have some kind of regularity differing from what one would expect.[22]

Boyle also suggests that the great irregularity observed in the weight of the atmosphere must derive 'for the most part from subterraneal steams'; the very variable and inconstant phenomenon of luminosity in the sea may also derive from 'some cosmical law or custom of the terrestrial globe'.[23] 'Some subterraneal changes' are also suspected as the cause of new epidemic diseases (he mentions syphilis and rickets) which 'invade whole countries (and sometimes greater portions of the earth)'.[24] Elsewhere, however, Boyle suggests that subterranean effluvia can make a region more healthy. 'For in some places', Boyle writes,

> the air is observed to be much more healthy, than the manifest qualities of it would make one expect: and in divers of these cases I see no cause, to which such a happy constitution may more probably be ascribed, than to friendly effluvia sent up from the soil into the air.[25]

The 'subterraneal parts of the globe' only really appear in Boyle's speculations, it seems, as a source for putative effluvia in the atmosphere which are held to have otherwise inexplicable effects. This brings us, therefore, to the second of Boyle's suggested influences on cosmical qualities, the atmosphere around us. On this topic the obvious source is Boyle's *Suspicions about the Hidden Qualities of the Air*. We can tell we are on the right track from the fact that Boyle opens with another apology for discussing occult qualities. Boyle tells us that this tract was originally intended for a collection on the natural history of the air but he now published it separately, even though

> if I had more consulted my own Reputation, I should least of all have suffer'd this Title to appear, there being none of the rest, that was not less conjectural.[26]

One of the most significant of these conjectures for our purposes is the suggestion that some bodies 'may be Receptacles, if not also Attractives of the Sydereal, and other Exotic Effluviums that rove up and down in our Air'.[27] Here we have the ingredients hinted at in *Cosmical Qualities*, exotic effluvia which operate in an occult way. It would seem that Boyle wished to prove the truth of his comment in *Cosmical Suspicions* that 'it [is] not time misspent to consider whether there may not be other, and even unobserved sorts of effluvia in the air'.[28]

The first such attractive to spring to Boyle's mind is the so-called 'Philosophical Magnet' of the alchemists, which is said to attract and corporify the Universal Spirit or the Spirit of the World. Boyle dismisses these ideas as 'abstrusities', but goes on to say that we may make use of the word 'Magnet' in this context 'without avowing the receiv'd Doctrine of Attraction'.[29] What follows is a typically Boylean piece of trying to have one's cake and eat it (or denying occult qualities and recurring to them). We are first of all told that when Boyle uses the word 'Magnet' he does not mean 'a body that can properly attract our foreign Effluviums', but merely a body that is 'fitted to detain and join with them, when . . . they happen'd to accost the magnet'. But the following paragraph throws all such caution to the wind:

> And, without receding from the Corpuscularian Principles, we may allow some of the bodies we speak of, a greater resemblance to Magnets, than what I have been mentioning. For not only such a Magnet may upon the bare account of adhesion by Juxtaposition or Contact, detain the Effluviums that would glide along it, but these may be the more firmly arrested by a kind of precipitating faculty that the Magnet may have in reference to such Effluviums; which, if I had time, I could illustrate by some Instances; nay I dare not deny it to be possible, but that in some Circumstances of time or place one of our Magnets may, as it were, fetch in such steams as would indeed pass near it, but would not otherwise come to touch it. On which occasion I remember, I have in certain cases been able to make some bodies, not all of them Electrical, attract (as they speak) without being excited by rubbing, &c. far less light bodies, than the Effluviums we are speaking of.[30]

Boyle concludes that he has hereby demonstrated that without meddling in 'the mystical theories of the chymists', 'one may discourse like a Naturalist about Magnets of Celestial and other Emanations that appear not to have been consider'd, not to say, thought of, either by the Scholastic, or even the Mechanical, Philosophers'.[31]

The main theme, then, of *Hidden Qualities of the Air* is that various phenomena (usually chemical) are actually brought about by the 'operations of the Air', or rather by some previously unremarked effluvia in the air which operate not merely by force of impact, as would be required by Cartesian dogma, but by some other means. In another example, Boyle writes:

> The Difficulty we find of keeping Flame and Fire alive, though but for a little time, without Air, makes me sometimes prone to suspect, that there may be dispers'd through the rest of the Atmosphere some odd substance, either of a Solar, or Astral, or some other exotic, nature, on whose account the Air is so necessary to the subsistence of Flame; which Necessity I have found to be greater, and less dependent upon the manifest Attributes of the air, than Naturalists seem to have observed.[32]

But we do not have to leave it there. Boyle returned time and again to experiments on the nature and usefulness of the air throughout his career, and in one of his very last works, *The General History of the Air* of 1692 (finally prepared for the press by John Locke after Boyle's death), Boyle presented another speculative account of the operations of the air. Moreover, because this essay deals with 'Celestial Influences or Effluviums in the Air' it brings us to the third of Boyle's candidate bodies for producing cosmical qualities, the stars and the aether.

The essay in question, however, which is explicitly described as an 'Apology for Astrology', was not written by Boyle himself. A Latin translation of this work has been recently discovered by Antonio Clericuzio among the Hartlib Papers, and it is clear from this that the author was in fact Boyle's friend and colleague, and the man who introduced Boyle to the pleasures and usefulness of natural philosophy, Benjamin Worsley.[33] We know from a letter of John Locke's to Boyle, written in October 1691, and from Locke's 'Advertisement of the Publisher to the Reader' prefixed to the *General History of the Air*, that Boyle himself compiled the materials which were to be included in this miscellany, but we still have to be circumspect in our assessment of the significance of this particular contribution to the work.[34] We cannot assume that Boyle wholeheartedly endorsed every detail of Worsley's account, but it seems clear from the fact that Boyle chose to include it that he was, at least for the most part, in sympathy with his late friend's attempt to explain the nature of celestial influences on the Earth. Other, much shorter, extracts from other works (by Pascal and Casati, for example) suggest that Boyle's *History of the Air* is not an indiscriminate hotchpotch of all things pertaining to the air, but a presentation only of those things which Boyle took to be particularly significant in attempts to understand the nature of the air. Boyle wrote in the preface of his conviction 'that scarce any Truth, whether Historical or Doctrinal, that relates to so important a subject as the Air, is unfit to be preserv'd', so it seems reasonable to conclude that Worsley's 'Apology for Astrology' seemed to Boyle to contain at least some important truths.[35] Indeed, in the preface Boyle went so far as to say that his 'collections of particulars' were intended to afford 'some specimens of what I should have thought requisite to do upon particular subjects, if I would have ventured upon such a task, as to write a natural history of the air'.[36] Moreover, the significance of Worsley's essay for Boyle would seem to be underscored by the fact that in 1691 he remembered the relevance to a natural history of the air of a speculative essay written by his former colleague in 1657.

Although it has to be admitted that the work is very unlike anything Boyle himself wrote, it is still possible to discern resonances with some of Boyle's own more speculative pieces. In particular, I believe it is as close as we can get to gleaning what Boyle had in mind when he spoke of the role of the stars and aether about them in the operations of cosmical qualities, and what he had in mind when he wrote of 'odd substance[s], either of a Solar, or Astral, or some other exotic, nature' dispersed through the atmosphere. In *Cosmical Suspicions* Boyle wrote that we have very little knowledge of 'what communication' the Earth has 'with the other Globes we call Stars, and with the Interstellar parts of Heaven', but he expressed an intention to 'elsewhere make it probable that there may be some Commerce or other'.[37] Boyle never did return to this topic, but perhaps it is true to say that, at the end of his life, he allowed Worsley to speak for him. Accordingly, we shall now turn to a brief look at the main themes of Worsley's posthumous contribution to Boyle's *General History of the Air*.

Worsley begins by insisting that 'it may . . . still be certain, that these celestial bodies . . . may have a power to cause such and such motions, changes, and alterations [on the Earth]' in spite of the

superstition and paganism incident to this kind of doctrine . . . the imposture,
ignorance and want of learning generally observed in the persons professing this
kind of knowledge, . . . the manifest mistakes and uncertainty that there is in
predictions of this nature, . . . and the inexplicableness of the way or manner they
come to affect one another.[38]

Predictably perhaps, Worsley's professed certainty is based upon 'undeniable
experiments' and 'undoubted observations', but he clearly believes there is scope for
development of these ideas, since he calls for a theory of the planets 'upon such grounds
as are indubitably demonstrable', without which 'it is impossible we should assert their
[the planets'] several aspects, and the mutual influences and virtues they have . . . one
upon another'.[39]

The principal means by which the planets and stars operate on the Earth is light, but
there are evidently many different kinds of light:

we say, that every planet hath its own proper light: and as the light of the sun is one
thing, the light of the moon another, so every planet hath its distinct light, differing
from all the other. Now we must either say, that this light is a bare quality and that
the utmost use and end of it is only to illuminate; or that there is no light but is
accompanied further with some power, virtue or tincture, that is proper to it.;
which if granted it will inform us then, that every light hath its own property, its
own tincture and colour, its own specific virtue and power.

It follows, therefore, that

those eminent stars and planets, that are in the heavens, are not to be considered by
us as sluggish, inenergetical bodies, or as if they were set only to be as bare candles
to us, but as bodies full of proper motion, of peculiar operation, and of life.

There is no suggestion of action at a distance in this, Worsley points out, because the
virtue or power of the planet is transmitted with 'its light, and is the real property of its
light'.[40]

Worsley's next move is to explain how these virtues affect the Earth. Firstly, the air
is affected, being 'moved, stirred, altered, and impressed by these properties, virtues
and lights, as penetrating each part of it'.[41] Secondly, various spirits which are held to
be constituents of all mixed bodies, are also affected:

as our spirits, and the spirits likewise of all mixed bodies, are really of an aerious,
etherial, luminous production and composition; these spirits therefore of ours, and
the spirits of all other bodies, must necessarily no less suffer an impression from the
same lights, and cannot be less subject to an alteration, motion, agitation, and
infection through them and by them, than the other, viz the air: but rather as our
spirits are more near and more analogous to the nature of light than the air, so they
must be more prone and easy to be impressed than it.[42]

These spirits now take on an indispensable role in the workings of the world, and are
said to be 'the only principles of energy, power, force and life, in all bodies wherein they
are', but these in their turn, as we have just seen, derive their properties from the truly

cosmic phenomenon of light. Here then we seem to have a clear statement of what Boyle only hinted at when he wrote of 'the impressions that a body may receive, or the power it may acquire from those vulgarly unknown or at least unheeded agents by which it is thus affected . . . by virtue of the general fabrick of the world'.[43] Magnetism, for example, can be seen as a cosmic quality because

> the earth, which is not only enlightened, warmed, cherished and fructified by the power, virtue, and influence of the sun, but hath its proper magnetical planetary virtue also fermented, stirred, agitated and awakened in it; . . . and together with this magnetic planetary property of the earth, which is stirred and raised by the sun, are awakened also the seminal dispositions, odours and ferments that are lodged in and proper unto, particular regions or places.[44]

Again, this reference to properties of the Earth which are seemingly confined to particular regions or places, echoes the concern in *Cosmical Suspicions* that, as well as laws of nature, we should also consider 'the customs of nature in this or that peculiar part of the world'.[45]

It is perhaps worth considering briefly some of the likely sources for Worsley's 'Celestial Influences or Effluviums in the Air'. One possible source is Francis Bacon's speculative philosophy, described by its modern rediscoverer, Graham Rees, as Bacon's semi-Paracelsian cosmology. It is a feature of this philosophy that all mixed bodies contain a spirit of an aerious, aetherial or luminous nature, and it is these spirits which are solely responsible for the activity and operations of those bodies, their matter being totally inert. Rees himself has argued that Bacon's speculative philosophy was almost entirely overlooked by Bacon's contemporaries and immediately succeeding generations, and it is possible that the similarities between Bacon and Worsley simply derive from a common interest in, and knowledge of, Paracelsian ideas. But, equally, it may be that Rees has been too hasty in reaching his conclusion.[46] *Thema coeli*, the work in which Bacon most clearly expounded these ideas, was published by Isaac Gruter in 1653, and it seems unlikely, given their admiration for Bacon, that Worsley and others in Hartlib's circle would not have seized the opportunity to read it.

The idea that the air is penetrated throughout by these celestial lights, and that these in turn affect our spirits, is also reminiscent of ideas in the medical tradition. William Harvey, for example, wrote in his *De generatione animalium* (1651) that there was in blood a spirit which corresponded to the element of the stars. 'The blood therefore is a spirit', he went on,

> by reason of its most excellent powers and virtues. It is also celestial because in that spirit is housed a nature that is the soul, analogous to the element of the stars, and this is something bearing an analogy with the heavens as being the instrument and deputy of heaven.[47]

If there is a similarity here, there is also a clear parallel with Boyle's talk of 'sydereal', 'celestial', 'solar, or astral' effluvia in the air in his *Hidden Qualities of the Air*.

Perhaps the most obvious source of Worsley's speculations in his 'Apology for Astrology' is the Neoplatonic tradition usually referred to as 'light metaphysics'. In this tradition light was regarded as the formal and efficient cause by which God brought about the Creation. The ability of light to diffuse itself instantaneously in all directions was invoked by light metaphysicians as the fundamental formative action in the universe. As light spread itself out from the centre it took with it the unformed *prima materia* which God had created. Since the diffusion of light defined the three dimensions of space, it also gave form to the unformed matter, so producing all the individual and separate bodies of the universe. Light, then, is the first corporeal form of the universe, from which all other forms are derived, and it is the original physical cause of all development and change in the universe. It seemed to follow, for those in this tradition, that the behaviour of light was the model of all causation in the physical universe. This belief gave rise to the rich tradition of geometrical optics and to the commonplace belief in the validity of astrological influence. The universal formative agency of the autodiffusive light ensured that all things in the universe were unified with everything else. Everything in the universe contains light within it, and this internal light is ultimately the origin and principle of all motion. Robert Grosseteste, a leading writer in this tradition, even described all qualitative change in terms of light within a body being emitted and light outside a body being absorbed. Of course, this internal 'essential' light is not the same as visible light. Sensible light was regarded as only one manifestation of the emanative power of *lux*. Al-Kindi, author of *De radiis stellarum*, one of the most influential works in the tradition (also known as *De theoria artium magicarum*), believed that everything in the world produced its own rays of influence like a star. Moreover, these rays were proper to each thing and different from all others, like Worsley's lights.[48]

Benjamin Worsley's essay of 'Celestial Influences or Effluviums in the Air' fits easily into the light metaphysical tradition, and Boyle himself may well have been sympathetic to this speculative tradition. Boyle pointed out in his *Aerial Noctiluca* (an essay about a luminescent body), for example, that 'light was the first corporeal thing the great Creator of the universe was pleased to make, and . . . he was pleased to allot the whole first day to the creation of light alone, without associating with it in that honour any other corporeal thing'.[49] Moreover, Boyle certainly entertained the possibility that the sun and planets might have specific influences 'here below' which are 'distinct from their heat and light': 'On which supposition', he wrote, 'it seems not absurd to me to suspect, that the subtil, but corporeal, emanations even of these bodies may . . . reach to our air, and mingle with those of our globe in that great receptacle or rendesvous of celestial and terrestrial influences, the atmosphere.'[50] Having said this, Boyle was quick to forestall any assumptions that he was proposing a simple mechanistic account of celestial influences:

> the very small knowledge we have of the structure and constitution of globes, so many thousands or hundreds of thousands of miles remote from us, and the great ignorance we must be in of the nature of the particular bodies, that may be presumed to be in these globes (as minerals and other bodies are in the earth) which in many things appear of kin to those that we inhabit, . . . this great imperfection, I say, of our knowledge may keep it from being unreasonable to imagine, that some,

if not many, of those bodies and their effluxions, may be of a nature quite differing from those we take notice of here about us, and consequently may operate after a very differing and peculiar manner.[51]

When Boyle wrote in *Cosmical Qualities* that there were three bodies in the world which, by unobserved operations, affect the cosmical qualities of things and so enable them to 'work . . . changes on other bodies' which they could not otherwise do, he had in mind the subterraneal parts of the globe, the atmosphere and the stars and aether around them. I have suggested that *Hidden Qualities of the Air* and Worsley's 'Celestial Influences or Effluviums in the Air' offer us our best chance of discovering Boyle's meaning. It would seem that Boyle wished to make a plea for the possibility that there were effluvia of an exotic nature, in the air, the heavens and the Earth, 'of peculiar operation', which were capable of altering Earthly bodies to give them relational properties such as gravity, magnetism, fermentation and other chemical properties. Accordingly, Boyle wished to infer, there are more laws of nature than just the three which Descartes enumerated and which (together with the derivative seven rules of impact), the influential French philosopher insisted, could explain all physical phenomena.[52] Moreover, although some of these laws may be general in their effects, there may also be local laws, or 'customs' as Boyle calls them, where phenomena which are far from universal may nonetheless be designated 'natural' and understood in terms of secondary causes operating in a regular way.

Cosmical Qualities was intended, therefore, to provide further support for Boyle's conviction that the natural world could not easily be understood in terms of the constant and undeviating operations of a few simple laws of nature. As Marie Boas Hall and Steven Shapin have both pointed out, Boyle sought to avoid conceptions of nature which were idealised and subject to mathematical abstraction.[53] The nature of the physical world was such, according to Boyle, that it could not properly be analysed in mathematical terms. The problem was not simply one of ensuring purity of samples, say, for the replication of chemical experiments, nor was it merely a recognition of the inevitability of experimenter error, or of discrepancies in instrumentation and apparatus. As Shapin has shown, Boyle believed that reality was structured in such a way that mathematical analysis was categorically inappropriate. The air, for example, was not 'a mere elementary body' but a 'great receptacle or rendesvous of celestial and terrestrial effluviums', whose precise composition varied from time to time and place to place. Similarly, Boyle believed that ores and minerals could differ in specific gravity and other qualities because 'neither nature nor art' gave 'all the productions, that bear the same name, a mathematical preciseness, either in gravity or in other qualities'.[54] Boas Hall and Shapin concentrated in their analyses on Boyle's chemical speculations, but in *Cosmical Suspicions* we see him extending his doubts about the regularity of nature to cosmology:

> when I likewise consider the fluidity of that vast interstellar part of the world wherein these globes swim, I cannot but suspect there may be less of accurateness, and of constant regularity, then we have been taught to believe, in the structure of

the universe, and a greater obnoxiousness to deviations than the schools, who were
taught by their master Aristotle to be great admirers of the imaginary perfections of
the celestial bodies, have allowed their disciples to think.[55]

Boyle's attitude to the concept of natural laws derives from two major preconceptions. First, Boyle's theology, as McGuire and others have shown, was
undeviatingly voluntarist.[56] For Boyle, it was God's supreme power and His arbitrary
will which were paramount. The idea that the complexity of the world system could be
reduced to a few simple laws of nature was associated, in Boyle's mind, with intellec-
tualist theologies in which, as far as Boyle and all other theological voluntarists were
concerned, God's power was circumscribed by the need to conform to the dictates of
man-made reason and the expectations of man's ethical assumptions. The rational
reconstruction of the system of nature, which Descartes no less than scholastic
philosophers indulged in, seemed to imply that God could not have done things
otherwise. Boyle's anti-rationalist experimental philosophy went hand in hand with his
theological voluntarism. Whatever regularity there was in the system of the world was
imposed by God arbitrarily by His absolute power, not as a result of God following the
dictates of reason. Accordingly, we could only discover the regularities of the world by
experience.

Secondly, as Steven Shapin has argued, the idealised, rationally reconstructed world
picture was liable, in Boyle's eyes, to be seen as a false authority for the truth of natural
philosophical claims.[57] The sceptical chemist, like many of his countrymen, was deeply
distrustful of reason. The trouble with reason, as Jeremy Taylor pointed out in his
Ductor dubitantium, was that it

> is such a boxe of quicksilver that it abides no where; it dwells in no settled mansion;
> it is like a doves neck, or a changeable taffata; it looks to me otherwise than it looks
> to you who do not stand in the same light that I doe: and if we inquire after the laws
> of Nature by the rules of our reason, we shall be uncertain as the discourses of the
> people, or the dreams of disturbed fancies.[58]

The claim to a philosophy or a theology based on 'reason' could be made by Hobbists
and other atheists, by Roman Catholics, or by some highly distasteful Protestant sect.
As Boyle himself wrote in his *Considerations about the Reconcileableness of Reason and
Religion*,

> there are very few conclusions, that we make, or opinions that we espouse, that are
> so much the pure results of our reason, that no personal disability, prejudice, or
> fault, has any interest in them, . . . the very body of mankind may be embued with
> prejudices, and errors, that from their childhood, and some also even from their
> birth, by which means they continue undiscerned, and consequently unreformed.[59]

The way to avoid bogus claims, based on reason, was to develop an epistemology in
which truth claims were sanctioned only by immediately accessible observations and
experiments. This is precisely what Boyle, to say nothing of a number of his colleagues
in the Royal Society, attempted to do.[60]

It should be realised, of course, that these two preoccupations of Boyle's have been expounded separately, and somewhat artificially, for the sake of elucidation. As far as Boyle was concerned, these two considerations were inextricably bound up with one another. The fact that voluntarist theology sanctioned, indeed demanded, an empiricist epistemology in which rationalist claims about the operation of a few laws of nature were illegitimate was hardly separable from the belief that rationalist arguments could be, and were, used by Roman Catholics and other groups (whether heterodox religious, or downright irreligious) merely to promote their own dubious claims. Whichever way you looked at it, as far as Boyle was concerned, rationalism was incompatible with sound religion.[61]

Whatever else may be said of Boyle's speculations in *Cosmical Qualities*, it seems clear that he was seeking to go beyond the bounds of the mechanical philosophy as it was set out by Descartes. Not only did he deny the competence of Descartes' laws in terms of the variability of things in themselves, and in terms of the variability of their inter-actions with other things to which they are 'manifestly related', but he also denied their ability to account for the variations brought about by the fact that all things are part of a system 'so constituted as our world is'. The world system as a whole is such, Boyle suggested, 'that there may be divers unheeded agents' operating 'by unperceived means' to directly affect bodies and to affect the way those bodies interact with other bodies.[62] The cosmical qualities themselves (whatever they might be) are to be attributed to the ordinary bodies of the world. These qualities are evidently not held to be inherent properties of the bodies in question, but relational properties, like the property of a key to open a lock, and may well be reducible to mechanistic terms.[63] Nevertheless, the 'unheeded agents' which produce the cosmical qualities in things do not seem to be so reducible. Particles of matter which operate only by impact cannot be held to affect the relational properties of other things in the way that Boyle seems to hint at. The answer, therefore, is to suppose, as Boyle does, that there must be 'peculiar sorts of corpuscles' which have 'peculiar faculties and ways of working'.[64]

One final question remains. If my reconstruction of the nature of the 'unheeded agents' capable of operating in peculiar ways to produce cosmical qualities is correct, why did Boyle not clearly say so in 1671? Why did he hold back his real thoughts on the matter? Why did he only hint obscurely at these ideas in 1671 and provide examples which, at least according to my reading, are extremely misleading because they deal with phenomena which seem to be easily explicable in terms of ordinary interactions or relationships between things, without invoking cosmic phenomena?

There are two or three closely connected reasons for Boyle's obscurity on this issue. First, Boyle did not want to be misunderstood. He did not want his readers in 1671 to think he was forsaking and rejecting the principles of the mechanical philosophy, in favour of a more occult philosophy. His point was that the mechanical philosophy was too restricted in the principles it allowed as legitimate. He did not deny the notion of natural law; he simply doubted that Descartes' three laws of nature could be used to explain all physical phenomena. By the same token, he doubted that all the complexity of interactions in the world could be explained in terms of the motions of particles of inert matter, colliding with one another and transferring their motions in accordance with Descartes' rules of impact. Nevertheless, as he wrote in *Hidden Qualities of the Air*, it should be possible to discourse like a naturalist about even such things as 'Magnets of Celestial Emanations'. Cosmical qualities should not be dismissed as chimerical, because they can be evinced by real experiments and physical phenomena.

It was important not to be misunderstood on this point, of course, because Boyle did not wish to disrupt the carefully forged links between the mechanical philosophy and Christian religion. The boast of the devout mechanical philosopher was that, by showing how all physical phenomena could be explained in terms of matter and motion, he simultaneously revealed that whatever cannot be explained in those terms cannot be a bodily or material phenomenon but must be spiritual. Any innovations in natural philosophy which threatened to blur the clear distinction between mechanical interactions and non-mechanical interactions (as the unheeded agents which were held to produce Boyle's cosmical qualities did) had to be very carefully managed. It is my belief that Boyle was not sure that his cosmical qualities, or rather the agents held to produce them, might not have been seized by irreligious thinkers to undermine the standard religious tropes of the mechanical philosophy, and he wished to be circumspect.

Boyle's concept of cosmical qualities was not merely an extension of the mechanistic ideas fully described in the *Origin of Formes and Qualities*, but was, in fact, an addendum to those ideas which Boyle regarded as essential to account for all the physical phenomena of nature. To interpret the cosmical qualities in such a way as to make them fit in with standard accounts of the mechanical philosophy is to see only part of the story. It is easy to see that the power of a key to open a lock does not have to correspond to any real or physical entity in the key, and Boyle's cosmical qualities were obviously intended to be seen in the same way. The qualities themselves, however, were produced by external agents operating in ways which were not allowed for in the standard contemporary accounts of the mechanical philosophy. It would seem that Robert Boyle might have said to Descartes or to scholastic philosophers what Hamlet said to Horatio:

> There are more things in heaven and earth . . .
> Than are dreamt of in your philosophy.[65]

Acknowledgements

I wish to thank the other contributors to the Stalbridge Boyle Symposium, all of whom made helpful comments on this paper, but I owe particular thanks to Antonio Clericuzio, Malcolm Oster and Michael Hunter for their comments upon an earlier draft.

NOTES

1 *Cosmical Qualities, Works*, iii, 306.
2 *Forms and Qualities, Works*, iii, 18.
3 *Ibid.*; Frederick J. O'Toole, 'Qualities and powers in the corpuscular philosophy of Robert Boyle', *Journal of the History of Philosophy* 12 (1974): 295–315, on pp. 308–9; Keith Hutchison, 'Dormitive virtues, scholastic qualities, and the new philosophies', *Hist. Sci.* 29 (1991): 245–78, on pp. 259–60.
4 *Cosmical Qualities, Works*, iii, 308, 309, 312, 313, 315.
5 *Ibid.* p. 307.
6 O'Toole, 'Qualities and powers' (n. 3), pp. 302–3.
7 *Cosmical Qualities, Works*, iii, 306. For a full account of what is meant here by powers and capacities, see O'Toole, 'Qualities and powers' (n. 3), pp. 310–15; and R. Harré, 'Powers', *British Journal for the Philosophy of Science* 21 (1970): 81–101.
8 O'Toole, 'Qualities and powers' (n. 3), p. 313, suggests that his analysis of Boyle's matter theory in terms of inherent non-relational and non-inherent relational properties provides us with a complete analysis of the ascriptions of powers and capacities to corporeal objects 'according to the corpuscular theory of matter'. I wish to suggest that Boyle himself would have wanted to insist that the category of non-inherent relational properties must be subdivided into those which O'Toole has discerned and others which, in Boyle's words (*Works*, iii, 306), 'are rather to be ascribed to some unheeded agents, which, by unperceived means, may have great operations'.
9 *Cosmical Qualities, Works*, iii, 306.
10 *Cosmical Qualities, Works*, iii, 316.
11 *Cosmical Qualities, Works*, iii, 307.
12 H. G. Alexander (ed.), *The Leibniz–Clarke Correspondence* (Manchester University Press, 1956), p. 94; Isaac Newton, *Opticks, or A Treatise of the Reflections, Refractions, Inflections and Colours of Light*, based on the 4th edn, London, 1730 (New York: Dover, 1952), p. 401. For a fuller discussion, see J. Henry, 'Occult qualities and the experimental philosophy: active principles in pre-Newtonian matter theory', *Hist. Sci.* 24 (1986): 335–81.
13 *Cosmical Suspicions, Works*, iii, 317–18.
14 *Cosmical Qualities, Works*, iii, 308.
15 *Cosmical Suspicions, Works*, iii, 316.
16 BP 9, fol. 40. I am very grateful to Prof. Steven Shapin for bringing this and the immediately following quotation to my attention. It may be objected to my interpretation that Boyle refers to these occult phenomena (gravity, fermentation, springiness, magnetism) as 'subordinate' and thereby implies that, ultimately, they are explicable in terms of matter in motion. I cannot answer this except by special pleading. In view of the fact that Boyle describes matter and motion as 'first and simplest principles', I believe that he might have meant to suggest that gravity, magnetism and the like were secondary and more complex explanatory principles. His use of the word subordinate was meant, therefore, to convey the notion only of an ordering of heuristic precepts without intending to imply dependence of one kind on the other. The only support I can give for this interpretation is to be found in the other quotations from Boyle's works which I discuss in this paper.
17 BP 8, fol. 166.
18 *Cosmical Suspicions, Works*, iii, 318.
19 *Ibid.* Descartes expounded his three laws of motion, or nature, and his seven rules of impact in his *Principia philosophiae* (Paris, 1644), Part II, sections 37–42 and 46–52. See R. Descartes, *Principles of Philosophy*, trans. V. R. Miller and R. P. Miller (Dordrecht: D. Reidel, 1983), pp. 59–62.
20 *Cosmical Qualities, Works*, iii, 307.
21 *Cosmical Qualities, Works*, iii, 319.
22 *Hidden Qualities of the Air, Works*, iv, 98.
23 *Cosmical Qualities, Works*, iii, 318, 319, 321.
24 *Cosmical Suspicions, Works*, iii, 321–2.
25 *Salubrity of the Air, Works*, v, 40.
26 *Hidden Qualities of Air, Works*, iv, 79.
27 *Ibid.* p. 95.
28 *Cosmical Suspicions, Works*, iii, 318.

29 *Hidden Qualities of Air, Works*, iv, 96.

30 *Ibid.*

31 *Ibid.* Note the similarity of expression at the end there with the quotation from *Cosmical Suspicious* at note 18 above. The fact that Boyle felt the same kind of defence of his position was required suggests that we are correct in linking these two discussions.

32 *Ibid.* p. 90.

33 'Of Celestial Influences or Effluviums in the Air', *Works*, v, 638–45. Worsley writes: 'You did not expect, I am sure, I should have adventured into so particular an Apology for Astrology', p. 642. The Latin translation of this essay appears under the somewhat nondescript heading, 'Exemplar literarum Benjamini Worslaei' in the Hartlib Papers in Sheffield University Library, Hartlib Papers, 42/1/18A–25B. Confirmation of Worsley's authorship is provided by a letter from Worsley to Hartlib, 20 October 1657, Hartlib Papers, 42/1/10A. I am extremely grateful to Dr Antonio Clericuzio for bringing this information to my attention. For a fuller treatment of Worsley's essays, together with related documents from the Hartlib archive, see A. Clericuzio, 'New light on Benjamin Worsley's natural philosophy', in *The Advancement of Learning in the Seventeenth Century: The World of Samuel Hartlib*, eds. M. Greengrass, M. Leslie and T. Raylor (Cambridge University Press, forthcoming). The fullest general account of the activities of Worsley and his associates in the 1640s and 1650s is to be found in Charles Webster, *The Great Instauration: Science, Medicine and Reform 1626–60* (London: Duckworth, 1975). Eight years Boyle's senior, Worsley was evidently the initiator of the 'Invisible College', of which, as far as we know, Boyle was the only other noted member. He died in 1677.

34 Letter from Locke to Boyle, 21 October 1691, in *The Correspondence of John Locke*, ed. E. S. De Beer, 8 vols (Oxford: Clarendon Press, 1976–89), iv, 320–2, also published in *Works*, vi, 543–4. The text appearing under the heading 'Of Celestial Influences or Effluviums in the Air' is clearly presented as a letter to Samuel Hartlib, but it is unsigned. There is nothing surreptitious about this; Locke apologises in his 'Advertisement' that 'The Negligence of Transcribers has let slip the Characters of Relators, and Names and Places of Authors from whom several of the Particulars in the following Papers were taken', *Works*, v, 609. In fact, the original of this letter in the Hartlib Papers (unlike the Latin translation discussed in the previous note), is unattributed (I owe this information to Dr Stephen Pumfrey), and it seems reasonable to suppose therefore that Boyle's copy was also unattributed.

35 *General History of Air, Works*, v, 611. For attributed extracts see, for example, Pascal 'in his small Tract, either De la Pesanteur de l'air, or in that Del'Equilibre des liqueurs', p. 694; 'Casati, Mechanicorum, lib. VIII, c. 5, p. 792, 793, &c.' (i.e. Paulo Casati, *Mechanicorum, libri octo*, Lyons, 1684), p. 616; 'Agricola, de re metallica, lib. duodec.', p. 634; and 'Jacobi Zabarelli *de Regionibus Aeris*, c. 8' (i.e. G. Zabarella, in his *De rebus naturalibus libri XXX*, Cologne, 1590), p. 695. For unattributed contributions, see, for example, letters from Mr J.T., pp. 624, 625, 634; or from 'Fort St. George', dated 23 January 1668, pp. 645–6.

36 *General History of the Air, Works*, v, 612.

37 *Cosmical Suspicions, Works*, iii, 318.

38 *General History of Air, Works*, v, 638–9.

39 *Ibid.* p. 638.

40 *Ibid.* p. 640.

41 *Ibid.* p. 641.

42 *Ibid.*

43 *Cosmical Qualities, Works*, iii, 307.

44 *General History of Air, Works*, v, 641.

45 *Cosmical Suspicions, Works*, iii, 318.

46 Graham Rees, 'Francis Bacon's semi-Paracelsian cosmology', *Ambix* 22 (1975): 81–101; 'Francis Bacon's semi-Paracelsian cosmology and the *Great Instauration*', *Ambix* 22 (1975): 161–73; 'Matter theory: a unifying factor in Bacon's natural philosophy', *Ambix* 24 (1977): 110–25; 'The fate of Bacon's cosmology in the seventeenth century', *Ambix* 24 (1977): 27–38.

47 William Harvey, *Disputations touching the Generation of Animals*, trans. Gweneth Whitteridge (Oxford: Blackwell, 1981), p. 378, see also pp. 374–83. For a full discussion of these ideas, see D. P. Walker, 'The astral body in Renaissance medicine', *Journal of the Warburg and Courtauld Institutes* 21 (1958): 199–33, reprinted in *Music, Spirit and Language in the Renaissance*, ed. P. Gouk (London: Variorum, 1985).

48 See A. C. Crombie, *Robert Grosseteste and the Origins of Experimental Science, 1100–1700* (Oxford: Clarendon Press, 1953); James McEvoy, *The Philosophy of Robert Grosseteste* (Oxford: Clarendon Press, 1982); David C. Lindberg, *Roger Bacon's Philosophy of Nature* (Oxford: Clarendon Press, 1983); and 'The genesis of Kepler's theory of light: light metaphysics from Plotinus to Kepler', *Osiris*, second series, 2 (1986): 5–42. On Al-Kindi, see Lynn Thorndike, *A History of Magic and Experimental Science*, 8 vols (New York: Columbia University Press, 1923–58), i, 642–6. For an indication of the role of light metaphysics in the medical tradition, see Walter Pagel, *New Light on William Harvey* (Basel: Karger, 1976), pp. 161–7.

49 *Aerial Noctiluca*, *Works*, iv, 384.

50 *Hidden Qualities of the Air*, *Works*, iv, 85.

51 *Ibid.* p. 86. For another example of Boyle's consideration of the possibility of astrological influences, see also p. 98.

52 Descartes, *Principles of Philosophy* (n. 19), part IV, sect. 199, pp. 282–3.

53 Marie Boas [Hall], *Robert Boyle and Seventeenth-Century Chemistry* (Cambridge University Press, 1958), p. 216; Steven Shapin, 'Robert Boyle and mathematics: reality, representation and experimental practice', *Science in Context* 2 (1988): 23–58.

54 Shapin, 'Boyle and mathematics' (n. 53), pp. 47–50. *Hidden Qualities of Air, Works*, iv, 91. See also *Two Essays concerning the Unsuccessfulness of Experiments*, in *CPE, Works*, i, 318–54. This aspect of Boyle's work is also discussed briefly in Paul B. Wood, 'Methodology and apologetics: Thomas Sprat's *History of the Royal Society*', *BJHS* 13 (1980): 1–26, on pp. 6–7; and Michael Hunter, 'Latitudinarianism and the "ideology" of the early Royal Society: Thomas Sprat's *History of the Royal Society* (1667) reconsidered', in *Establishing the New Science: The Experience of the Early Royal Society* (Woodbridge, Suffolk: Boydell Press, 1989), pp. 45–71, esp. pp. 53, 62, 68–9, where Hunter publishes a manuscript draft by Henry Oldenburg from which Sprat evidently drew for pp. 243–5 of his *History*.

55 *Cosmical Suspicions, Works*, iii, 322. Boyle's use of the word 'obnoxiousness' here should be taken in the sense of the fourth meaning given for 'obnoxious' by the *Oxford English Dictionary*: 'with *to*: Exposed to the (physical) action or influence of, liable to be affected by . . . '. Hence, the planetary and stellar bodies are in a more pronounced state of being liable to be influenced by variations than the schools would have us believe. See also Hunter, 'Latitudinarianism and the Society's "ideology"' (n. 54), p. 68, where Oldenburg talks of 'Experiments [being] very obnoxious to contingencies'. It is my belief, though only speculative, that Boyle closely supervised this document of Oldenburg's. Boyle does introduce non-chemical considerations elsewhere when discussing this theme: in his *Two Essays concerning the Unsuccessfulness of Experiments* (1661), Boyle had previously mentioned the variation of the compass and some anatomical observations. But most of his examples, as he mentioned in the full title were 'chiefly Chymical'; *Works*, i, 318, 343, 345.

56 J. E. McGuire, 'Boyle's conception of nature', *JHI* 33 (1972): 523–42; Eugene M. Klaaren, *Religious Origins of Modern Science: Belief in Creation in Seventeenth-Century Thought* (Grand Rapids, Mich.: William B. Eerdmans, 1977); J. Henry, 'Henry More versus Robert Boyle: the Spirit of Nature and the nature of providence', in *Henry More (1614–87): Tercentenary Studies*, ed. Sarah Hutton (Dordrecht: Kluwer Academic, 1990), pp. 55–76.

57 Shapin, 'Boyle and mathematics' (n. 53), pp. 50–2. See also Shapin and Schaffer, *Leviathan and the Air-Pump*, for a fuller account of Boyle's concerns with the authority for natural philosophical pronouncements.

58 Jeremy Taylor, *Works*, ed. R. Heber, 15 vols (London, 1828), xi, 356.

59 *Works*, iv, 164. For fuller discussions of the suspicion of 'reason' among English thinkers at this time, see H. G. van Leeuwen, *The Problem of Certainty in English Thought, 1630–1690* (The Hague: Martinus Nijhoff, 1963); Barbara J. Shapiro, *Probability and Certainty in Seventeenth-Century England* (Princeton University Press, 1983); Christopher Hill, ' "Reason" and "reasonableness" ', in *Change and Continuity in Seventeenth-Century England* (London: Weidenfeld and Nicolson, 1974), pp. 103–26; Lotte Mulligan, ' "Reason", "right reason", and "revelation" in mid-seventeenth-century England', in *Occult and Scientific Mentalities in the Renaissance*, ed. Brian Vickers (Cambridge University Press, 1984), pp. 375–401; J. Henry, 'The Scientific Revolution in England', in *The Scientific Revolution in National Context*, eds. R. Porter and M. Teich (Cambridge University Press, 1992), pp. 178–209.

60 See Shapin and Schaffer, *Leviathan and the Air-Pump*; Henry, 'Occult qualities and the experimental philosophy' (n. 12); *ibid.* 'Scientific revolution in England' (n. 59).

61 For another treatment of this theme from a different perspective, see Peter Dear, 'Miracles, experiments and the ordinary course of nature', *Isis* 81 (1990): 663–83.

62 *Cosmical Qualities, Works*, iii, 306.

63 Here again I am referring to the analysis of O'Toole, 'Qualities and powers' (n. 3), in which he distinguishes between non-relational inherent properties of bodies and non-inherent relational properties. See notes 6 and 8 above.

64 *Cosmical Suspicions, Works*, iii, 316. My conclusion should be taken to endorse and support, from a non-chemical perspective, Antonio Clericuzio's recent attempt to redefine Boyle's corpuscular philosophy in terms of his chemical philosophy. See A. Clericuzio, 'A redefinition of Boyle's chemistry and corpuscular philosophy', *Ann. Sci.* 47 (1990): 561–89. See also Henry, 'Occult qualities' (n. 12).

65 *Hamlet*, 1. 5. 166–7.

The theological context of Boyle's
Things above Reason

JAN W. WOJCIK

Scholars have long been aware of Robert Boyle's status as a lay theologian.[1] To date, however, no one has analysed any of Boyle's theological writings in the specific context of the theological controversies of his day. On the face of it, there is good reason for this neglect. These writings appear to be non-polemical in nature, and when this fact is considered along with Boyle's known distaste for controversy it is easy to assume that his theological writings represent simply Boyle's desire to show that a dedicated natural philosopher could be an even more dedicated Christian.

A closer look at Boyle's 1681 *Discourse of Things above Reason*, however, indicates that when he wrote it Boyle was indeed engaging in a religious controversy – one which has important implications for his philosophical epistemology. Specifically, he was addressing the question of the extent to which human reason can and should be applied to the mysteries of Christianity. In this controversy the Anglican Latitudinarian, Joseph Glanvill, in arguing reason's competence, had pitted himself against certain nonconformist ministers (those refusing to conform to the Church of England after the Restoration of Charles II) who had argued – and would continue to argue – that human reason is not capable of understanding (at least some of) the mysteries of Christianity. Further, we find that Boyle was prompted to publish his views by a particular controversy among *other* nonconformists over predestination. This controversy was occasioned by the publication of John Howe's *Reconcileableness of God's Prescience of the Sins of Men, with the Wisdom and Sincerity of . . . whatsoever Means He uses to prevent them* (1677), a work which Howe, a nonconformist minister himself, had written at Boyle's request. Boyle's *Things above Reason* should be read both as his response to the controversy generated by Howe's work and as a systematisation and extension of the previous arguments of those nonconformists who had emphasised the limits of human reason.

Robert Boyle's *Things above Reason*

Before looking more specifically at the context within which Boyle formulated his views concerning reason's limits, I shall briefly summarise those views as expressed, primarily, in *Things above Reason* and its accompanying *Advices in Judging Things above Reason* (1681) and in *Reflections upon a Theological Distinction* (published with *Christian Virtuoso*, 1690). Boyle's main purpose is to assert that, where both theology and natural philosophy are concerned, one of the most important roles of human reason is to ascertain and acknowledge its limits. Boyle, aware that he could be accused of setting reason 'awork to degrade itself', argues that he is instead exercising its 'most noble and genuine exercise'; he is showing that it is but a 'limited faculty . . . [and that it is] an injury to the author of it, to think man's understanding infinite like his'.[2]

In theology, things that are above reason are in the category of 'such notions and propositions, as mere reason, that is, reason unassisted by supernatural revelation, would never have discovered to us'.[3] Even after these notions and propositions are revealed, we cannot understand fully some of them for one of three (not mutually exclusive)[4] reasons: first, we cannot form adequate ideas of the thing revealed; secondly, we cannot explain how certain effects are caused; or thirdly, the revelation contradicts something else which we know to be true. Boyle calls the first category truths that are 'incomprehensible', the second category truths that are 'inexplicable', and the third category truths that are 'unsociable'.[5] (Boyle occasionally substitutes the term 'asymmetrical' for 'unsociable', and 'inconceivable' or 'supra-rational' for things above reason generally [and sometimes for the subcategory of 'incomprehensible' things in particular].)[6] As a group, the things that are above reason are called 'privileged things',[7] and are 'above reason' at least partly because they involve notions of which human beings can have no 'clear and distinct' conception.[8]

God's nature is an example of one of the things that is incomprehensible to us.[9] How God made the world out of nothing, or how He unites an immaterial soul to a human body and maintains that union, are among those things that Boyle terms 'inexplicable'.[10] Boyle's most frequent example of two 'unsociable' (or contradictory) truths is that both the proposition that God has foreknowledge of future contingents and the proposition that man has free will are true;[11] in addition he argues that various of God's attributes may appear contradictory to human beings.[12] It is this third category of things above reason which is most relevant to Howe's *Reconcileableness*, as we shall see when we examine that work.

Boyle goes on to explain just how it is that these 'privileged things' are exempt from the rules of reasoning. He does this by dividing the rules of reasoning into two categories. The first includes those axioms which are applied to particular fields and involves 'physical propositions' which are but 'limited, collected truths, gathered from the phenomena of nature'. Boyle terms these 'gradual' or 'emergent' truths and offers, as an example, the principle *ex nihilo nihil fit*.[13] Privileged things are 'of so heteroclite a nature [that they] may challenge an exemption from some of the rules employed about common things'.[14] These 'physical propositions' which are abstracted from natural things should not be applied to supernatural things.

The second category of rules of reasoning includes those which spring from 'the rational faculty itself, furnished with the light, that accompanies it, when it is rightly disposed and informed', such as that snow is white, that from truth nothing but truth can be deduced and that contradictory propositions cannot both be true.[15] The light which accompanies this rational faculty 'enable[s] a man, that will make a free and industrious use of it . . . to pass a right judgment of the extent of those very dictates that are commonly taken for rules of reason'.[16] In considering the objection that nothing is more unreasonable than to embrace opinions that violate the very rules of reasoning, Boyle argues that even the principles of logic should be considered to be but 'gradual' or 'emergent' truths from God's point of view because our 'inbred or easily acquired ideas and primitive axioms . . . do not extend to all knowable objects whatsoever, but reach only to such . . . as God thought fit to allow our minds in their present . . . condition'.[17] Boyle will not commit himself as to the exact nature of reason's inability to comprehend fully all truths,[18] but he most often appeals to a comparison of man's finite reason with God's infinite reason to support his argument.[19] He dares not, however, deny that there may be other factors as well, such as the way in which God has united individual souls with individual bodies, or Adam's fall.[20] In any event, God has helped us to ascertain the limits which He has set on human understanding by furnishing us with 'a domestick monitor, or a kind of internal criterium' which allows us to discern that some matters are quite beyond our reach.[21]

To argue that the content of revelation is above reason, and something that human beings cannot fully understand, does not seem to be an unusual position to take. But to argue that the content of revelation can be *contradictory* to reason (i.e. that revelation can violate the law of contradiction) poses far more problems. That any conclusion whatsoever follows from contradictory premises was well known in Boyle's day.[22] Boyle himself notes the difficulty involved when he observes that the rational soul

> cannot avoid admitting some consequences as true and good, which she is not able to reconcile to some other manifest truth or acknowledged proposition. And whereas other truths are so harmonious, that there is no disagreement between any two of them, the heteroclite truths appear not symmetrical with the rest of the body of truths, and we see not how we can at once embrace these and the rest, without admitting that grand absurdity, which subverts the very foundation of our reasonings, that contradictories may both be true.[23]

Strictly speaking, Boyle avoids violating the law of contradiction by arguing that God, in His infinite reason, perceives the harmony among all truths.[24] And yet Boyle does 'subvert the very foundation of our reasonings' when he asserts that two propositions which appear to human beings to be truly contradictory may both be true. What would cause Boyle to take such an extreme position? If we look at Boyle the natural philosopher we might think that he was simply trying to protect the mysteries of religion from being challenged by any findings of the new natural philosophy, and indeed this was probably what he had in mind when he said that a revealed truth might appear contradictory to those 'emergent' or 'gradual' truths that are gathered from the phenomena of nature. But when Boyle went on to argue that some revelations are exempt even from the laws of logic he had something else in mind.

The context of the question of *Things above Reason*

When we look at the history of the argument concerning things above and contrary to reason in the 1650s, 1660s and 1670s in England we see just what it was that Boyle had in mind. He was entering into a controversy which, despite the fact that it had significant philosophical implications, was essentially theological in nature.

During the Reformation one of the stock Protestant arguments had been that some Roman Catholic doctrines (such as Transubstantiation) should be rejected on the grounds that they are contrary to reason. For example, in advocating Scripture alone as the Rule of Faith, William Chillingworth observed in 1638 that although nothing in *Scripture* is contrary to reason, some doctrines of *Catholicism* require men to believe things which are impossible. As Chillingworth expressed it,

> following your [Roman Catholic] Church, I must hold many things not only above reason, but against it . . . whereas, following the Scripture, I shall believe many mysteries, but no impossibilities; many things above reason, but nothing against it; many things, which had they not been reveal'd, reason could never have discovered, but nothing which by true reason may be confuted; many things, which reason cannot comprehend how they can be, but nothing which reason can comprehend that it cannot be.[25]

In time, as Protestants splintered into sects, some argued that a particular doctrine held by an opposing sect was contrary to reason; Arminians, for example, often argued that the Calvinist doctrine of absolute predestination made God, Who is infinitely good, the author of sin. The Socinians, a sect of radical reformers known primarily for their denial of the orthodox interpretation of the Trinity, carried such thinking to its logical conclusion, and explicitly argued that consonance to human reason should be the criterion against which differing interpretations of Scriptural passages should be judged.[26] They argued that, if a passage of Scripture appears to violate the laws of logic, then that passage is being misinterpreted and should be studied more carefully in terms of its original language and historical context. In denying the Trinity, for example, the Socinians first quoted Scriptural passages in support of their views. Then they argued that any other passages in Scripture which appear to support Trinitarianism should not be so interpreted because such an interpretation is repugnant to reason, on the grounds that two substances imbued with opposite qualities (mortality and immortality, to have a beginning and to be eternal, to be mutable and to be immutable) cannot combine into one person. John i, 14, for example, traditionally invoked in support of Trinitarian views, should be translated 'The Word was flesh', rather than 'The Word became flesh'.[27]

Socinian writings became widely available in England during the Interregnum breakdown in censorship, and this availability prompted the first scholarly refutation of Socinian doctrines. In the full title of his 1655 *Vindiciæ Evangelicæ; or, The Mystery of the Gospell Vindicated, and Socinianisme Examined*,[28] the Calvinist John Owen made it clear that the work was written in response to the Socinian John Biddle's 1654 *Scripture Catechisme*, and to 'the Catechisme of Valentinus Smalcius, commonly called the *Racovian Catechisme*'.[29] Although Owen notes and rejects the 'high *pretences* to *Reason*' of the Socinians,[30] it is obvious that he believes that human reason is competent to combat their heterodox views and that the Socinians should be attacked on their own grounds. 'Let not any *attempt dealing* with these men', Owen urges, 'that is not in some *good measure* furnished with those kinds of *literature*, and those *common Arts*, wherein they excell: as first, the *knowledge* of the *Tongues*, wherein the *scripture* is *written* . . . [as well as] *Logick* and *Rhetorick*'.[31] Owen goes on to devote over 800 pages to a point-by-point repudiation of Socinian doctrines.

Richard Baxter also responded to the Socinian threat in 1655 in his *Arrogancy of Reason against Divine Revelations, Repressed*,[32] but his approach was totally different from Owen's. In *The Arrogancy*, Baxter attacks the Socinians on methodological grounds, denying that human reason is competent to judge the mysteries of Christianity. Using John, iii, 9, as his Scriptural text (in which Nicodemus had asked Christ 'How can these things be?'), Baxter argues that

> the corrupt nature of man is more prone to question the truth of Gods Word, then to see and confess their own ignorance and incapacity; and ready to doubt, whether the things that Christ revealeth are true, when they themselves do not know the nature, cause, and reason of them.[33]

In the process of complaining that some individuals reject Christianity because the Christian revelation contains apparent contradictions, and that yet other individuals discard one revealed truth because of its apparent inconsistency with another truth, Baxter, in a passage which foreshadows Boyle's later writings on apparent contradictions in Scripture, claims that

> no man can know Gods truths perfectly, till he see them all as in one Scheam or Body, with one view, . . . not onely to see that there is no contradiction, but how every Truth doth fortify the rest. All this therefore is exceeding desirable, but it is not every mans lot to attain it, nor any mans in this world perfectly, or near to a perfection.[34]

Both Baxter and Owen were among those who would refuse to conform to the reestablished Church of England after the Restoration of Charles II in 1660. Although the two men represented different doctrinal views, both can be classified as 'Puritan' in outlook, and their writings illustrate a curious tension in the Puritan attitude toward reasoning and learning. Although Puritans tended to argue that human reason was debased by the fall, and is incompetent to judge revelation, they were also known for their scholarly Scriptural exegesis.[35] Owen's *Vindiciæ* is an excellent example of such Puritan scholarship and its role in sectarian controversies, yet in 1676, as we shall see, he would also argue for the limits of reason in much the same way as had Baxter in *The Arrogancy*, and he would do so without acknowledging (or perhaps even recognising) the apparent inconsistency involved in applying human reason to specific controversies about those very things which he argued human reason cannot understand. Baxter, too, reflects this tension in Puritan thought. Despite his argument in *The Arrogancy* that debased reason is incompetent to understand fully the content of revelation, Baxter was one of the most prolific contributors to the sectarian controversies of his day.[36] It may have been individuals such as Owen and Baxter who prompted the nonconformist Benjamin Calamy's complaint that 'it were much to be wished that men had never gone about to explain those things which were mysterious, and to give an account of those things which themselves acknowledg'd incomprehensible'.[37] Certainly one of Robert Boyle's concerns in *Things above Reason* is to point out the inconsistency involved when one argues confidently about those very things one has also argued one cannot understand.

In 1670 Joseph Glanvill aimed his *Seasonable Recommendation* at divines who were arguing views similar to those expressed in Baxter's *The Arrogancy* from the pulpit. Perhaps most often remembered for his urging that the Royal Society investigate the evidence of witches in order to establish the existence of an immaterial realm of spirits, Glanvill was also an Anglican divine and Latitudinarian. In addition, he was significantly influenced by the Cambridge Platonists, and as such saw human reason as one of the ways in which human nature is most like the divine.[38] In *A Seasonable Recommendation* Glanvill complains that some ministers 'set up a *loud* cry against *Reason* . . . under the misapplied names of *Vain Philosophy*, *Carnal Reasoning*, and the *Wisdom of this World*'.[39] Although aware that some of his contemporaries claim that ''tis Socinianism to plead for Reason in the affairs of Faith, and Religion', Glanvill responds that it is absurd to suppose that the Socinians are 'the only *rational* men'.[40]

In his argument for the competency of human reason, Glanvill makes certain distinctions which provide the framework for subsequent discussions of the question. He defines exactly what he means by 'reason', 'religion' and 'Christianity'. In discussing the revealed doctrinal articles of Christianity, Glanvill divides them into those involving 'mixt' and those involving 'pure' revelations; 'mixt' revelations are those Scriptural revelations which can also be discovered by human reason, and 'pure' revelations are those which reason can neither discover nor understand *how* they can be true. So far, Glanvill's argument seems to be simply a more sophisticated version of Baxter's views, but Glanvill minimises the impact of the inability of reason to comprehend 'how these things can be' when he argues that reason cannot even comprehend the ways of acting of the most ordinary and obvious natural things, such as how the parts of matter cohere and how the human soul is united to the human body. When some of the propositions of Christianity 'are said to be *Above Reason*', Glanvill claims, 'no more is meant, than that *Reason* cannot *conceive how those things are*; and in *that* sense many of the affairs of *nature* are *above it* too'.[41]

That Glanvill is arguing for the competency of human reason rather than reason's limits is most obvious in an inference which he draws from his argument as a whole. 'No Principle of Reason', Glanvill claims, 'contradicts any Articles of Faith.' Both the principles of reason and the articles of faith are gifts of God, and, as such, both are certainly true. They may sometimes seem to clash, but in those cases 'either something is taken for *Faith*, that is but a *Phancy*; or something for *Reason*, that is but *Sophistry*; or the *supposed contradiction* is an *error*, and *mistake*'. 'When any thing is pretended from Reason, against any Article of Faith', Glanvill continues, 'we ought not to cut the knot, by denying reason; but indeavour to unty it by answering the Argument, and 'tis certain it may be fairly answered'.[42] Nowhere does Glanvill suggest that the apparent contradiction in fact poses a real contradiction for debased human understanding.

In 1675, Robert Ferguson, a nonconformist who was serving at the time as Owen's personal assistant at Owen's church in Leadenhall Street,[43] entered the controversy with his *Interest of Reason in Religion*. Ferguson's primary purpose in this 665-page work is to argue that Christ's union with the believer during the sacrament of Communion is an immediate union with individual believers and not a union mediated by the Church, as claimed by Anglicans urging conformity to the Church of England such as William Sherlock,[44] against whom the work was written. In the process, however, he identifies himself as the target of Glanvill's *A Seasonable Recommendation* and devotes the first 274 pages of *Interest* to defending his views concerning the relationship between reason and revelation from Glanvill's attack. In doing so, Ferguson reveals the polemical use of his views, for he emphasises that there are some revelations which cannot be rationally attacked or defended, among them the nature of Christ's union with individual believers. Ferguson accepts Glanvill's division of revelations into those that are 'pure' and those that are 'mixt', but goes on to distinguish between 'pure' revelations which, once revealed, can be supported by reason, and 'pure' revelations which, even after they are revealed, reason cannot comprehend. In one sense Ferguson is extending reason's sphere into the content of revelation even further than Glanvill when he assigns reason a role in supporting some 'pure' revelations. He is doing this, however, in order to support the Calvinist interpretation of Christ's Satisfaction; he places Christ's Satisfaction in the category of 'pure' revelations which reason can support. Concerning such Scripturally revealed doctrines, Ferguson borrows Baxter's argument that we cannot tell 'the manner and way' in which these mysteries can be. We are simply to believe that they are true, but their mode (or how they are true), Ferguson argues, is no article of the creed.[45] Where these revelations are concerned it is reason's role, first, to show that we cannot comprehend them; secondly, to judge any argument against such a revelation to be a sophism, even if we cannot detect the fallacy; and, thirdly, to answer the objection by showing that it is based on principles which have been extended beyond their sphere, for 'to urge them beyond their bounds, is to contradict Reason, which tells us that they hold only so far, and no farther'.[46] In discussing the third role of reason where 'pure' revelations are concerned, Ferguson makes a connection between the Cartesian criterion of clear and distinct ideas and the arguments of the Socinians (with which he shows considerable familiarity). The problem is that too many things are being urged as principles of reason (that is, as clear and distinct perceptions) which are in fact not clear and distinct, for human reason has been debased by the fall.[47]

As had Glanvill, Ferguson denies that any Christian belief is contrary to any maxim of reason, but Ferguson's explanation of what he means by this is less than clear. He states that when he says no Christian doctrine is contrary to reason he means 'universal' or 'abstract' reason, not the 'concrete' reason which is exercised by individuals and is debased by the fall.[48] At the same time, he claims that all of our maxims of reason (whether logical, moral, physical or mathematical) have their root in God and are discerned by the human mind.[49] This seeming contradiction in Ferguson's discussion can perhaps be resolved by his statement that revelation perfects our intellectual powers;[50] apparently Ferguson means that if one perceives the revelation correctly (as, in his opinion, the Calvinists do) then nothing in revelation is contrary to reason.

With his *Nature of Apostasie* (1676), John Owen, perhaps influenced by the views of his assistant, Robert Ferguson, joined the ranks of his fellow nonconformists in arguing for reason's inability to comprehend the content of revelation. In *The Nature of Apostasie*, Owen complains that reason, which had been rightly emancipated from the Roman Catholic 'bond of Superstition & Tradition', has recently 'enlarge[d] its Teritories'.[51] Owen blames this extension of reason's 'Teritories' on the Socinians. The Socinians have been so successful in introducing their doctrines, Owen claims, because, by arguing that human reason is competent to comprehend revelation, they have managed to convince their contemporaries that whatever reason cannot comprehend is contrary to reason.[52] According to Owen, the Socinians have wrestled revelations concerning the attributes of God, His eternal decrees, the office and mediation of Jesus Christ, the doctrine of justification by Christ's righteousness and the role of grace in the conversion of sinners into doctrines which human reason can understand,[53] and in doing so have encouraged the defection of the Church of England from what Owen considers to be the true Reformation doctrine (i.e. Calvinism).[54] The purpose of Owen's discussion is to argue that there are indeed revelations which reason cannot comprehend. Owen goes much further than either Baxter or Ferguson, however, when he claims, not only that some of these revelations are above reason insofar as reason is finite, but also unambiguously states that others are contrary to human reason insofar as reason is depraved.[55] Owen's work, like Ferguson's, is polemical in nature; he blames the anti-predestinarianism predominant in the reestablished Church on reason's having become 'the sole and *absolute Judge* of what is Divinely proposed unto it'.[56]

By 1676, then, Owen, Baxter and Ferguson had introduced all of the basic concepts that Boyle would use in his *Things above Reason*, and the views of these nonconformists on the limits of reason had been opposed by Joseph Glanvill, who had contributed the framework for subsequent discussions of the issue by Ferguson (and Boyle). Further, by 1676 some nonconformists were not only arguing the limits of reason in opposition to the claims of the Socinians but were also using those arguments to protect certain Calvinistic doctrines from the rational scrutiny of their Anglican opponents. In *Things above Reason*, Boyle, himself an Anglican who wished to make room in the salvation process for man's free will, consolidated, explicated and argued for the position of the nonconformists arguing the limits of reason, but he did so in such a way as to show that, although it is rational for one to argue that some revealed matters are impervious to human understanding, it is not rational to *also* argue that one actually understands those same revelations. Before continuing this discussion of *Things above Reason*, however, I shall first examine the specific controversy which prompted Boyle to publish his views.

John Howe's *Reconcileablensss*

One of the most heated of all the theological issues in seventeenth-century England was the question of predestination. The Protestant doctrine of predestination can be interpreted in one of two general ways: either predestination is absolute (the Calvinistic position), or it is conditional (the more moderate position). In absolute predestination, God, without consideration of His foreknowledge of any individual's faith or works, decided to save some particular souls and damn others. In this interpretation the emphasis is on the absolute role of God's grace in the salvation process. In the conditional predestination favoured by the Arminians of the Laudian era and most of the Anglican divines in the post-Restoration Church, God's salvation of some souls and His damnation of others is based upon His foreknowledge of who will have faith and who will not; His offer of salvation extends to all individuals, even though only some will choose to respond. In this interpretation the emphasis is on man's free will. There are a variety of different interpretations of both the absolute predestination and the conditional predestination doctrines, but such subtleties need not concern us here.[57] It is, however, important to note that arguments concerning the nature of predestination were often based on the grounds that the view being opposed was contrary to something we know about God's nature. Those believing in conditional predestination, for example, argued that a good God would not damn individuals for sins they are unable to avoid, and Calvinists argued that the more moderate conditional predestination position contradicts God's omnipotence, for it makes God dependent on human contingencies. The Socinians, who shared the Arminian and Anglican emphasis on man's free will, went so far as to deny God's prescience (or foreknowledge) of those future contingencies which depend upon man's free will, partly on the ground that if He foresaw particular sins, then God, in His exhortations to men not to sin, would be attempting to deter the unavoidable; this, they argued, would be contradictory to God's wisdom.[58] John Howe's *Reconcileableness of God's Prescience of the Sins of Men, with the Wisdom and Sincerity of his Counsels, Exhortations, and whatsoever*

Means He uses to prevent them (London, 1677)[59] was but one of many works in which the author tried to show that his own doctrinal views were more reconcilable to what we know about God's nature than his opponents' views.

Howe was a nonconformist minister whose position on predestination was essentially a moderate conditional one. His work is in the form of a letter to Robert Boyle. In his dedicatory epistle, Howe specifically notes that he had written the work 'at the request of Mr. Boyle'. Basically, Howe follows the argument that had been presented by the Anglican Henry Hammond in 1660 to Robert Sanderson, a divine who had, in the late 1650s, revised his Calvinistic views in favour of acknowledging the role of free will in the salvation process. In 1660, Hammond had published his Χάρις καὶ ἐιρήνη; *or, A Pacifick Discourse of Gods Grace and Decrees: In a Letter, of full Accordance written to the Reverend, and most learned, Dr: Robert Sanderson* in order to make public Sanderson's change of heart.[60] The gist of Hammond's (and Howe's) argument is that God's prescience of future contingents does not necessitate those events. Both Hammond and Howe attempt to reconcile their conditional predestination with Calvinist absolute predestination by arguing that God may, on occasion, effect the salvation of an individual by overwhelming that individual with His grace. Thus God's grace is offered to all individuals (which affirms God's wisdom in exhorting individuals not to sin), even though the salvation of some individuals who resist God's offer may be effected by a 'super-effluency' of grace.

It is particularly important to note two things about Howe's work. First, although early in his argument Howe warns against assuming the sufficiency of human reason to resolve all doctrinal difficulties, Howe's emphasis is on removing as many difficulties as possible by showing his position to be a rational one. Howe sounds more like Glanvill than his fellow nonconformists when he argues that if we ascribe 'inconsistencies' to or 'give a self-repugnant notion' of God, we 'gratifie atheistical minds'.[61] Even though Howe urges that when we encounter problems we should ask first whether the difficulty is in our own understanding,[62] and should be careful not to mistake a dictate of our depraved nature for a common notion,[63] he also warns against using the limits of human understanding as an excuse not to attempt to resolve the apparent difficulty or inconsistency.[64] And, in his *Reconcileableness*, Howe proceeds to resolve the apparent inconsistency implied by God's warning individuals not to do those very things which God foresees that they will do. However, Howe failed to note (or was perhaps unaware of the fact) that Hammond, in 1660, had concluded his similar argument with the statement that Christians must assent to both God's prescience and man's free will, and that, if someone were to judge Hammond's reconciliation of these two truths inadequate, then that person should still assent to both truths in the 'assurance that God can reconcile his own contradictions'.[65]

Secondly, in his argument Howe included a lengthy digression against the Calvinist doctrine of absolute predestination on the grounds that such a doctrine makes God, Who is infinitely good, the author of sin.[66] Howe's claim was certainly not original; this was not the first time that Calvinists had encountered such an argument. But it did renew sectarian debate on the question. That same year the nonconformist

Calvinist Theophilus Gale attacked Howe in *Part IV* of his *Court of the Gentiles*. In *Reconcileableness* Howe had argued that God concurs only in the good actions of individuals.[67] In *Court of the Gentiles* Gale objects to this on the grounds that no action of any human being is unmixed with sin, and because of this God cannot concur only in the good part of an act without concurring in the sinful aspect at the same time.[68] Howe responded immediately with a *Postscript* in which he defended his position and reiterated that he had originally written about the subject at Boyle's request.[69] In the *Postscript* Howe distinguished between 'immediate' concurrence and 'predeterminative' concurrence, stating that he does not believe that God immediately concurs in all actions but that he denies that God predeterminatively concurs in wicked actions.[70]

The following year another nonconformist Calvinist, Thomas Danson, attacked Howe in his *De Causa Dei*. In this work Danson asserts that 'God does determine, or move all Creatures to all and each of their actions', and argues that, by making human beings the cause of some actions and displacing God as the first cause, Howe borders 'as near upon *Arminianism* as *Scotland* does upon *England*'.[71] The poet, Andrew Marvell, immediately answered Danson and defended Howe in his *Remarks upon a Discourse*, in which he argued that many divines 'out of a vain affectation of Learning, have been tempted into Enquiries too curious after those things which the Wisdom of God hath left impervious to Humane Understanding'.[72] In addition, Marvell argues that, no matter how much strict Calvinists might deny it, their doctrine of absolute predestination does indeed make God the author of sin.[73] Howe's work had generated quite a controversy between the Calvinist nonconformists and the more moderate nonconformists, and he had made it clear that he had written that work at Boyle's request.

Boyle's contribution to the controversy

Boyle's *Things above Reason* should be read as a response to Howe and to the controversy that Howe's work generated. In it Boyle assumes that God's prescience and man's free will are in fact not reconcilable by human beings, and that both are true; he bases the truth of prescience on revelation (specifically, fulfilled prophecies) and on the grounds that foreknowledge is a perfection, and he bases the truth of free will on the fact that we have 'felt it' in ourselves.[74] Arguing that finite human reason cannot perceive the harmony among all truths, Boyle, like Hammond, was willing to let God resolve His own contradictions.

In *Things above Reason*, Boyle developed his point that finite human reason is unable to penetrate the mysteries of Christianity, and in doing so he drew upon the writings of those who had engaged in controversies concerning the limits of reason over the previous three decades. It is impossible to know which, if any, of the specific works I have discussed Boyle had read. That he had read at least some of them (or had encountered similar arguments elsewhere) is shown by his comment in *Reflections upon a Theological Distinction* (1690), where he notes (concerning the distinction concerning things 'above reason' and things 'contrary to reason') that

> as far as I can discern by the authors, wherein I have met with it, (for I pretend not to judge of any others) there are divers that employ this distinction, [but] few who have attempted to explain it, (and that I fear not sufficiently,) and none that has taken care to justify it.[75]

Although the framework of his argument corresponds to the framework originated by the Anglican divine, Glanvill, in arguing that human reason is not competent to judge the content of God's revelation, Boyle aligned himself with what was predominantly a nonconformist position. Boyle's category of 'inexplicable' things above reason echoes Baxter's (and Ferguson's) argument that human beings cannot understand 'how these things can be'. Boyle's category of 'incomprehensible' things echoes Ferguson's discussion concerning the error of assuming that the content of revelation includes only ideas that can be 'clearly and distinctly' perceived.

Boyle, in his discussion of 'emergent' or 'gradual' truths, further echoes Ferguson, who had argued that principles should not be extended beyond their sphere, and extends Ferguson's comments to include explicitly even the principles of logic, which, Boyle argues, should be considered 'emergent' or 'gradual' truths from God's perspective. In doing this Boyle provides a rationale for the category of 'unsociable' things, a category which had only been hinted at earlier when Baxter had argued that human beings cannot perceive the harmony among all truths, when Hammond had stated that God can reconcile His own contradictions and when Owen had stated that some revelations may indeed be contrary to reason.

What did Boyle hope to accomplish in his writings about things that are above reason? Unfortunately there do not seem to be any surviving records of communications between Boyle and Howe. I have been unable to find any such communication among the Boyle Papers and Letters at the Royal Society, and Howe's papers were burned, at his request, shortly before his death.[76] After 1675 Howe resided in London; perhaps his contacts with Boyle were oral rather than written. In any event, it seems likely that Boyle had suggested that Howe expand Hammond's argument and publish it in an attempt to bring about a reconciliation between the conditional predestinarians and the Calvinists; as I noted above, Howe's (and Hammond's) emphasis on man's free will accommodates the conditional predestinarian position by allowing for the possible salvation of all souls, and Howe's (and Hammond's) explanation that by a 'super-effluent grace' God miraculously saves some particular souls allows for the Calvinist position. It is possible that Boyle's original intent was that Howe should write only about Hammond's statement that 'God can reconcile his own contradictions' (or, in other words, that Boyle originally intended Howe to say what Boyle would eventually say in *Things above Reason*), but it seems more likely that, when Howe's attempted reconciliation was vehemently rejected by the Calvinists, Boyle decided that what needed to be emphasised was the fact that these matters are above reason. At the same time, he made an implicit statement of his own conditional predestinarian views by using God's prescience and man's free will as his primary example of two contradictory propositions which are both true; Boyle *could* have used absolute predestination and man's free will as his example, but did not do so. That he used God's prescience rather than God's absolute decrees as his example indicates his rejection of the Calvinist position, especially since he did so in the wake of the objections of strict Calvinists to Howe's *Reconcileableness*. Further, by stressing reason's limits in *Things above Reason*, Boyle was able to exempt the mysteries of Christianity from rational scrutiny and analysis, including not only the mysteries inherent in the process of salvation, but also other mysteries, such as that of the Trinity,[77] the six-day creation, Christ's nature as both divine and human and the virgin birth.[78] His major point, however, was irenical in nature: If finite human reason is unable to comprehend the mysteries of Christianity, then the claim that any individual or sect has achieved a uniquely correct interpretation of those mysteries is unfounded.

'A moderate and well-temper'd man', Boyle noted elsewhere, 'in those controversies in Divinity where the Love of truth dos not engage him to spake, will let the Love of peace keep him silent'.[79] In *Things above Reason* the love of truth engaged Boyle implicitly to state his conditional predestinarian views by using God's prescience and man's free will as his example, but the love of peace combined with his convictions concerning the limits of human reason contributed to his silence where the Calvinist doctrine of predestination was concerned. Boyle's purpose in *Things above Reason* was neither to engage in anti-Calvinist polemics nor to argue for the correctness of his own beliefs, but rather to persuade his fellow Christians that the inability of human reason to understand God's revelations completely should prohibit their speaking confidently and dogmatically about those revelations.

Conclusion

My purpose has been to show how the theological controversies of his day contributed to Boyle's views on the limits of human reason. What is the significance of this for our understanding of Boyle's philosophy as a whole?[80]

First of all we should revise our view of Boyle as being one of those Christian virtuosi from whom 'deism, the religion of reason, steps full grown'.[81] If one of the roots of the religion of reason was the emphasis placed on natural religion by the Christian virtuosi, an equally important source was the insistence of the Socinians (and some other Christians) on 'de-mystifying' Christianity, and this was a development that Boyle opposed.

Secondly, we should extend our notion of Boyle's voluntarism to include the fact that, in Boyle's view, God freely chose to impose limits on human understanding. There are a number of passages in Boyle's published works in which Boyle makes this point,[82] but he makes it most explicitly in an unpublished passage in which he states that just as God 'freely establisht the . . . Laws of Motion by which the universe was framed, & doth act, so he freely constituted the Reason of Man'.[83] In Boyle's view, human beings are capable of understanding just as much as God intended and no more.

Thirdly, we should recognise the significance of Boyle's views on the limits of reason where the development of so-called 'British' empiricism is concerned,[84] for Boyle thought that the limits God set on human reason affect the ability to know natural as well as revealed truths. In fact, Boyle explicitly stated that *Things above Reason* should be considered a philosophical work rather than a theological work;[85] his purpose was to show that there are things both in nature and in theology which reason cannot comprehend. Although God could have constituted human reason in such a way that we would have perfect knowledge, He did not do so. 'We know perfectly and fully no truth', Boyle states, 'because we know not all its Connections and respects'.[86]

Finally, and most importantly, we should recognise the importance of seeing the unity of Boyle's thought, and in doing so we shall come to understand the extent to which Boyle's theological views affected his thought as a whole. James E. Force has argued that scholars who study Isaac Newton's theology, science and politics as separate units fail to recognise the extent to which Newton's voluntarism 'underlies all his work'.[87] The same thing is true where Boyle studies are concerned; Boyle's view of God is the starting point from which all of his other views follow. In recent years scholars have become increasingly aware of the extent to which theological views helped to shape the modern mind,[88] and I hope that this study of Boyle's thought will help to reinforce that awareness.

Acknowledgements

I am grateful to James E. Force and Michael Hunter for their extensive and extremely helpful comments on an earlier draft of this essay.

NOTES

1 In addition to the relevant essays in this volume, see, *inter alia*, the items in the Bibliography by E. B. Davis, J. T. Harwood, M. Hunter, E. M. Klaaren, J. E. McGuire, M. J. Osler, T. Shanahan and R. S. Westfall.
2 *Things above Reason, Works*, iv, 411.
3 *Theological Distinction, Works*, v, 542.
4 *Things above Reason, Works*, iv, 423.
5 *Ibid.* p. 407.
6 See, e.g., *ibid.* p. 422 and *Theological Distinction, Works*, v, 543.
7 *Things above Reason, Works*, iv, 409.
8 *Ibid.* pp. 420–2.
9 *Ibid.* p. 407.
10 *Ibid.* p. 423.
11 See, e.g., *ibid.* pp. 408–9; *Advices, Works*, iv, 466; *Reason and Religion, Works*, iv, 176.
12 *Things above Reason, Works*, iv, 415; *Christian Virtuoso I, Appendix, Works*, vi, 697–8.
13 *Advices, Works*, iv, 462–3.
14 *Ibid.* p. 451.

15 *Ibid.* p. 458.

16 *Ibid.* pp. 461–2.

17 *Things above Reason, Works,* iv, 445. Note that the text is mispaginated, skipping from p. 424 to p. 445; nevertheless, the content of the work is complete.

18 *Ibid.* p. 413.

19 See, e.g., *Christian Virtuoso I, Works,* v, 515; *Reason and Religion, Works,* iv, 161.

20 *Things above Reason, Works,* iv, 413, 445.

21 *Ibid.* p. 446.

22 William Kneale and Martha Kneale, *The Development of Logic* (New York: Oxford University Press, 1962), p. 314.

23 *Things above Reason, Works,* iv, 423.

24 *Advices, Works,* iv, 450, 463; *Theological Distinction, Works,* v, 547; *Excellency of Theology, Works,* iv, 32.

25 William Chillingworth, *The Religion of Protestants A Safe Way to Salvation; or, An Answer to a Booke Entituled Mercy and Truth, or, Charity maintain'd by Catholiques, Which pretends to prove the Contrary* (Oxford, 1638), p. 77.

26 See, e.g., Joachim Stegmann, *Brevis disquisitio . . .* (Amsterdam, 1633); this work was later translated by John Biddle under the title *Brevis Disquisitio; or, a Brief Enquiry Touching a Better way Then is commonly made use of, to refute Papists, and reduce Protestants to certainty and Unity in Religion* (London, 1653). For background information on the Socinians, see George Huntston Williams, *The Radical Reformation* (Philadelphia: Westminster Press, 1962); *ibid.* (ed., trans.), *The Polish Brethren: Documentation of the History and Thought of Unitarianism in the Polish–Lithuanian Commonwealth and in the Diaspora, 1601–1685,* 2 vols (Missoula, Mn.: Scholars Press, 1980). For general information concerning Socinianism in England, see H. John McLachlan, *Socinianism in Seventeenth-Century England* (London: Oxford University Press, 1951). For studies concerning the influence of Socinianism on the rise and growth of English deism, see Zbigniew Ogonowski, 'Le "Christianisme sans mystères" selon John Toland et les sociniens', *Archiwum historii filozofii i mysli spolecznej* 12 (1966): 205–23; Robert E. Sullivan, *John Toland and the Deist Controversy* (Harvard University Press, 1982); and J. A. I. Champion, *The Pillars of Priestcraft Shaken: The Church of England and its Enemies 1660–1730* (Cambridge University Press, 1992).

27 Faustus Socinus *et al., The Racovian Catechisme: Wherein you have the substance of the Confession of those Churches, which in the Kingdom of Poland, and Great Dukedome of Lithuania, and other Provinces appertaining to that kingdom, do affirm, that no other save the Father of our Lord Jesus Christ, is that one God of Israel, and that the man Jesus of Nazareth, who was born of the virgin, and no other besides, or before him, is the onely begotten Sonne of God,* trans. John Biddle (Amsterledam (sic) [London], 1652), pp. 28, 53.

28 John Owen, *Vindiciæ Evangelicæ; or, The Mystery of the gospell vindicated, and Socinianisme Examined, In the Consideration, and Confutation of A Catechisme, called A Scripture Catechisme, Written by J. Biddle M.A. And the Catechisme of Valentinus Smalcius, commonly called the Racovian Catechisme. With The Vindication of the Testimonies of Scripture, concerning the Deity and Satisfaction of Jesus Christ, from the Perverse Expositions, and Interpretations of them, by Hugo Grotius and his Annotations on the Bible. Also an Appendix, in Vindication of some things formerly written about the Death of Christ, & the fruits thereof, from the Animadversions of Mr. R[ichard] B[axter]* (Oxford, 1655).

29 The *Racovian Catechism* (n. 27) was first published in 1605 at Raków and was re-issued in at least fifteen editions or reprints (some expanded and revised) during the seventeenth century. Biddle's *Scripture Catechism* referred to is *A Twofold Catechism: The One simply called A Scripture-Catechism; The Other, a brief Scripture-Catechism for Children* (London, 1654).

30 Owen, *Vindiciæ* (n. 28), separately paginated preface, p. 63.

31 *Ibid.* pp. 65–6.

32 Richard Baxter, *The Arrogancy of Reason against Divine Revelations, Repressed; or, Proud Ignorance the cause of Infidelity, and of Mens Quarrelling with the Word of God* was one of four parts of his *Unreasonableness of Infidelity* (London, 1655); *The Arrogancy* was the part aimed at 'staying . . . the present Course of Apostasie' (title page of *Unreasonableness*). *The Arrogancy* was also published as a separate work that same year; the pagination in the two editions is the same. The timing of publication and the fact that this part is aimed at those who made human reason the judge of revelation make it reasonable to assume that the Socinians were (at least among) those whom Baxter had in mind.

33 *Ibid.* p. 13.

34 *Ibid.* pp. 21–3, on p. 23.

35 John Morgan, *Godly Learning: Puritan Attitudes towards Reason, Learning, and Education, 1560–1640* (Cambridge University Press, 1986), esp. ch. 6.

36 Edward Arber (ed.), *The Term Catalogues, 1668–1709, Vol. I* (London, 1903), introduction, p. xiv.

37 Benjamin Calamy, *A Sermon Preached before the Right Honourable . . . Lord Mayor . . . upon the 13th of July, 1673* (London, 1673), p. 15.

38 Jackson I. Cope, *Joseph Glanvill: Anglican Apologist* (St Louis: Washington University Studies, 1956), pp. 85–6, ch. 6.

39 Joseph Glanvill, *ΛΟΓΟΥ ΘΡΗΣΚΕΙΑ; or, A Seasonable Recommendation, and Defence of Reason, in the Affairs of Religion; Against Infidelity, Scepticism, and Fanaticisms of all sorts* (London 1670), p. 1. *A Seasonable Recommendation* is an early version of Glanvill's 'Agreement of Reason and Religion' published in his *Essays on Several Important Subjects in Philosophy and Religion* (London, 1676).

40 Glanvill, *A Seasonable Recommendation* (n. 39), p. 22.

41 *Ibid.* p. 14.

42 *Ibid.* p. 25.

43 Peter Toon, *God's Statesman: The Life and Work of John Owen, Pastor, Educator, Theologian* (Exeter: Paternoster Press, 1971), pp. 156–7.

44 William Sherlock, *A Defense and Continuation of the Discourse concerning the knowledge of Jesus Christ, and our Union and Communion with Him. With a particular respect to the Doctrine of the Church of England, and the Charge of Socinianism and Pelagianism* (London, 1675).

45 Robert Ferguson, *The Interest of Reason in Religion* (London, 1675), p. 226.

46 *Ibid.* pp. 228–30.

47 *Ibid.* pp. 266–7.

48 *Ibid.* pp. 14–19.

49 *Ibid.* pp. 22–3.

50 *Ibid.* p. 20.

51 John Owen, *The Nature of Apostasie* (London, 1676), p. 7.

52 *Ibid.* pp. 9–11.

53 *Ibid.* p. 297.

54 *Ibid.* pp. 12–14, 301.

55 *Ibid.* pp. 9, 284–97.

56 *Ibid.* p. 278.

57 For a contemporary account of these subtleties, see J[ohn] H[umphrey], *The Middle-Way in one Paper of Election & Redemption. With Indifferency between the Arminian & Calvinist* (London, 1673).

58 Examples of such arguments include: the Calvinist Peter Sterry's denial of free will on the grounds that it contradicts God's omniscience, *A Discourse on the Freedom of the Will* (London, 1675); the Arminian Samuel Hoard's denial of predestination on the grounds that the doctrine is inconsistent with God's exhortations against sin, *Gods Love to Mankind*, 2nd edn (London, 1673, first published 1633), esp. pp. 157–70. For the Socinian claim, see *The Racovian Catechisme* (n. 27), p. 146.

59 John Howe, *The Reconcileableness of God's Prescience of the Sins of Men, with the Wisdom and Sincerity of his Counsels, Exhortations, and whatsoever Means He uses to prevent them, in a Letter to the Honourable Robert Boyle Esq.* (London, 1677) was originally published under Howe's 'penultimate letters' ('H.W.'). My citations come from the edition of the same year, *To which is added a Post-Script in London 1677, Defence of the said Letter*; the *Post-Script* is separately paginated. Because the original publication of *Reconcileableness* precipitated a controversy, Howe considered it more 'becoming' for the subsequent edition to be published under his full name (*Postscript*, pp. 2–3). Unless otherwise noted, my citations to this work refer to the body of the original text rather than to the *Postscript*.

60 About the time that Hammond published *A Pacifick Discourse*, Boyle was subsidising Sanderson's writings concerning cases of conscience, using Thomas Barlow, later Bishop of Lincoln, as an intermediary; see Michael Hunter, 'Casuistry in action: Robert Boyle's confessional interviews with Gilbert Burnet and Edward Stillingfleet, 1691', *Journal of Ecclesiastical History* 44 (1993): 82–3.

61 Howe, *Reconcileableness* (n. 59), p. 6.

62 *Ibid.* p. 7.

63 *Ibid.* p. 15.

64 *Ibid.* p. 10. (The text is mispaginated; there are two pages numbered '10'. My reference is to the first.)
65 Henry Hammond, Χάρις καὶ εἰρήνη; *or, A Pacifick Discourse of Gods Grace and Decrees: In a Letter, or full Accordance written to the Reverend, and most learned, Dr. Robert Sanderson, to which are annexed the Extracts of three Letters concerning Gods Prescience reconciled with Liberty and Contingency* (London, 1660), p. 160.
66 Howe, *Reconcileableness* (n. 59), pp. 31–50.
67 *Ibid.* p. 44.
68 Theophilus Gale, *The Court of the Gentiles, Part IV, Of Reformed Philosophie, wherein Plato's Moral, and Metaphysic, or prime Philosophie is reduced to an useful Forme and Method* (London, 1677), p. 523.
69 Howe, *Reconcileableness, Postscript* (n. 59), p. 2.
70 *Ibid.* pp. 28–9.
71 T[homas] D[anson], *De Causa Dei; or, A Vindication of the Common Doctrine of Protestant Divines, concerning Predetermination: (i.e., the Interest of God as the first Cause in all the Actions, as such, of all Rational Creatures:) from the invidious Consequences with which it is burdened by Mr. John Howe, in a late Letter and Postscript, of God's Prescience* (London, 1678), pp. 3, 121.
72 [Andrew Marvell], *Remarks upon a Discourse writ by one T.D., under the pretence of De Causa Dei, and of Answering Mr. John Howe's Letter and Postscript of God's Prescience, Affirming, as the Protestant's Doctrine, That God doth by Efficacious Influence universally move and determine Men to all their Actions, even to those that are most Wicked* (London, 1678), p. 3.
73 *Ibid.* p. 6.
74 *Advices, Works*, iv, 466.
75 *Theological Distinction, Works*, v, 542.
76 Henry Rogers, *The Life and Character of John Howe, M.A., with an Analysis of His Writings* (London, 1836), p. 2.
77 *Advices, Works*, v, 455; *Christian Virtuoso I, Appendix*, vi, 695.
78 *Theological Distinction, Works*, vi, 543.
79 BP 5, fol. 32v.
80 I explore the following and other implications of Boyle's thought more fully in 'Robert Boyle and the limits of reason: a study in the relationship between science and religion in seventeenth-century England', Ph.D. dissertation, University of Kentucky, 1992.
81 Richard S. Westfall, *Science and Religion in Seventeenth-Century England* (New Haven: Yale University Press, 1958), p. 141.
82 See, e.g., *Christian Virtuoso I, Appendix*, vi, 694, 697; *Advices, Works*, iv, 466.
83 BP 36, fol. 46v.
84 For an explanation of why the term 'British empiricism' is misleading, see David Fate Norton, 'The myth of British empiricism', *History of European Ideas* 1 (1981): 331–44.
85 *Things above Reason, Works*, iv, 418, 423.
86 BP 25, fol. 297.
87 James E. Force, 'Newton's God of dominion: the unity of Newton's theological, scientific, and political thought', in Force and Richard H. Popkin, *Essays on the Context, Nature, and Influence of Isaac Newton's Theology* (Dordrecht: Kluwer Academic Publishers, 1990), pp. 75–102, on p. 95.
88 In addition to the items in the Bibliography to this volume cited in note 1, see, e.g., Force and Popkin, *Essays* (n. 87); Peter Dear, 'Miracles, experiments and the ordinary course of nature', *Isis* 81 (1990): 663–83; Alan Gabbey, 'Henry More and the limits of mechanism', in *Henry More (1614–1687): Tercentenary Studies*, ed. Sarah Hutton (Dordrecht: Kluwer Academic Publishers, 1990), pp. 19–35; Michael Hunter, *Science and Society in Restoration England* (Cambridge University Press, 1981); Paul Russell, 'Scepticism and natural religion in Hume's *Treatise*', *JHI* 49 (1988): 247–65.

'Parcere nominibus': Boyle, Hooke and the rhetorical interpretation of Descartes

EDWARD B. DAVIS

It was not uncommon for Robert Boyle, who feared plagiarists almost as much as atheists, to entrust his precious drafts to Henry Oldenburg, who was certainly neither, and who could if Boyle desired publish them properly in the *Philosophical Transactions*. The draft for one of his greatest works, *A Disquisition about the Final Causes of Natural Things*, was no exception. Dedicated to Oldenburg, who had 'pressed me for it', the body of the treatise had been written 'many years' before its actual publication in 1688, according to Boyle's own account in the preface. Oldenburg certainly knew of its existence, and probably saw a draft early in 1677, but upon his death later that year the unpublished draft was 'thrown by',[1] not to see the light of day for more than a decade. By that time Boyle had dredged up for publication another draft he had thrown by – that of *A Free Inquiry into the Vulgarly Received Notion of Nature*, originally written in 1666, but not printed until 1686.

The central problem Boyle confronted in these works is one that dogged him throughout his scholarly life: how to reconcile the mechanical philosophy with Christian theology. Boyle chose the high road to harmony, arguing that the new science was not only acceptable to Christians, but actually more amenable to Christianity than less mechanistic alternatives. In staking out his position, he set himself squarely in between the opposite extremes of Epicurean atomism, with its absolute denial of divine providence, and a form of Christian natural philosophy that transformed nature into God's intelligent agent whose anthropocentric purposes were ever so clear.

Hand in hand with this intellectual problem was the social problem posed by the fact that some of his opponents were personal friends, fellow Christians who detected in Descartes the faint odour of atheism. How could they be persuaded to give Cartesianism a fair hearing? Manuscript evidence I have recently discovered shows that, at the very time Boyle was putting these treatises into final form in the early 1680s, considerations about the best rhetorical strategy for dealing with Henry More and other Christian opponents were uppermost in his mind. In the process, Robert Hooke, Boyle's one-time laboratory assistant, stands revealed as Boyle's confidential counsel on matters Cartesian, and gives his approval to Boyle's strategy for dealing with Cartesianism and its adherents.

The crucial piece of evidence lies buried in the morass known as the Boyle Papers, housed at the Royal Society, an archive which until very recently has been essentially uncatalogued. With the aid of Michael Hunter's new finding list,[2] I located and identified a single folio sheet containing Robert Hooke's notes on a draft of *Final Causes*. It is to the best of my knowledge the only document of its kind in the entire collection of some 20 000 pages. Nowhere else can we read for ourselves the advice that Boyle received in confidence about the draft of any of his treatises, let alone one as significant as *Final Causes*. In the following analysis of this manuscript and its context, several questions will feature prominently: How did Robert Boyle prefer to understand Cartesianism, especially Descartes' position on final causes? How does this fit in with Boyle's general rhetorical strategy for dealing with intellectual opponents, especially when they shared Boyle's Christian faith? What role did Robert Hooke play in all of this? And, finally, how then should we understand the relationship between Boyle and his former assistant?

Boyle, Hooke and the issue of final causes

It was clear to me almost as soon as I saw the manuscript that the key to its identity was the presence of several underlined words and phrases, apparently taken from another document, not named. The subjects mentioned in the notes led me to believe that a draft of *Final Causes* was most likely to be that document. A quick check revealed that the underlined words and phrases matched almost perfectly with passages in the published version of the treatise. (For the text of the notes, annotated with the printed texts to which they refer, see the Appendix to this chapter.) The few minor differences are easily explained as revisions Boyle made prior to publication; indeed, their very presence is tangible proof that the notes were comments on a draft rather than a printed text – comments, as we shall see, that Boyle did not ignore.

The earliest possible date for the manuscript can be established from the very last line, which cites an edition of Descartes printed in 1677, the year of Oldenburg's death. The most likely date, however, is the early 1680s. Although *Final Causes* was not printed until 1688, manuscript evidence proves definitively that Boyle made his final revisions in 1682–3. In the preface Boyle says that 'the contagious boldness of some baptized Epicureans engaged me to dwell much longer on the third proposition of the fourth section than I at first intended'.[3] Drafts of this very part of the treatise are found in commonplace books used mostly by Hugh Greg, who began working for Boyle about 1680.[4] The same notebooks contain draft material for several other works printed in 1684–5: *Experiments about the Porosity of Bodies*, *Memoirs for the Natural History of Humane Blood*, *Of the Reconcileableness of Specifick Medicines to the Corpuscular Philosophy*, and *An Essay of the Great Effects of even Languid and Unheeded Motion*, placing a firm upper limit on the date for these additions to *Final Causes*. There are also excerpts from *A Free Inquiry into the Vulgarly Received Notion of Nature* in both note-books,[5] supporting my claim (repeated below) that it too was completed in 1682–3. Of course, the notebooks themselves were not sent to Hooke, the printers, or anyone else. In accord with Boyle's usual practice, the passages they contain would first have been copied out cleanly onto folio sheets, none of which are still extant.[6] I know of just one other surviving draft of *Final Causes*, several folio pages penned by John Warr in the 1670s, when the treatise was first dictated.[7] They show that Boyle added a few passages and rearranged parts of his argument before arriving at the version Hooke saw, which was very close to what was actually printed. Indeed one of the passages Hooke singles out (marked 'c' in his notes) is missing in this draft, which covers the relevant part of section I.

Hooke's comments fall into three general categories: editorial suggestions (a, b, k, p and q); observations on certain astronomical details Boyle calls upon to illustrate his argument (h, i, l, m, n and o); and remarks about Descartes and the Cartesians, with particular attention to final causes (b, c, d, e, f, g and r). It is entirely possible that originally there were more notes, on other aspects of the book; the last line is right at the bottom of the paper. But there is no indication of this on the single page we have.

Although the notes are not signed, several clues point clearly to Robert Hooke as the source. The author had a thorough knowledge of astronomy, which Boyle respected as he (correctly) deferred to him on the frequency of eclipses. He was also very familiar with the works of Descartes and the Cartesians, and those of Boyle himself, as evident from notes (i) and, as we shall see, (f). The juxtaposition of remarks like these with comments on natural theology, especially the phrase about the eye of a fly, immediately led me to consider Robert Hooke. My suspicion was quickly confirmed by Michael Hunter, whose familiarity with the handwriting of manuscripts associated with the early members of the Royal Society is second to none. I have myself compared the manuscript under discussion with several samples of Hooke's hand from the 1670s and early 1680s that are known to be authentic, and have found it to be a very close match.[8]

The fact of Hooke's authorship is of course interesting in itself, and I will give it due consideration later. For the moment I want to examine the way in which Boyle interprets Descartes' position on final causes, and the way in which Hooke responds to this interpretation. My first point is simply to note that Hooke makes this very matter the primary focus of his comments, which suggests that he was responding to a specific request from Boyle to do so – or, at very least, that he knew implicitly that Boyle would want his comments on matters Cartesian. Indeed, Boyle had every reason to consult Hooke about Descartes, for it was with Hooke's help that Boyle had first really studied Descartes: a fact that scholars have not appreciated. The crucial information comes from John Aubrey, who tells us that Dr Thomas Willis recommended Hooke 'to the Hon[oura]ble Robert Boyle, Esqre, to be usefull to him in his Chymicall operations. Mr. Hooke then read to him (R.B., esqre) Euclid's Elements, and made him understand Descartes' Philosophy.'[9]

The meaning of this remark is more fully realised when it is coupled with something Boyle himself said in an unpublished fragment on generation, evidently dating from the early or mid-1650s, before Hooke began working for Boyle in 1658. There Boyle observes that Descartes' *Passions* is 'the only book of his which I remember my selfe to have read over'.[10] The clear implication is that Robert Hooke helped Boyle work through the bulk of Descartes' philosophical works during the four years that he remained in Boyle's employ. This enables us to correct Marie Boas Hall's view[11] that Boyle was only feigning a lack of familiarity with Descartes' *Principles of Philosophy* when he said (in 'A Proëmial Essay' to *Certain Physiological Essays*) that he had 'purposely refrained, though not altogether from transiently consulting about a few particulars, yet from seriously and orderly reading over'[12] that work and certain others. She would have us read this as 'deliberate (and Baconian) propaganda' rather than as an accurate assessment of the facts. I am hardly about to deny, in the midst of an essay on Boyle's rhetorical strategy, that Boyle was wont to hold his cards close to his chest. Nonetheless it seems that he was no more than frank when he proceeded to say that he was just 'beginning now to allow myself to read those excellent books'.[13] Written about 1658–60, at precisely the time when Hooke was 'ma[king] him understand Descartes' Philosophy', it only confirms what Aubrey claims.

Hooke's notes deal with precisely those works of Descartes that he and Boyle must have studied together. They certainly zero in on a topic of mutual interest, natural theology. The most interesting examples are those Hooke labelled (h) and (i), which call attention to a phrase in Boyle's draft that read, 'the eye of a Fly is a more curious peece of workmanship than the Body of the Sun'. The eye of a fly – how could Hooke possibly fail to notice this allusion to one of the most stunning diagrams from his own *Micrographia*?[14] He also picked up on Boyle's view here that the organic creation, above all the smaller creatures within it, exhibits design more clearly and obviously than the inorganic world. Hooke is less sure of this, as an accomplished astronomer might be, and Boyle takes his advice to treat more cautiously.[15] But otherwise both men agreed about the great value of the least of God's creatures to illustrate divine wisdom, a point John Harwood has recently emphasised.[16] 'Nor are the smallest and most despicable productions of nature so barren', Boyle had said in *The Usefulness of Natural Philosophy*, 'but that they are capable both to invite our speculations, and to recompense them'.[17] This work, which was drafted in the 1650s and printed in 1663 (just one year after Hooke left Boyle's laboratory for the Royal Society), was surely well known to Hooke, perhaps even in manuscript. No doubt Hooke echoed these words deliberately in *Micrographia*, when he proclaimed in a discussion of flying insects 'that the Wisdom and Providence of the All-wise Creator, is not less shewn in these small despicable creatures, Flies and Moths, which we have branded with the name of ignominy, calling them Vermine, then in those greater and more remarkable animate bodies, Birds'.[18] Likewise for moss, which Hooke reckoned like Solomon to rank 'with the tall Cedar of Lebanon'.[19] Here again he followed Boyle, who had argued that the wisest of natural historians 'did not only treat of the tall cedars of Lebanon, but that despicable plant . . . that grows out of the wall'.[20]

We must not allow the fact that *Micrographia* was written under the auspices of the Royal Society to please certain virtuosi, Boyle among them, to cause us to conclude that

Hooke's many invocations of the design argument were in any way insincere. Nothing we know supports this.[21] If anything, we must conclude that the architect and mechanic Robert Hooke saw in nature the work of the supreme architect and mechanic. Just as Hooke sought Boyle's approval by implicitly citing Boyle's arguments in *Micrographia*, so Boyle sought Hooke's approval by asking him to review *Final Causes* prior to publication.

After publication, Boyle sent Hooke a copy of the book,[22] perhaps in gratitude for Hooke's contribution, which we can now see as part of an ongoing conversation about Descartes and final causes that had begun when Hooke first 'made him understand Descartes' Philosophy'. The thrust of Hooke's comments is to support Boyle's overly generous reading of Descartes' position on final causes, the clearest statements of which are just those that Hooke cited in note (b): in the *Principles of Philosophy*, part III, articles 1–3, and in the sixth paragraph of the fourth Meditation. In the first place Descartes warns his readers not to underestimate the power of God, and not to overestimate ourselves, by assuming that God made everything solely for our use. 'It would clearly be ridiculous', he says, to appeal to such purposes to support physical reasoning, since there are some things we have never seen or known about. In the *Meditations* he put it even more strongly, stating that God's purposes are frankly inscrutable to finite creatures, and thus 'the kind of cause known as final is useless in physical things'.[23] This would seem to leave little room for manoeuvring, yet Boyle found some, and Hooke apparently concurred. A comparison of Hooke's notes with the associated passages in Boyle shows that they chose to interpret Descartes very liberally, to make him appear to allow the possibility that we humans might comprehend at least some of God's purposes in some of God's works, though an exhaustive knowledge of final causes would always exceed our grasp. The only allowance Descartes had made was his concession that, 'from a moral perspective, it might be pious to believe that God made all things for us, because this might move us even more to love him and give him thanks for abundant blessings'.[24] But here, as elsewhere, Descartes had pushed for a separation between the concerns of the pious and those of the natural philosopher; turning Augustine on his head, he had argued that what was good for the faith was irrelevant, if not absurd, in physics.

It was this very attitude more than anything else that Boyle objected to in Descartes. Piety in Boyle's view was no less worthy a concern for the naturalist than for the theologian. As he wrote in *Final Causes*,

> it seems injurious to God, as well as unwarrantable in itself, to banish from natural philosophy the consideration of final causes; from which chiefly, if not only, I cannot but think (though some learned men do otherwise) that God must reap the honour, that is due to those glorious attributes, his wisdom, and his goodness. And I confess that I sometimes wonder, that the Cartesians, who have generally, and some of them skilfully, maintained the existence of a Deity, should endeavour to make men throw away an argument, which the experience of all ages shews to have been the most successful (and in some cases the only prevalent one) to establish, among philosophers, the belief and veneration of God.[25]

Indeed veneration, even more than belief, was what Boyle wanted to encourage in his contemporaries. Although Boyle acknowledged the legitimacy of the ontological argument, which Descartes had placed at the centre of his philosophy, he remained lukewarm about it, always preferring to infer God's existence and attributes from the evidence of design in nature, which was much more likely to produce a real sense of piety and devotion. It was his desire, he stated near the end of *Final Causes*, 'that my reader should not barely observe the wisdom of God, but be, in some measure, *affectively convinced of it*'.[26] The ontological argument simply did not go far enough to produce genuine reverence. 'It is true', Boyle admitted, that

> in the idea of a Being infinitely perfect, boundless wisdom is one of the attributes that is included: but, for my part, I take leave to think, that this general and indefinite idea of the divine wisdom, will not give us so great a wonder and veneration for it, as may be produced in our own minds, by knowing and considering the admirable contrivance of the particular productions of that immense wisdom, and their exquisite fitness for those ends and uses, to which they appear to have been destinated.[27]

With his own way of considering the world manifestly a product of both power and wisdom, Boyle believed, 'men may be brought, upon the same account, both to acknowledge God, to admire Him, and to thank Him'.[28]

Boyle's desire to encourage vigorous piety and active devotion – in short, to make his contemporaries over in his own image – motivated his use of teleological rather than ontological arguments, as well as his rejection of Cartesianism pure and simple. Certainly his deep suspicion of pure reason and his strong commitment to theological voluntarism also contributed here. A free and sovereign God, Boyle believed, reveals Himself more clearly, more fully, and with more religious effect in His written word and in His freely created, exquisitely crafted works than in seeds planted in the human mind. It is worth noting in passing that Isaac Newton shared Boyle's disenchantment with the more abstract Cartesian approach, and for similar reasons.[29]

Going little by names: Boyle's rhetorical strategy

If Boyle sought to distinguish himself from Descartes and his followers on the necessity of including final causes within natural philosophy, nevertheless he sought equally to ally himself with Descartes, and if possible his followers also, on the more fundamental issue of Christian theism. Noting in the preface to *Final Causes* that both the Epicureans and the Cartesians were 'my adversaries, not my assistants', Boyle tried not so subtly to make the Cartesians at least friendly adversaries if not also his assistants. The Epicureans and the Cartesians were after all 'opposite sects', he observed, the one denying that final causes even exist (since there is no God), the other simply asserting our presumption in trying to know them (since our minds are finite and God's is not).[30] That left Boyle some middle ground which he promptly seized. 'It is not without trouble', he wrote, 'that I find myself obliged . . . to oppose . . . some sentiments of M. Des Cartes, for whom otherwise I have great esteem, and from whom I am not forward to dissent'. 'This I the rather declare to you', he hastened to add, 'because I am not of their mind, that think M. Des Cartes a favourer of atheism, which, to my apprehension, would subvert the very foundation of those tenets of mechanical philosophy that are particularly his'.[31]

No doubt this was said for the benefit of those who gave ear to Henry More, the Cambridge philosopher and divine with whom Boyle maintained an amicable, if not always harmonious, relationship.[32] More had long been ambivalent about Descartes' philosophy, even sharing some of his concerns in letters sent to the Frenchman during the last couple of years of his life in 1648–9. It was in another letter – this one to Boyle on 4 December 1671 – that More revealed most clearly his differences with Boyle, and the reasons for Boyle's very careful treatment of the matter of Descartes' theism.[33] More had learned from his colleague Ezekiel Foxcroft, that Boyle disapproved of those parts of More's recently published *Enchiridion metaphysicum* (1671) which drew on Boyle's hydrostatical experiments to criticise mechanists. Having talked face to face with Boyle about his forthcoming book earlier in the year, he wrote to Boyle as much to smooth ruffled feathers as to defend his view of Descartes. 'When I was with you', More observed, 'you seemed not to be concerned for yourself, but for *Des Cartes*. I shall now briefly answer as touching both.' After noting that some 'mere scoffers at religion, and atheistical' Hollanders had 'professed themselves *Cartesians*', More reiterated his opinion that Cartesianism was in fact 'prejudicial . . . to the belief of a God'. For his efforts to oppose the Cartesians, More told Boyle, he 'expected, that all, whose hearts are seriously set upon God and religion, would give me hearty thanks for my pains'. If he had misunderstood Boyle's experiments, he had meant no offence, attributing any errors to 'a career of heat and zeal for the main points I labour for', and expressing the hope that 'your great candour and Christianity would easily pardon it'.

Not himself much given to zeal – his piety ran deep and strong, but not hot – Boyle replied with a similar graciousness, as we shall shortly see. For the moment, however, I want only to connect this exchange with Boyle's rhetorical stance in *Final Causes*, where he certainly saw More and his Cambridge colleagues as among his friendlier adversaries (though clearly not of the Cartesian variety).[34] By refusing even to suggest that Cartesianism was atheistic, Boyle left wide open the door for dialogue. However important the issue of final causes was for Boyle, he wanted his readers to know that, so far as he was concerned, his disagreement with Descartes was between Christians, an internal argument that Boyle would not poison with spurious charges of atheism. Nor would he allow anyone else to create unnecessary conflict by reading Descartes too rigidly so as to disallow his own very liberal interpretation of Descartes' position on teleology.

This is why he carefully separated Descartes from his interpreters when he spelled out the two extreme positions he would avoid in the early part of the first section of *Final Causes*. After distinguishing four different ways of understanding purpose in the universe, Boyle indicated his intent to 'dissent from both the opposite opinions; theirs, that, with the vulgar of learned men, will take no notice of final causes but those we have stiled human ones; and theirs, that (as they think with Descartes) reject final causes altogether'.[35] I believe that Boyle added the parenthetical expression after seeing Hooke's note (c), which (again, in a liberal reading of Descartes) pointed out Descartes' endorsement of one of Boyle's four types of purpose, universal ends. The use of such expressions to refine his meaning is absolutely commonplace in Boyle's prose. It should hardly be surprising that they figured prominently in the process of revision. Of the changes Boyle made in the second (1669) edition of *Certain Physiological Essays*, the work he altered between editions more than any other, almost one-fifth amounted to the simple addition or deletion of words in parentheses.[36] In at least two other places in *Final Causes* Boyle appears to have added parenthetical qualifiers after reading Hooke's suggestions – I have in mind those parts of the text that relate to notes (h) and (l). In any case, the effect of the printed text is to show the self-styled Cartesians that they have exceeded their authority, and thus subtly to enlist their support in what follows.

In keeping with his congenial tone, Boyle tried very hard to understand, if not readily to allow, Descartes' objections to teleological explanations. Nowhere is this clearer than in the middle of the first section of *Final Causes*, where he speculates on Descartes' motive, which he regards with as much sympathy as he can. There Boyle suggests that what Descartes really opposed was the explication of final causes at the expense of efficient causes, the understanding of which ought to be the goal of physics. (I wonder at this point whether Boyle has had Bacon for breakfast, as this comes right out of *The Advancement of Learning*.)[37] Boyle's reply is accommodating but firm: 'I readily admit, that in physics we should indeed ground all things upon as solid reasons as may be had; but I see no necessity, that those reasons should be always precisely physical' and never metaphysical. What matters 'in physics, or any other discipline', Boyle continued, is not whether 'a thing be proved by the peculiar principles of that science or discipline, provided it be firmly proved by the common grounds of reason'.[38] A few lines later, after restating Descartes' apparently uncompromising refusal to employ final causes, Boyle tried again to appreciate his motive: 'since I writ the foregoing part of this treatise, I lighted upon a passage of his, wherein he seems to speak more cautiously or reservedly, opposing his reasoning to their opinion, who teach, that God hath no other end, in making the world, but that of being praised by men'.[39] The hesitation is almost audible, yet Boyle cannot quite bring himself to endorse this, adding immediately that this same passage contains 'two or three other things, wherein I cannot acquiesce'. But to ensure that there might be no mistaking his respect for Descartes, Boyle closes the final section of his treatise by affirming a proposition guaranteed to warm the heart of any Cartesian (or Baconian, for that matter): 'That the naturalist should not suffer the search, or the discovery of a Final Cause of Nature's works, to make him undervalue or neglect the studious indignation of their efficient causes',[40] a statement echoed in the last lines of the conclusion.

Clearly, then, Boyle had a deliberate rhetorical strategy for dealing with Descartes: to state his differences honestly, but fairly, making every attempt to understand him, and taking pains not to question the Frenchman's Christianity or to engage in any sort of personal attack. In fact it was his general strategy, applied not only to Descartes but to others as well. Evidently Hooke realised this, for he refers to it in the most interesting of the notes he gave Boyle – the second part of note (f), which comes at the bottom of the first column, after all but one of his comments on Descartes and final causes. Together with the comments in the second column, it appears to have been written separately from what comes before, as the colour of the ink changes quite noticeably from dark to light just at this point. If so, then it represents an afterthought, which would make sense given Hooke's phrasing: 'upon the whole matter therefore were it not more Charity at least, if not justice, &c in this case Parcere Nominibus – or to use names no more then needs must & then too with tenderness &c'.

This was a direct reference to the preface of Boyle's *Hydrostatical Discourse*, written in 1671–2 as a reply to More's *Enchiridion metaphysicum*.[41] There, after apologising for going against a private promise he had made some time before not to write any more books 'in answer to any that should come forth against mine', Boyle proceeded to set the ground rules for his defence against More's objections to Boyle's mechanistic theory of the air. First Boyle acknowledged his appreciation for the tone of More's attack, 'whose civility I would not leave unanswered'. For this reason, Boyle said, 'I have restrained myself to the defensive part, forbearing to attack anything in his Enchiridium Metaphysicum, save the two chapters, wherein I was particularly invaded'. Nevertheless Boyle could not overlook

> that part of his preface, which falls afoul upon Monsieur *des Cartes*, and his philosophy. For though I have often wished, that learned gentleman had ascribed to the divine Author of nature a more particular and immediate efficacy and guidance, in contriving the parts of the universal matter into that great engine we call the world; and though I am still of opinion that he might have ascribed more than he has to the supreme Cause, in the first origin and production of things corporeal, without the least injury to truth, and without much, if any, prejudice to his own philosophy; and though not confining myself to any sect, I do not profess myself to be of the Cartesian: yet I cannot but have too much value for so great a wit as the founder of it, and too good an opinion of his sincerity in asserting the existence of a deity, to approve so severe a censure, as the doctor is pleased to give of him. For I have long thought, that in tenets about religion, though it be very just to charge the ill consequences of men's opinions upon the opinions themselves; *yet it is not just, or at least not charitable, to charge such consequences upon the persons, if we have no pregnant cause to think they discern them, though they disclaim them* ... And I am the more averse from so harsh an opinion of a gentleman, whose way of writing, even in his private letters, tempts me very little to it, because I cannot think him an atheist, and an hypocrite, without thinking him (what Dr. *More* has too much celebrated him) to call him a weak head, and almost as bad a philosopher as a man.[42]

'*Parcere nominibus*', to go little by names. With his choice of words Hooke was clearly signalling his agreement with what he obviously knew to be Boyle's standard policy of taking the rough edges off an argument by avoiding needless *ad hominem* remarks, being fair with opponents and toning down absolute statements whenever possible by adding qualifiers, often in parentheses. It was a policy Boyle had proclaimed years before in the 'Proëmial Essay' that opens *Certain Physiological Essays*, as Steven Shapin and Simon Schaffer have forcefully pointed out.[43] What I want to stress here is the explicit theological justification behind the policy, which made it all the more imperative that Boyle apply it in disputes with More or Christian Cartesians. 'I love to speak of persons with civility, though of things with freedom', Boyle told Pyrophilus (his nephew Richard Jones, for whom Boyle pretended to write the work), for 'a quarrel-some and injurious way of writing does very much misbecome both a philosopher and a Christian'. It was one thing civilly to reason a man out of his opinions, quite another bitterly to oppose his errors. In the latter case, Boyle noted in appropriately ecclesiastical language, 'it is very uneasy to make a proselyte of him, that is not only a dissenter from us, but an enemy to us'.[44]

Going little by names and much by arguments was precisely the policy Boyle employed often in *Final Causes*, where his opponents were frequently Christians. Boyle carefully avoided attacking, or even citing, his contemporaries by name lest he make enemies needlessly. For example, he spoke cryptically in the preface of 'the contagious boldness of some baptized Epicureans' that had caused him to expand that part of his treatise most directly concerned with inferring a deity, where he then lamented 'this libertine age; wherein too many men, that have more wit than philosophy or piety, have, upon Epicurean (and some also even upon Cartesian) principles, laboured to depreciate the wisdom of God'.[45] In the second section he chastised 'a late ingenious author [who] did causlessly reflect upon me, for having mentioned the enlightning of the earth, and the service of men, among the ends of God, which he thought undiscoverable by us'. Then he admonished 'those Cartesians, that, being divines, admit the authority of holy scripture' to pay heed to biblical references to final causes. In the same place he referred to 'those undervaluers of their own species, that are divines', who thought that God would do nothing simply for the good of humanity, a group whom Boyle labels 'those moderns' a bit later.[46] Those of the opposite opinion, 'the school-philosophers, and many other learned men', assumed that all things had been made purely for human good, thus meriting Boyle's description 'the vulgar of learned men'.[47] The list could easily be much longer.

Boyle used an identical strategy in another unpublished document from the early 1680s, a putative 'Post-Script' to the *Vulgarly Received Notion of Nature* that was also intended to calm the waters of controversy among Christians.[48] Full details about the content and dating of the manuscript will be part of a separate study that Michael Hunter and I are preparing of all draft material related to the *Notion of Nature*. Here I will focus only on what it reveals about Boyle's rhetorical strategy for dealing with his Christian opponents. He starts by refusing once again to use names: 'I have so much respect & kindness for some of the persons I am in this paper to opose that I forbare expressly to name them (tho tis like that by the mention of their opinions & arguments you will guess who they are).'[49] What Boyle wants most from his critics is the same respect and kindness, the same fair play and discernment he accords them – and not only for himself, but also for other adherents of the new science. It would be a good start if they would take seriously Descartes' opposition to atheism, which Boyle sees as sincere if not particularly strident. More than this, Boyle desires that critics 'would be henceforth less forward to impute to all Mechanicall Philosophers in generall those opinions unfavourable to Religion that divers of them disowne & that they would take some care to discriminate such from the rest'.[50] Boyle readily grants that, if he and those of like mind had 'indiscriminately swallowed' the Epicurean variety of mechanical philosophy, 'the objection of our sensourers would be of great weight'. But 'we thinke it not just to have imputed to us those errors of the Antient Attomists that we utterly disclaime'.[51] The 'impious errours of those that excluded God from the formation of the world', Boyle added, 'are not to be imputed to the mechanical principles themselves but to the Personall arrogance of those Philosophers that rashly & unskillfully misapply'd those innocent principles by straineing them to do a work for which they were insuffitient'.[52] Having done his best to defend the mechanical philosophy against charges of atheism, Boyle bends over backward in a conciliatory gesture to those who wrongly oppose him. '[B]ut as for those learned men we descent from', he concludes, 'the knowledge I have of some of them which inclines me to favourable thoughts of the rest makes me willing to thinke that the severe censures are but the effects of their zeal to maintaine Religion against Atheisme'. Boyle finds this 'so laudible a cause' that he admits he is 'more influenc'd by my finding them heartily ingag'd in the main quarrel with me against irreligion than by the fault they find with the weapons I employ'.[53]

Once again, it can only be the zeal of Henry More and his circle whom Boyle has in mind. This confirms what John Henry suspects: that a primary target of Boyle's *Notion of Nature* was More, whose notion of a 'Spirit of Nature' as God's vicegerent arose from his rationalist theology of creation. Boyle's commitment to a voluntarist theology of creation made him wish, in Henry's words, 'to repudiate the very basis of More's philosophical theology'.[54] He did just this, though without using names, in both the *Notion of Nature* and *Final Causes*, which have a unity of purpose that has hitherto gone unnoticed. In each case his audience included the same group of divines who misunderstood the new science and failed to see its advantages for the faith. As with the Cartesians, Boyle treated these opponents as a group of potential allies in the real battle against atheism. If properly courted, they could be won over. To that end Boyle styled himself like Galileo before him as a true champion of religion, whose defence of the new science was motivated by his conviction that it is true and therefore not contrary to the faith. In doing so he sought to steer a middle course between the Epicurean atheists (some of whom he thought to be divines), who had done violence to the new science by making mathematical explanations do too much, and More and his like, who stubbornly clung to a notion of nature that Boyle found less harmonious with Christianity than the new one.

Boyle's refusal to accept either extreme, his denial of a simple contradiction between Christianity and the mechanical philosophy, his insistence on sticking to what he thought were the crucial issues separating belief from unbelief, was rooted in his solid grasp of the dialectical nature of Christian theology. This understanding is evident in other aspects of his thought – the dialogue of divine will and reason that underlies both his rejection of rationalist natural philosophies and his belief that the created order is nevertheless intelligible; his outspoken advocacy of a clockwork universe tempered by his conviction that God has not abandoned the world He has made; and his view that Scripture is both the Word of God and the words of men.[55] A world view based on the subtlety of theological dialectic cries out for a rhetorical strategy that seeks clearly defined positions and depth of understanding, not the premature dismissal of apparently opposing views. Such a strategy would only facilitate Boyle's desire to make allies of his Christian opponents and thereby to present a united front against the spectre of atheism, a desire ultimately fulfilled by the lectures established in his will, which called for sermons against unbelief that would rise above conflicts among Christians.

And this is exactly what Boyle himself had tried to do in the decade following Henry Oldenburg's death: to rise above conflicts among Christians. As he prepared for publication two of his greatest treatises, he kept that irenical purpose in mind. The positions he opposed lay on both sides: on the one hand, Epicurean atomism, which Boyle saw as genuinely atheistic; on the other hand, a needlessly anthropocentric form of Christian natural philosophy advocated by certain divines, among them some personal friends. In answering these opponents Boyle staked out his position on the middle ground between them, all the while making overtures to both the Christian Cartesians on the one hand and Henry More on the other, hoping in the end to win their support against the common foe of atheism.

Robert Hooke and Robert Boyle: reconsidered

Having done with Boyle and the Cartesians, I want finally to consider the question of Hooke's role, and what this implies about the relationship between the two men. I will do this by way of a commentary on Steven Shapin's provocative treatment of Hooke and Boyle in a recent article. Hooke is said to have been 'recognised as a person dependent upon others, a person of at best compromised freedom of action, of ambiguous autonomy, and of doubtful integrity. This is to say, his contemporaries might not generally recognise Hooke as a gentleman.' Nor was he a paragon of Christian virtue. Robert Boyle, by contrast, 'modelled the experimental philosopher on the recognised patterns of the devout Christian and the English gentleman'. He was, in a few words, the epitome of his own Christian virtuoso: possessing commitment to truth, humility, honour, devotion to God and independence. This contrast between the two men and the social roles they filled, in Shapin's opinion, led to an ambiguity in Hooke's identity as an experimental philosopher which gave him problems in his work. And the personal relations between Hooke and Boyle 'reflect their relative public standing'. Though Hooke often dined with Boyle at Lady Ranelagh's house, Boyle almost never crossed Hooke's threshold. 'Hooke and the philosophical world came to Boyle; Boyle and the philosophical world did not come to Hooke.'[56]

I find Shapin's description of Boyle quite convincing, even eloquent. To comment further on just one point, the quality of humility that is integral to Shapin's reading of Boyle's character as the Christian virtuoso was central to his dealings with Cartesian opponents, and would have been an appropriate theme to stress earlier in this essay. But I am much less pleased with his picture of Hooke, or at least his picture of how Hooke was perceived by his contemporaries. To cite a particularly glaring example: if Hooke was really thought to lack integrity and autonomy, then why were his services as an architect in demand among gentlemen? Simply because Hooke was not a gentleman, which he was not, is no reason in itself why gentlemen should not have valued his abilities and sought his opinions, trusting Hooke's integrity enough to believe that he could keep confidences and his autonomy enough to render independent judgements.

Let us consider the case of one particular gentleman, Robert Boyle, who knew Mr Hooke very well indeed – well enough to have weathered a few arguments and well enough, no doubt, to be aware of Hooke's rather unvirtuous nocturnal life. The fact that Hooke was neither a gentleman nor a model of the Christian virtuoso does not seem to have prevented Boyle from having great respect for Hooke. Boyle thought highly enough of Hooke's abilities to recommend the best mechanic he would ever have to the Royal Society. He valued his conversation enough to invite Hooke constantly to share his table and to correspond with Hooke when not living in London. Very likely, he trusted Hooke enough to suggest to his dearest sister Katherine that Hooke be employed to supervise construction at Chiswick. And he esteemed Hooke enough to leave him, by that time a man of independent means, his best microscope in his will. Above all, Boyle did what Shapin implies he could not have done: Boyle took his philosophical world to Hooke. Hooke's comments on the draft of *Final Causes* have a confident tone that belies any difference in social status. He is quick to correct the esteemed gentleman, and to suggest improvements in his style and argument. At no time does he assume an air of deference or inferiority. This is not the work of a mere mechanic or even a trusted servant; it is rather the private opinion of a fellow natural philosopher, the former laboratory assistant who had first made Boyle understand Descartes' philosophy some twenty years before.

Acknowledgements

I am happy to state that this paper is based upon work supported by the National Science Foundation under grant number DIR-8821955, without which my work with the Boyle Papers at the Royal Society would have been impossible. Ongoing support from Messiah College has been no less important, and the final draft of this paper was completed with the aid of a Mellon Fellowship in the Humanities at the University of Pennsylvania. The comments of John Henry and Michael Hunter, and the expert help so courteously provided by Mrs Sheila Edwards and her staff at the library of the Royal Society, were invaluable. I thank the President and Council of the Royal Society for permission to cite and to publish transcripts of certain manuscripts in their possession.

Appendix
Royal Society, Boyle Papers 10, fol. 116

The anonymous author is Robert Hooke. In two columns, formerly folded down the middle; entries a–f in column 1, g–r in column 2. The comments in curly brackets are mine; editorial additions are in square brackets. In the notes I have indicated, among other things, the locations of the underlined phrases as found in *Works*, vol. v. The symbol '???' denotes a word I cannot decipher. Any words crossed out are also crossed out here.

    (a) del <u>For</u>. because used twice in 8 words.[57]

    (b) r[ead]. <u>Des-Cartes</u> instead of Cartesius,[58] which name is only used in Latin, & himselfe was not pleased with it, as <u>I find</u> by one of his Epistles.[59]

        I do not remember that Des-Cartes doth discourse of Final Causes in more then two places, viz. towards the beginning of the 3d part of his Principles, & in his 4th Meditation, not far from the beginning also. in both which he seemes to speake of them but incidently, & yet with much Caution & ~~upon the wholl matter~~ in the former especially, which was writt last, he seemes to assert no more concerning them than severall of the particulars, in this your excellent discourse will perhaps amount unto.

    (c) <u>reject final causes altogether</u>[60]
        Des-Cartes in the 3d pt Princ. Philos [Articles] 2 & 3 doth expressly allow your Universal Ends, if not some of the 3 other sorts. there, & in his Epistles &c.

    (d) Tis indeed a part of his hypothesis, & so in this That he suppose, all the parts of matter were at first equall, because both hypotheses did seeme very simple; but he sayes expressly He could have proceeded upon an hypothesis different from the second, & that he had thoughts once of doing so. <u>& perhaps</u> he he might have said as much at the first, ~~??? ???~~ at least the grounds on which he argues for it, seeme not so cogent, as you very well observe.[61]

    (e) <u>Duty</u>,[62] which Des-Cartes acknowledges in the fore cited place. viz [*Principles of Philosophy*, Part III] Art. 3.

    (f) I do not know any Cartesian who denys this.[63]
        ~~quere whether not should be transposed viz ??? not the curious & will contradict him~~[64]
{From here on the same hand continues, but in a much lighter ink.}
        upon the whole matter therefore were it not ~~better in this~~ more Charity at least, if not justice, &c in this case Parcere Nominibus – or to use names no more then needs must & then too with tenderness &c
{The rest of column 1 is left blank; point g begins column 2.}

    (g) <u>for the service of the earth & its inhabitants</u>[65]
        all this is perfectly Cartesian. see again Prins Phil. Pt III. [Art] 3.

    (h) read (if you please) <u>there is more admirable contrivance in a mans muscles</u>, than <u>in</u> (what we know of) <u>the Coelestial orbs. & that the eye of a Fly is a more curious peece of workmanship</u> (quod nos at least as in what appears to us) than the body of the Sun.[66] St Augustin affirms some such thing, but those that come after him would not suffer him to run away were it unattackd.

        The business of Chance is admirably well discourst & illustrated.[67]

(i) The Sun, which is a Nobler Body than the earth.[68] We Plebeans are not so fit to judge of Degrees of Nobility: or els if the eye of a Fly be a more curious peece of workmanship than the Body of the Sun, the assertion may be disputed. were it not therefore better that both Assertions were delivered in Termes a little more warily qualified according to your usuall prudent practice in other cases?

(k) r[ead]. third Question (viz whether & in what sense &c[69]

(l) r[ead]. & (if the Earth stand still) their scarce conceivable rapidity (as before).[70]

(m) That light Milky Way seemes to me i.e. to my ~~understanding~~ Intellect, the darkest part of heaven, notwithstanding that the Telescopes have discoverd it to be a congeries of lesser starrs, & that the new starrs have all of them ~~appeard~~ shoon in it, & may prove one Phainomenon to enlighten our Understandings.[71]

{This entire point m is enclosed in a large parenthesis that was added after it had been written.}

(n) yet this perhaps may be worth noting viz That the Countryes lying neer the poles do have a very small proportion in respect of those lying neer the Line; that is, only according to the respective Segments of the Axis of the Earth (as is demonstrated by Archimedes de Sphaera & Cylind, or at least my be easily deduced from what he there delivers) viz that the Teraqueous Globe within the Latitude of 30 deg. on either side of the Aequator (which is but 6 1/2 degrees more then the Torrid Zone) is equall to all the rest &c. and still the ??? the Poles the lesser the proportion.[72]

but these notes (m) & (n) only obiter.

(o) dele almost for there are every yeare at least 2 Eclipses, tho in no yeare above 8.[73]

(p) The II Proposition is rather a Permission then a Caution, which Name you give to all the five.[74]

(q) displac'd by dislocation que & displace one of those two words.[75]

(r) Dioptricks, & posthumus Tract De Homine from p. 72v De Amst[derdam]. 1677.[76]

NOTES

1 *Works*, v, 393. An essay called 'Whether and how far a Naturalist should consider Finall causes' appears as item nine in the second part of Oldenburg's own list of papers that 'were mentioned to me by the Hon[oura]ble Robert Boyle March. 26. 1677'. See BP 36, fols. 88–9. Oldenburg died on 5 September 1677.

2 Michael Hunter, *Letters and Papers of Robert Boyle: a Guide to the Manuscripts and Microfilm* (Bethesda, Md.: University Publications of America, 1992). I used an earlier version of this work to hone in on certain parts of the massive archive, especially those manuscripts lacking precise identifications that seemed to be drafts of published treatises. In many such cases the final identification given in the printed catalogue is mine.

3 *Works*, v, 394. The section Boyle cites is *Works*, v, 428–39.

4 See R.S. MSS 198, fols 68v–72r, 113, 114; 199, fols 16r–17r, 124r, 129–30, 132. All are in Greg's hand, with corrections in Boyle's hand on 198, fols 68v–72r. On the dates of Boyle's various amenuenses, see the introduction to Hunter, *Letters and Papers* (n. 3); on Greg specifically, see *ibid*. p. xxxii.

5 See R.S. MSS 198, fols 1–8; 199, fols 147–50.

6 Readers are cautioned not to be fooled by BP 9, fols 35–41, Robin Bacon's clean copies of material from two manuscript notebooks: R.S. MSS 185, fols 29–31; 191, fols 110–16. These were intended for an Appendix to *Final Causes* that was never printed, and were clearly written after that treatise had already appeared. MS 191 starts with a list of Boyle's papers (fols 1–2r) dated 22 July 1688, the year in which *Final Causes* was printed.

7 See BP 10, fols 42–6, a draft of the first section of the treatise. Although some other manuscripts

relate to *Final Causes*, including a virtually complete Latin translation by Thomas Ramsay that was never printed (BP 14, fols 86–107; 17, fols 126–41; 24, pp. 419–555), none contains further draft material.

8   See, for example, the following collections at the Royal Society: Classified Papers xx (for a good set of samples covering many years); Classified Papers xxiv (especially item 88, astronomical notes from 30 December 1680); and Early Letters H.3 (note, however, that some letters were written by Hooke's amenuenses).

9   The words 'and made him understand' replace Aubrey's original phrase 'and taught him'. The revised wording puts greater emphasis on Hooke's role as an expositor of Descartes. This remark, from Aubrey's *Brief Lives*, ed. Andrew Clark, 2 vols (Oxford: The Clarendon Press, 1898), i, 411, is not found in the modern reprint editions, which otherwise print the section on Hooke without alteration – a puzzling omission. Michael Hunter called this statement to my attention.

10   BP 28, p. 270.

11   Marie Boas [Hall], *Robert Boyle and Seventeenth-Century Chemistry* (Cambridge University Press, 1958), pp. 26–7.

12   *Works*, i, 302.

13   *Ibid.*

14   See Robert Hooke, *Micrographia* (London, 1665), scheme XXIV (facing p. 175). I am using the modern facsimile (New York: Dover, 1961) of the original.

15   See note 68.

16   See John Harwood's fascinating essay, 'Rhetoric and graphics in *Micrographia*', in *Robert Hooke: New Studies*, eds Michael Hunter and Simon Schaffer (Woodbridge: The Boydell Press, 1989), pp. 119–48. The rest of this paragraph borrows shamelessly from Harwood.

17   *Works*, ii, 12. Cf. pp. 2, 14 and 24ff., all passages which Harwood cites. For a longer discussion that places Boyle's natural theology (including his special admiration for the smallest parts of the Creation) in the context of larger theological issues, see the fourth chapter of my doctoral dissertation, 'Creation, contingency, and early modern science: the impact of voluntaristic theology on seventeenth century natural philosophy', completed in 1984 at Indiana University.

18   Hooke, *Micrographia* (n. 14), p. 198. I first noticed this passage in Steven Shapin, 'Who was Robert Hooke?', in *Robert Hooke: New Studies* (n. 16), pp. 253–85, on p. 277.

19   Hooke, *Micrographia* (n. 14), p. 134.

20   *Usefulness II, sect. 1, Works*, ii, 142.

21   See Harwood, 'Rhetoric and graphics' (n. 16) for a comprehensive review of the circumstances surrounding the publication of *Micrographia*, and support for the sincerity of Hooke's interest in natural theology.

22   A torn scrap of paper bearing the names of those to whom Boyle sent copies of *Final Causes* survives as R.S. MS 190, fol. 174r. In addition to Hooke, recipients included (among others) 'My L[ady]. Ranelaugh', Nehemiah Grew and Bishop Edward Stillingfleet. John Evelyn's copy, now in the British Library, was also a gift from the author, although his name is not on this list.

23   Charles Adam and Paul Tannery (eds), *Oeuvres de Descartes*, 12 vols (Paris, 1897–1910), citing, respectively, from viiiA, 81 and vii, 55.

24   Again citing Descartes, *Principles of Philosophy*, iii. 2: Adam and Tannery (eds), *Oeuvres de Descartes* (n. 23), viiiA, 81.

25   *Works*, v, 401.

26   *Ibid.* p. 439; my italics.

27   *Ibid.*

28   *Ibid.* p. 402.

29   On the significance of voluntarist elements in the thought of Boyle and Newton, see Davis, 'Creation, contingency, and early modern science' (n. 17), ch. 4, 5, and references cited therein; 'Newton's rejection of the "Newtonian world view": the role of divine will in Newton's natural philosophy', *Fides et Historia* 22 (1990): 6–20. An excellent overall discussion of the various theological themes in both thinkers is found in John Hedley Brooke, *Science and Religion: Some Historical Perspectives* (Cambridge University Press, 1991), pp. 130–51.

30   *Works*, v, 393.

31   *Ibid.* p. 401.

32 On Boyle and More, see John Henry, 'Henry More versus Robert Boyle: the spirit of nature and the nature of providence', in *Henry More (1614–87): Tercentenary Studies*, ed. S. Hutton (Dordrecht: Kluwer Academic Publishers, 1990), pp. 55–76; Shapin and Schaffer, *Leviathan and the Air-Pump*, pp. 75–6, 155–6, 207–24, who show that More's colleagues Simon Patrick and Ralph Cudworth were two of his allies; Simon Schaffer, 'Godly men and mechanical philosophers: souls and spirits in Restoration natural philosophy', *Science in Context* 1 (1987): 55–85, esp. 74–6; Maddison, *Life*, p. 135; and Louis Trenchard More, *The Life and Works of the Honourable Robert Boyle* (New York: Oxford University Press, 1944), p. 178, where the passage just quoted is explicitly connected with Henry More.

33 The letter is printed in *Works*, vi, 513–15.

34 I want to stress my own words: More and his Cambridge colleagues were *among* his adversaries. Boyle's strategy of avoiding names makes it devilishly difficult, at times, to identify with real confidence just whom he is arguing against. In seeing More as one specific target here and elsewhere in this paper, I am not in any way ruling out other possible targets. Boyle's rhetorical cannons, as John Henry has put it, 'were loaded with grapeshot'. See Henry, 'Henry More versus Robert Boyle' (n. 32), p. 68).

35 *Works*, v, 396.

36 I generated a list of these changes in the process of preparing with Michael Hunter a new edition of Boyle's works, to be published by Pickering & Chatto.

37 See Francis Bacon, *The Advancement of Learning*, Book II, section vii. 7, ed. G. W. Kitchin (London: J. M. Dent, 1973), p. 97: 'For the handling of final causes mixed with the rest in physical inquiries, hath intercepted the severe and diligent inquiry of all real and physical causes, and given men the occasion to stay upon these satisfactory and specious causes, to the great arrest and prejudice of further discovery.'

38 *Works*, v, 399ff.

39 *Ibid.* p. 400. Boyle quotes this passage in French in the margin, but fails to say that it comes from Descartes' letter to 'Hyperaspistes' from August 1641. There Descartes had railed against the childishness and absurdity of claiming that 'God, like some vainglorious human being, had no other purpose in making the universe than to win men's praise; or that the sun, which is many times larger than the earth, was created for no other purpose than to give light to man, who occupies a very small part of the earth'. Quoting from the translation by Anthony Kenny, *Descartes: Philosophical Letters* (Oxford: Clarendon Press, 1970), pp. 117ff. The original letter is printed in Adam and Tannery (eds), *Oeuvres de Descartes* (n. 23), iii, 422ff.

40 *Works*, v, 443.

41 Part of *Tracts written by the Honourable Robert Boyle* (London, 1672).

42 Quoting throughout from *Works*, iii, 596–8; my italics.

43 See Shapin and Schaffer, *Leviathan and the Air-Pump*, pp. 72–9, especially their comment (p. 73) that 'The *ad hominem* style must at all costs be avoided, for the risk was that of making foes out of mere dissenters.' I will be citing some of the same passages they also cite.

44 *Works*, i, 312.

45 *Works*, v, 394, 439.

46 *Ibid.* pp. 411 (twice), 412, 415.

47 *Ibid.* pp. 395, 396.

48 See BP 7, fols 186–93, written first by an anonymous amenuensis and revised by Boyle himself. Robin Bacon's clean copy of the opening lines survives as BP 36, fol. 22.

49 BP 7, fol. 186r.

50 *Ibid.* fol. 189v.

51 *Ibid.* fol. 191r/v.

52 *Ibid.* fol. 192v.

53 *Ibid.* fol. 193r.

54 Henry, 'Henry More versus Robert Boyle' (n. 32), p. 69. Henry notes (p. 76, note 65) that J. E. McGuire 'is one of the few historians to have noticed that More is one of Boyle's targets'. See McGuire, 'Boyle's conception of nature', *JHI* 33 (1972): 523–42.

55 On all of these points, see Davis, 'Creation, contingency, and early modern science' (n. 17). For a lucid discussion of the dialectic features of Boyle's subtly formulated theology of nature, see Timothy

Shanahan, 'God and nature in the thought of Robert Boyle', *Journal of the History of Philosophy* 26 (1988): 547–69.

56 Shapin, 'Who was Robert Hooke?' (n. 18), pp. 256, 260, 269–72, 282.

57 This may refer to the fourth sentence in the preface (at the top of p. 393): 'For Epicurus, and most of his followers (for I except some few late ones, especially the learned Gassendus) banish . . .'. Here Boyle opens with the word 'For', which is capitalised in Hooke's notes, and then uses it exactly eight words later, as Hooke says. If this is correct, then clearly Boyle did not take this bit of advice to delete the first word in the sentence.

58 I assume this comment accompanied Boyle's first reference to Descartes, which could have been in the fifth sentence of the preface (immediately after the sentence I have quoted in note a), in the second sentence of Section I of the treatise (p. 395), or anywhere else, depending on the draft Hooke was given. Boyle did use the name 'Cartesius' at least once in the published version, on p. 403.

59 I do not know which letter this was.

60 Boyle uses this wording in the ninth paragraph of Section I (p. 396), where he wants to distance himself from those 'that (as they think with Des Cartes) reject final causes altogether . . .'. Here I think Boyle added the parenthetical expression to stress his polemical point that those who rejected final causes altogether were erroneous in their belief that Descartes had done the same.

61 This whole comment would seem to refer to Boyle's discussion of Descartes' account of the conservation of motion and the Creation of the world, which begins in the tenth paragraph of Section I (pp. 396ff.). Relevant passages in the *Principles of Philosophy* include articles II. 36–42, III. 44–7, and IV. 1.

62 See the twelfth paragraph of Section I (p. 397), where Boyle says, 'But to pretend to know God's ends in the former sense, is not a presumption, but rather to take notice of them is a duty.'

63 This could refer to almost anything in Boyle's draft. Perhaps it refers to paragraph thirteen of Section I (p. 398), which begins, 'It is not to be denied, that He may have more uses for it than one, and perhaps such uses as we cannot divine . . .'.

64 Although Boyle uses the words 'curious' and 'contradict' several times in Section I, no one place stands out as the obvious referent here.

65 Boyle uses this phrase in the third paragraph of Section II (p. 403), where he laments the narrow self-centredness of humans: 'I am apt to fear too, that men are wont, with greater confidence than evidence, to assign the symmetrical ends and uses of the coelestial bodies, and to conclude them to be made and moved only for the service of the earth and its inhabitants.'

66 Boyle says this at the end of the third paragraph in Section II (p. 403): 'And, for my part, I am apt to think, there is more of admirable contrivance in a man's muscles, than in (what we yet know of) the coelestial orbs; and that the eye of a fly is (at least as far as it appears to us) a more curious piece of workmanship than the body of the sun.' I believe that the two parenthetical expressions were added by Boyle, at the suggestion of Hooke, to qualify his argument somewhat.

67 Obviously refers to the main argument of Section II.

68 This phrase is not found in the treatise, at least not in Section II. I believe this is because Boyle deleted it from his draft in response to Hooke's comments. In the draft, I think, it followed the phrase quoted in the previous note, thus forming the conclusion to that sentence and thereby the end of the third paragraph.

69 This refers to the question that opens Section III (at the bottom of p. 412): 'Viz. Whither, and in what sense, the acting for ends may be ascribed to an unintelligent (and even inanimate) body.' Boyle opens the body of the *Disquisition* by directly stating the questions he will address in each of the four sections of the treatise. He does not repeat the first question at the start of Section I, for the obvious reason that he has just finished stating the questions he will answer. He does, however, repeat the second and fourth questions verbatim at the start of those two sections. But he forgot to repeat the third question at the beginning of Section III, a fact that Hooke points out to him. Boyle corrected this oversight by repeating that question not in the text itself, but in a note.

70 Boyle opens the second paragraph to Proposition I of Section IV as follows (p. 420): 'I will not only allow you, but encourage you, to take a rise from the contemplation of the coelestial part of the world, and the shining globes that adorn it, and especially the sun and moon, to admire the stupendous power and wisdom of Him, that was able to frame such immense bodies, and, notwithstanding their vast bulk, and (if the earth stand still) scarce conceivable rapidity, keep them, for so many ages, so constant,

both to the lines and paces of their motion, without justling or interfering with one another.' Earlier, three paragraphs before the end of Section II (pp. 410ff.), he had said, 'I think the coelestial bodies do abundantly declare God's power and greatness, by the immobility of their bulk, and (if the earth stand still) the celerity of their motions, and also argue his wisdom and general providence as to them . . .'. Apparently Hooke was advising Boyle to repeat the parenthetical expression he had used earlier to clarify the point that the heavens would be moving with great speed only if the earth were at rest. It is interesting to note that Boyle was willing to sidestep the issue of the earth's motion, at least in this place, perhaps because he wanted not to alienate Catholics and other readers who would be inclined to reject a geodynamic cosmology – the very sort of cosmology that Descartes himself pretended to believe in the *Principles of Philosophy*. Actually by the 1680s Boyle harboured no doubts about the truth of what he customarily referred to as the Copernican 'hypothesis'. See his discussion of this near the end of the fourth section of the *Free Inquiry*. For a particularly interesting unpublished document comparing the Ptolemaic, Tychonic and Copernican hypotheses, see BP 7, fols 253–60.

71 This information, which Boyle did not use, is related to his discussion of the Milky Way in Proposition I of Section IV (p. 421).

72 Comments pertaining to Boyle's remarks about the polar regions, in Proposition I of Section IV (pp. 421ff.).

73 Near the end of Proposition I in Section IV (p. 424), Boyle says that God 'did not think fit to alter the tracts or lines of motion, that He assigned the luminaries, to avoid the eclipses, that must yearly ensue upon their moving in such lines'. Obviously his draft had used the wording, 'must almost yearly ensue', which Boyle altered to get the astronomical facts correct.

74 Section IV of the *Disquisition* (p. 419) considers the question, 'With what cautions Final Causes are to be considered by the naturalist?' Hooke is remarking about the different tenor of the second proposition in this section, which states (p. 424) that 'In the bodies of animals it is oftentimes allowable for a naturalist, from the manifest and apposite uses of the parts, to collect some of the particular ends, to which nature destinated them. And in some cases we may, from the known natures, as well as from the structure, of the parts, ground probable conjectures (both affirmative and negative) about the particular offices of the parts.' Apparently Boyle saw no reason to change this proposition, which is in fact similar in tone to the third proposition.

75 In the fourth paragraph of Proposition II in Section IV (p. 425) Boyle says, 'For if so much as a finger be made bigger by tumours, or displaced by being put out of joint', which reflects the change of wording suggested by Hooke.

76 In the fifth paragraph of Proposition II in Section IV (p. 426), while discussing the exquisite design of the parts of the eye, Boyle praises 'the industrious Scheiner's Oculus, and Des Cartes's excellent Dioptrics'. Hooke makes reference to that part of the 1677 Latin edition of Descartes that begins with article 36, about the construction of the eye. It may be that, in his draft, Boyle had cited only Scheiner; then, after seeing Hooke's comments, he added the reference to Descartes in order to enlist the Frenchman once again as an ally on the question of final causes.

Teleological reasoning in Boyle's
Disquisition about Final Causes

TIMOTHY SHANAHAN

Ex rerum Causis Supremum noscere Causam[1]

Introduction

Robert Boyle's *Disquisition about the Final Causes of Natural Things* is the most detailed treatment of final causality to appear in the seventeenth century. Remarkably, given the importance of the doctrine of final causes for understanding the dramatic shift in philosophies of nature during this period, this work has received scant attention from historians and philosophers of science.[2] To date there has been only one sustained scholarly examination of this treatise. In an article published a decade ago, James Lennox argues that Boyle's primary aim in this work was to provide a defence of teleological inference in experimental science.[3] In Lennox's view, the *Disquisition* should be viewed as an early exercise in the analysis of teleological explanation in natural science, a project that has proven popular in recent years.[4]

I believe that Lennox's interpretation presents a distorted picture of Boyle's thinking, and that a much more satisfactory account of his aims in this treatise is possible. To this end, I undertake a fresh examination of this work, paying particular attention to (i) the adversarial context in which it was written, (ii) the kinds of questions Boyle sought to answer, (iii) the technical distinctions deployed in his arguments and (iv) the reasoning undergirding his conclusions. What emerges is a kind of topographic map of the argumentative contours of the *Disquisition*.[5]

I then use this account to assess Lennox's interpretation. I argue that viewing Boyle's thought through the lens of contemporary philosophy of science, as Lennox does, badly distorts it. Rather than appropriating Boyle for contemporary defenders of teleological explanation in science, he should be viewed in the context of the intellectual challenges he perceived as he drafted this treatise. When this is done, he emerges not as a champion of the scientific virtues of considering final causes (though he did not dispute this), but rather as a pious Christian who wishes to preserve 'an argument, which the experience of all ages shews to have been the most successful . . . to establish, among philosophers, the belief and generation of God'. The result of this thorough reconsideration of the *Disquisition* is a quite different picture of Robert Boyle the defender of final causes than the one that Lennox proposes – one which, I believe, provides a better understanding of his aims and concerns in writing it, and hence of the intellectual context in which this treatise was published.

Boyle's friendly adversaries

The full title of Boyle's work on final causes is *A Disquisition about the Final Causes of Natural Things: Wherein it is inquired, Whether, and (if at all) with what Cautions, a Naturalist should admit them?* This treatise was initially written before the death in 1677 of Henry Oldenburg, first Secretary of the Royal Society, apparently at Oldenburg's request, but was not published until 1688, just a few years before Boyle's death. Robert Hooke read with approval an account of this work to the members of the Royal Society on 23 November 1687.[6]

If we take him at his word in the preface to the *Disquisition*, Boyle wrote this treatise as a response – and rebuke – to certain factions of the mechanical philosophy, which in his view had drawn incorrect (and unfortunate) conclusions from the basic principles of this approach to nature. On the one hand, there are the Epicureans, who (except for Gassendi), 'banish the consideration of the ends of things; because the world being, according to them, made by chance, no ends of any thing can be supposed to have been intended' (393). (Just who these Epicureans are is not clear from Boyle's remarks.) Such philosophers 'think the world was the production of atoms and chance, without any intervention of a Deity; and that consequently it is improper and vain to seek for final causes in the effects of chance' (395). Epicurean philosophers challenge the very *existence* of final causes in nature. They present an ontological challenge.

On the other hand, there are the Cartesians, who 'suppose all the ends of God in things corporeal to be so sublime, that it were mere presumption in man to think his reason can extend to discover them' (393). The Cartesians (including Descartes himself) *accept* the reality of final causes in nature, but judge that, 'God being an omniscient agent, it is rash and presumptuous for men to think, that they know, or can investigate, what ends He proposed to Himself in his actings about his creatures' (395). Descartes did not reject the *existence* of final causes; rather, he rejected the claim that we can *know* the final causes of natural things. Descartes (and those who follow him), therefore, present an epistemological challenge to the doctrine of final causes.

'According to these opposite sects', Boyle writes, 'it is either impertinent for us to seek after final causes [because none can be supposed to exist], or presumptuous to think we may find them [because our reason is not powerful enough for the task]' (393). Final causes either do not exist or, if they do, then we cannot discover them. Boyle finds himself unable to accept either of these views, nor at the same time is he willing to side with the Aristotelians, who place final causes on a par with (or even above) efficient and material causes. Consequently, he sets about a thoroughgoing analysis of the role of final causes in the context of the mechanical philosophy. Against the Epicureans, he tries to show that the postulation of final causes is not only permissible, but is in fact essential for understanding the phenomena of nature. Against the Cartesians he tries to show that we can, at least in some cases, know with a high degree of probability the final causes of natural things.

It is worth noting that for Boyle this was an 'in-house' dispute among advocates of the mechanical philosophy. Despite the differences between Epicureans, Cartesians, and Boyle himself, they all represent 'sects' of one fundamental approach to nature. Although he considers the Epicureans and Cartesians his 'adversaries' (393) in this treatise, they are nonetheless *friendly* adversaries, in the sense that they are not opposed to the basic approach to nature that he advocates, but only to one of its more problematic features. Aristotelians, on the other hand, could be expected to assent to Boyle's general claims about final causes, but to do so on grounds that he finds unacceptable. Despite their common endorsement of final causality in nature, there is nonetheless a greater divide between Boyle and the Aristotelians than between him and the other sects of the mechanical philosophy which reject the study of final causes. Hence he is concerned in this treatise more with implications than with basic principles.

The argumentative structure of the *Disquisition*

Boyle's treatment of final causality in the *Disquisition* is organised around four key questions:

  I.   Whether, generally or indefinitely speaking, there be any final causes of things corporeal, knowable by naturalists?
  II. Whether, if the first question be resolved in the affirmative, we may consider final causes in all sorts of bodies, or only in some peculiarly qualified ones?
  III. Whether, or in what sense, the acting for ends may be ascribed to an unintelligent (and even inanimate) body?
  IV. And lastly, how far, and with what cautions, arguments may be framed upon the supposition of final causes? (394–5)

The *Disquisition* is divided into four sections, each of which is ostensibly devoted to answering one of these four questions. In a final concluding section Boyle states explicitly and succinctly the results of his inquiry concerning each question. We shall return to his conclusions shortly, but first it is important to introduce the technical conceptual apparatus he develops for addressing the issue of final causes in nature.

Boyle begins Section I by noting that perhaps the chief thing which 'alienated' Descartes from allowing the consideration of final causes in nature was 'that the school-philosophers, and many other learned men, are wont to propose it too unwarily, as if there were no creature in the world that was not solely, or at least chiefly, designed for the service or benefit of man' (395). This view can be attributed directly (and fairly) to Aristotle. In the *Politics* Aristotle had claimed that, 'If then we are right in believing that nature makes nothing without some end in view, nothing to no purpose, it must be that nature has made all things specifically for the sake of man.'[7] As his remarks make clear, Boyle was opposed to such an exclusively anthropocentric view. In order to forestall the objections that may easily be framed on so restricted a conception of final causes, he proposes an important distinction (which 'ought neither to be overlooked nor confounded'), which he hopes will forestall potential misunderstandings with regard to his own treatise.

When we speak of the ends which the Author of nature is said to have in corporeal things, one of four things may be signified:

(1) *Universal ends* concern the creation of the whole universe, whose purpose is the 'displaying of the Creator's immense power and admirable wisdom, [and] the communication of his goodness', so that intelligent creatures may express the 'admiration and thanks due to Him' (395).

(2) *Cosmical* (or *systematical*) *ends* concern 'the ends designed in the number, fabric, placing, and ways of moving the great masses of matter, that, for their bulks or qualities, are considerable parts of the world' (396). Cosmical ends are further divided into:
      (a) *celestial ends*, which concern the sun, moon and fixed stars; and
      (b) *sublunary ends*, which concern the Earth and the sea and all the components thereof.
      Such bodies 'were so framed and placed, as not only to be capable of persevering in their own present state, but also as was most conducive to the universal ends of the creation, and the good of the whole world, whereof they are notable parts' (396). Cosmical ends concern those bodies that, because of their size and importance, constitute the gross structure of the universe.

(3) *Animal ends* concern the ends that 'the particular parts of animals are destinated to, and for the welfare of the whole animal himself, as he is an entire and distinct system of organized parts, destinated to preserve himself and propagate his species' (396). Plants, too, are included under this category.

(4) *Human ends* concern those ends 'aimed at by nature, where she is said to frame . . . her productions, for the use of man'. Human ends may be further distinguished into:
      (a) *mental ends*, which relate to man's mind; and
      (b) *corporeal ends*, which relate to man's body not only *qua* animal, but also as he is framed for dominion over the other works of nature to make use of them for his service and benefit (395–6).

Note that only 'human ends', strictly speaking, are directly concerned with securing the welfare of mankind.

At the beginning of Section IV Boyle makes another distinction, the want of which, he thinks, has contributed to the obscurity of his subject. He observes that there are two ways of *reasoning* from the final causes of natural things that ought not to be confounded:

(1) *Physico-theological* or *metaphysical reasoning* begins with the *uses* of things, and conclusions are then drawn concerning the Author of nature and the general ends He is supposed to have intended in corporeal things.
(2) *Physical reasoning* begins with the supposed *ends* of things, and arguments are then framed concerning the peculiar nature of the things themselves. In
 (a) *affirmative* physical arguments, one begins with the supposed ends of things and one argues that a particular characteristic of a natural body or part ought to be *affirmed* or granted; and in
 (b) *negative* physical arguments one begins with the supposed ends of things and one argues that a particular characteristic of a natural body or part ought to be *denied* (420).

When one considers that each of the different kinds of *ends* Boyle distinguishes can be conjoined with each of the different kinds of *reasoning*, the subtlety (and complexity) of his discussion of final causality becomes apparent (and somewhat overwhelming). In the *Disquisition* he has something to say about most of the different possible kinds of teleological arguments (for example, sublunary cosmical–physical arguments, mental human–metaphysical arguments, etc.), but makes no attempt to treat these various kinds of arguments in a systematic manner. He does, however, attempt at the end of this work to provide an explicit answer to each of the four questions he posed at the beginning of his treatise.

With regard to his first question, whether there are any final causes of corporeal things, and whether they can be known by naturalists, Boyle concludes:

> That all consideration of final causes is not to be banished from natural philosophy;
> but that it is rather allowable, and in some cases commendable, to observe and argue
> from the manifest uses of things, that the author of nature pre-ordained those ends
> and uses. (444)

In other words, his first conclusion is that 'physico-theological' (or 'metaphysical') reasoning from the uses of things in nature to the Creator of these things is often permissible, and sometimes laudatory as well. Not only does he think (contra the Epicureans) that there *are* final causes in nature, but also (contra the Cartesians) that we can *discover* at least some of them.

Two distinct sets of arguments (each utilising different considerations) are used to support this claim. The first set of arguments constitutes a direct attack on the Cartesian prohibition against the consideration of final causes in nature, using Descartes' own principles against him (and showing, incidentally, Boyle's own philosophical sophistication). To begin with, Descartes had claimed that there is always the same 'quantity' of motion in the world, which is identical to that originally introduced by God at the creation of the universe. Having put a determinate quantity of motion into the universe, there is no reason why He should later have need to augment or decrease it.[8] But to argue thus, Boyle points out, is to suppose that one is capable of judging God's ends in designing the universe – precisely what Descartes denies is possible!

> For, without a supposition, that they know what God designed in setting matter a–moving, it is hard for them to show, that His design could not be such, as might be best accomplished by sometimes adding to, and sometimes taking from, the quantity of motion He communicated to matter at first. (396)

In a sense, the Cartesians are even more presumptuous than most philosophers in judging of God's designs, because they assume that God could only have achieved His ends for the universe in one way, whereas Boyle thinks that God, being very wise, has 'various ways of compassing his several ends', and that we may be able to discover at least *some* of these ends by observing nature.

Besides, Boyle asks, how does Descartes know that among the many ends for which God has designed natural things, one of them was not that we as intelligent creatures should discern the beauty and goodness of the world, which discernment is only possible if we know the ends for which God has designed natural things (397)? In other words, Descartes claims that we cannot know any of the ends for which God has designed the world, but at the same time assumes that God did not design the world in such a way that humans should discern the final causes of natural things. Given his own principles, Boyle suggests, he has no right to assume this.

Finally, Descartes objected that it was presumptuous for man to suppose that he could know God's ends in the universe. But it is important to make a distinction here. One may claim to know only *some* of God's ends in *some* of His works, or one may claim to know *all* of God's ends in *all* of His works. The latter kind of claim *is* presumptuous, Boyle thinks, but the former is not necessarily so. Indeed, he thinks that the Cartesians avoid intellectual presumption only by incurring the greater expense of intellectual blindness.

> For there are some things in nature so curiously contrived, and so exquisitely fitted for certain operations and uses, that it seems little less than blindness in him, that acknowledges, with the Cartesians, a most wise Author of things, not to conclude, that, though they may have been designed for other (and perhaps higher) uses, yet they were designed for this use. (396)

The Epicureans have a kind of excuse for not recognising God's ends in nature, believing as they do that the world was produced by the casual collision of atoms without the intervention of any intelligent being; but other philosophers, and especially the followers of Descartes, Boyle thinks, have no such excuse.

Descartes assumed that all of God's ends in nature 'lie equally hid in the abyss of the divine Wisdom', and are thus equally unknowable. But this, Boyle thinks, is simply not true. The second set of arguments he develops against Descartes, therefore, depends on a careful *empirical* examination of natural phenomena to determine to what extent God's ends in nature are *in fact* knowable. Such arguments constitute the bulk of the *Disquisition*.

Boyle's strongest support for the conclusion that some of God's ends in nature *are* knowable comes from the legitimacy of 'animal-metaphysical' arguments. Such arguments take as their premises certain observations concerning the uses and structures of the parts of animals, and then an inference is made from these to the ends that God has intended with respect to the animal's welfare. Boyle obviously had a great interest in such arguments, because significant portions of the *Disquisition* are devoted to giving examples of the fitting of structures to ends in various species of animals. (He gives no less than sixty such examples.) He makes the reason for his extended discussion of animal–metaphysical arguments clear: 'There is no part of nature known to us, wherein the consideration of final causes may so justly take place, as in the structure of the bodies of animals' (424).

Of the structures of the bodies of animals, no structure intrigued Boyle more than the vertebrate eye.[9] He gives the following argument in defence of the claim that we can *know*, with a reasonable degree of certitude, that the eye was not only *designed*, but designed for *seeing*:

> The eye (to single out again that part for an instance) is so little fitted for almost any other use in the body, and is so exquisitely adapted for the use of seeing, and that use is so necessary for the welfare of the animal, that it may well be doubted, whether any considering man can really think, that it was not destinated to that use. (425)

Of course, some considering men, namely the Epicureans, *did* doubt that the eye was destinated to that use, or destinated to *any* use for that matter. Boyle's argument, here as elsewhere, is directed against both them and the Cartesians, since if we can *know* the end of a natural thing, then *a fortiori* such an end must *exist*.

Earlier Boyle had admitted that the Epicureans 'may have a kind of excuse' for not recognising the specific purpose of a bodily part, since on their supposition a man's eyes are made by chance and have no relation to a designing agent. But, he says, when one takes account of how exquisitely fitted the eye is to be an organ of sight one can hardly doubt, 'that an artificer, who is too intelligent either to do things by chance, or to make a curious piece of workmanship, without knowing what uses it is fit for, should not design it for an use, to which it is most fit' (398).

This kind of argument is thus equally directed against the Cartesians, who admit final causes, but who deny that we can *know* any of them. Boyle wants to show, on the contrary, that in *some* cases, for example with the eye, it is just plain obstinacy to deny our knowledge of what seems to be the obvious purpose of an anatomical part. Hence, there *are*, he concludes, final causes of corporeal things that can be known by naturalists.

With regard to the second and third questions, namely whether final causes may be considered in *all* sorts of bodies, or only in some peculiarly qualified ones, and whether the acting for ends may be ascribed to unintelligent and even inanimate bodies, Boyle concludes:

> That the sun, moon, and other coelestial bodies, excellently declare the power and glory of God; and were some of them, among other purposes, made to be service-able to man. (444)

Thus he thinks that, in addition to animal–metaphysical arguments, celestial cosmical–metaphysical arguments are also warranted, and so are at least some human–metaphysical ones framed with respect to the celestial spheres.

Celestial cosmical–metaphysical arguments take as premises certain facts concerning the symmetry and construction of the heavenly bodies, and an inference is then made from these to their designed ends. That Boyle considers *one* of the designed ends of the visible cosmos to be to inspire belief in God is evident:

> And to this use, the distance and vastness of the fixed stars, the immensity of the heavens, and the regular motion of the superior planets, (supposing they can bring man no other advantage) may do him good service, since they afford him rational and solid grounds to believe, admire, adore, and obey the Deity. (416)

Still, he insists that celestial cosmical–metaphysical arguments are not nearly so probable as animal–metaphysical arguments because 'the situation of the coelestial bodies do not afford by far so clear and cogent arguments of the wisdom and design of the Author of the world, as do the bodies of animals and plants' (403). In a pair of memorable phrases, he writes that in his view, 'the eye of a fly is . . . a more curious piece of workmanship than the body of the sun' and 'there is incomparably more art expressed in the structure of a dog's foot, than in that of the famous clock at *Strasburg*' (403–4).

With regard to his fourth question, namely how far, and with what cautions, arguments may be framed upon the supposition of final causes, Boyle concludes:

> That from the supposed ends of inanimate bodies, whether coelestial or sublunary, it is very unsafe to draw arguments to prove the particular nature of those bodies, or the true system of the universe. [But] as to animals, and the more perfect sorts of vegetables, it is warrantable, not presumptuous to say, that such and such parts were pre-ordained to such and such uses, relating to the welfare of the animal (or plant) itself, or the species it belongs to. (444)

He thinks that it is a very risky business to infer the nature of inanimate bodies from our guesses about the purposes of such things. One must not assume, for instance,

> that the sun and other vast globes of light ought to be in perpetual motion to shine upon the earth; because they fancy, it is more convenient for man, that those distant bodies, than the earth, which is his habitation, should be kept in motion. (420–1)

The reason is that we know too little of the particular purposes of inanimate bodies, and we cannot be sure that they exist solely for the benefit of humans. Although such arguments are sometimes permissible, they may easily lead one astray.

Animal–physical arguments, on the other hand, can be a good deal more probable. Such arguments take as premises certain supposed ends which have been designed in the parts or systems of animals, and from these one infers the natures of particular parts or systems. For example, from the use of the eye for seeing one may make judgements about the structure and function of various parts of the eye. We may, from the known ends of nature, as well as from the structure of the parts, 'ground probable conjectures, both affirmative and negative, about the particular offices of the parts' (427).

Boyle refers to Scheiner's work on optics to illustrate this point. Although anatomical and optical writers had for centuries located the principal seat of vision in the crystalline humour,

> Yet the industrious *Scheiner*, in his useful tract entitled *Oculus*, does justly enough reject that received opinion, by shewing, that it suits not with the skill and providence of nature, to make that part the seat (or chief organ) of vision, for which it wants divers requisite qualifications, especially most of these being to be found in the retina.[10] (427)

Boyle argues that the supposition of purposeful design in the organs of animals allows Scheiner to reject a certain structure of the eye for a particular use, since (i) that structure does not, and (ii) another structure does, perform that use most efficiently. Without the design assumption, Scheiner would have had no good reason to suppose that there was some connection between the physical characteristics of the structures of the eye and their particular uses. Recall that according to the *Epicurean* hypothesis, the structure of the eye, as well as the structure of all other material bodies, is the product of *chance*. There would be no reason to suppose, according to *this* hypothesis, that we could identify the particular use of a structure from that structure's suitability to perform that use. The design assumption is crucial.

The above example is one that Boyle would term a *negative* animal–physical argument, since from the supposition of design it was *denied* that a certain structure performed a particular use. To this same purpose he also considers an *affirmative* animal–physical argument. Here again he identifies two desiderata for such arguments: (i) the particular use that the part is supposed to perform should be a *necessary* one for the welfare of the animal, and (ii) there must be no other part that is as suited for this use as this part is. He appeals to the example of Harvey to illustrate his point with regard to this kind of argument. Upon asking Harvey what made him think of a circulation of the blood,

> he answered me, that when he took notice, that the valves in the veins of so many parts of the body were so placed, that they gave free passage to the blood towards the heart, but opposed the passage of the venal blood the contrary way; he was invited to imagine, that so provident a cause as nature had not so placed so many valves, without design; and no design seemed more probable, than that since the blood could not well, because of the interposing valves, be sent by the veins to the

limbs, it should be sent through the arteries, and return through the veins, whose valves did not oppose its course that way. (427)

Again, the affirmative form of the argument can be seen to depend upon the assumption of design in the construction of the animal. Harvey begins by noticing certain structures of the circulatory system, and from the supposition of design he infers the particular *uses* of these structures.[11] Boyle stresses that the assumption of *design* was essential for discovering the true uses of the valves:

> For I consider, that even in man himself, though there be numerous valves found in his veins, yet for those many ages, that the true uses of them lay hid, an *Asclepiades*, or some other bold Epicurean physician, might have thought himself well grounded to look upon as superfluous parts, which, now that the circulation of the blood is discovered, they are acknowledged to be far from being. (437–8)

The Epicureans, or those who deny design, have no reason to suppose that every structure in the body of an animal *has* a use, so they are not encouraged to ask, and perhaps to *discover*, what the particular use of a part is. Boyle is convinced that identifying the particular use of a part of the body is greatly facilitated by having good reasons for supposing that *every* part of the body has been purposely *designed* to perform some particular use. Those who, like Harvey, embrace such a supposition in their anatomical researches, have a decisive advantage over those who do not. He is insistent, however, that the arguments framed on the supposed purposes of bodily parts may easily deceive one, since correctly identifying the most efficient *means* to a certain end is difficult. Nature often takes various routes to achieve the same ends, and discovering which of these routes is the *primary* one in any given case may be impossible. Since anatomists are too often over-hasty in framing arguments for the particular uses of the parts of animals from their supposed ends, he thinks that 'there is less need of my increasing them, than of my proceeding to give you a caution about them . . . ' (428). With this 'caution' in mind, we now consider Lennox's interpretation of Boyle's arguments.

A critique of Lennox's interpretation

As noted above, I believe that Lennox's interpretation of the *Disquisition* as a defence of teleological explanation in science does not provide a satisfactory account of Boyle's discussion of final causes. His interpretation depends for its plausibility upon the cogency of more specific claims, as follows:

    (1) *On the content of the* Disquisition: 'The bulk of the *Disquisition* is a defense of teleological inference in the subjects of [experimental] anatomy and physiology.'[12]

    (2) *On the aim of Boyle's discussion of final causes*: 'Boyle . . . was attempting to establish just those minimal conditions which contemporary philosophy of

science insists must be met if teleological explanations are to be scientifically "respectable".'[13]

(3) *On Boyle's critique of Epicureans and Cartesians*: 'Boyle's critique of the rejection of teleology by two factions of the mechanical philosophy . . . is designed to disclose that mechanical and teleological explanations of biological phenomena are compatible, and that pursuing each mode of explanation has heuristic value for the other.'[14]

In Lennox's view, the *Disquisition* is a valuable document because it anticipates contemporary defences of teleological explanations in biology. According to Lennox, Boyle was trying to do exactly what contemporary defenders of teleological explanation in science are trying to do, namely to define the precise conditions under which teleological explanations may be said to have scientific justification. Because experimental anatomy and physiology are the disciplines where teleological explanations are most obviously appropriate, Boyle fills the *Disquisition* with examples from these disciplines. Finally, in Lennox's view, Boyle is not concerned to show that either Epicureans or Cartesians are mistaken, but rather he wants to show again how teleological and mechanical explanations are perfectly compatible.

Each of Lennox's claims may be evaluated in turn. Consider, first, the claim that, 'The bulk of the *Disquisition* is a defense of teleological inference in the subjects of [experimental] anatomy and physiology.'[15] Unfortunately, this claim does not square with what we find in Boyle's treatise. In the preface Boyle expressly declares that the subjects of anatomy and physiology will *not* be his chief concerns in this discourse, and warns readers not to expect a detailed discussion of these subjects:

> Anatomists indeed, and some physicians, have done laudably upon the uses of the parts of the human body; which I take this occasion to declare, that it may not be suspected, that I do in the least undervalue their happy industry, because I transcribe not passages out of their books: the reasons of which omission are, not only, that I had not any one book of anatomy at hand, when I was writing, but that the uses of the parts of man's body related but to a small part of my discourse. (393)

Note that Boyle does not deprecate the work of anatomists and physiologists; he had the highest admiration for the accomplishments of men like Scheiner and Harvey. He is merely stating at the outset that such work will not have a prominent place in this discourse. Indeed, he warns the reader that,

> Those, that relish no books in natural philosophy, but such as abound in experiments, are seasonably advertised, that I do not invite them to read this treatise; wherein I thought it much more suitable to the nature of my subject and design, to declare the works of God, than of men; and consequently to deliver rather observations, than artificial experiments. (393)

His orientation to the potential reader of the *Disquisition* is that he should not look for any scientific experiments in the following pages of the sort that the reader of some of his other treatises might naturally expect. Neither the *subject* nor the *purpose* of this treatise, he forewarns, has to do with *experimental* science.

True to his word, Boyle devotes little of his attention to experimental science in the remainder of the treatise. By far the greatest part of the *Disquisition* is devoted to delivering observations drawn from nature in order to support the legitimacy of animal–metaphysical arguments. Indeed, his *only* discussion of experimental anatomy and physiology in the *Disquisition* consists of the two adjoining paragraphs in which he discusses the discoveries of Scheiner and Harvey. These two paragraphs, taken together, constitute less than two per cent of the entire treatise, and thus can hardly be said to constitute the 'bulk of the *Disquisition*'.

Of course, to point out that the subjects of experimental anatomy and physiology occupy only a small part of the physical space of the *Disquisition* is not to demonstrate their *unimportance* in the context of the work. Here I am merely interested in achieving a more *realistic* picture of this work. Much of the plausibility of Lennox's presentation depends on taking for granted that the *Disquisition* is a treatise composed chiefly of this concern with experimental anatomy and physiology. Boyle does discuss these subjects. But unless this discussion is seen in the context of the whole work, its importance can easily be overemphasised, and the nature of his thought distorted.

Next, consider Lennox's claim that in the *Disquisition*, 'Boyle . . . was attempting to establish just those minimal conditions which contemporary philosophy of science insists must be met if teleological explanations are to be scientifically "respectable"'.[16] In Lennox's view, Boyle's critique of the Epicureans and Cartesians should be seen as a defence of the *scientific* value of teleology.

A number of features of the *Disquisition*, however, make clear that, whereas this might have been *part* of Boyle's aim, it could not have been his *primary* concern in writing this work. In the opening sentence of the *Disquisition* he writes that, 'There are not many subjects, in the whole compass of natural philosophy, that better deserve to be inquired into by Christian philosophers, than that which is discoursed of in the following essay' (392). He goes on to say that Christian philosophers ought to see whether there are any final causes to be considered in the works of nature,

> since if we neglect this inquiry, we live in danger of being ungrateful, in overlooking those uses of things, that may give us just cause of admiring and thanking the Author of them, and of losing the benefits, relating as well to philosophy as piety, that the knowledge of them may afford us. (392)

As these remarks suggest, his foremost concern is to strengthen the case for acknowledging the existence of an Author of nature, and to stimulate devotion to such an Author. His critiques of the Epicureans and Cartesians are integral to this endeavour. With respect to the Epicureans, he admits that the 'contagious boldness of some baptized Epicureans' compelled him to spend a disproportionate amount of his treatise on establishing the claim that the ends or uses of many things in nature 'were framed or ordained thereunto by an intelligent and designing agent' (394, 428). Later he states that he has said many things in this treatise against that part of the Epicurean hypothesis, 'which ascribes the origin of things to chance, and rejects the interests of a deity, and the designing of ends, in the production and management of natural things' (428). His reason for arguing against the Epicureans is apparently *not* because of the ill effects of the Epicurean hypothesis on natural science,

> But because I observe, not without grief, that of late years too many, otherwise perhaps ingenious men, have with the innocent opinions of *Epicurus* embraced those irreligious ones, wherein (as I was saying) the deity and providence are quite excluded from having any influence upon the motion of matter, all of whose productions are referred to the casual concourse of atoms: for this reason, I say, I thought it a part of my duty, as well to the most wise author of things, as to their excellent contrivance and mutual subserviency, to say something, though but briefly, yet distinctly and expressly to shew, that, at least in the structure and nature of animals, there are things, that argue a far higher and nobler principle, than is blind chance. (428)

This 'higher and nobler principle' is, of course, God's continual providence in the universe.[17] This explains why he devotes so much of the *Disquisition* to constructing animal–metaphysical arguments: they provide the strongest arguments against the Epicureans for the existence of the Deity.

If the Epicureans posed one problem for acknowledging God's providential activity, then Descartes posed another, 'the Cartesian opinion having of late made it requisite to handle the formerly difficult question, about the consideration of final causes, after a new manner' (394). Boyle knows that Descartes was certainly no atheist, yet his 'rejection of final causes from the consideration of naturalists, tends much to weaken . . . if not quite to deprive us of one of the best and most successful arguments to convince men that there is a God, and that they ought to admire, praise, and thank Him' (401). This is all the more unfortunate since,

> the excellent contrivance of the great system of the world, and especially the curious fabric of the bodies of animals, and the uses of their sensories, and the other parts, have been made the great motives, that in all ages and nations induced philosophers to acknowledge a Deity, as the author of these admirable structures. (401)

Indeed, he thinks that it is actually *injurious* to God to banish the consideration of final causes, and expressed wonder

> that the Cartesians, who have generally, and some of them skilfully, maintained the existence of a Deity, should endeavour to make men throw away an argument, which the experience of all ages shews to have been the most successful . . . to establish, among philosophers, the belief and veneration of God. (401)

Granted, Descartes had done his part in arguing for God's existence with his ontological proof. But this is no reason to throw away an argument that has proven effective in all ages, especially when such an argument displays a decisive *advantage* over the Cartesian proofs for God's existence:

> For, whereas a *Cartesian* does but shew, that God is admirably wise, upon the
> supposition of his existence; in our way the same thing is manifested by the effect
> of a wisdom, as well as power, that cannot reasonably be ascribed to any other than
> a most intelligent and potent Being; so that by this way men may be brought, upon
> the same account, both to acknowledge God, to admire Him, and to thank Him.
> (402)

Earlier in the *Disquisition* he writes:

> It is true, that, in the idea of a Being infinitely perfect, boundless wisdom is one
> of the attributes that is included: but, for my part, I shall take leave to think, that
> this general and indefinite idea of the divine wisdom, will not give us so great a
> wonder and veneration for it, as may be produced in our minds, by knowing and
> considering the admirable contrivance of the particular productions of that
> immense wisdom, and their exquisite fitness for those ends and uses, to which they
> appear to be destinated. (439)

Later he amplifies this idea. He explains that he devoted so much space in this treatise
to recounting instances of design in nature because 'too many men, that have more wit
than philosophy or piety, have, upon Epicurean (and some also even upon Cartesian)
principles, laboured to depreciate the wisdom of God'. His purpose in writing, he adds,
was that his reader 'should not barely observe the wisdom of God, but [should] be, in
some measure, affectively convinced of it' (439).

Finally, recall that, for Lennox, Boyle's critique of the Epicureans and Cartesians
'is designed to disclose that mechanical and teleological explanations of biological
phenomena are compatible, and that pursuing each mode of explanation has heuristic
value for the other'.[18] If the account given so far is correct, however, then it is clear
that Boyle's critique of the Epicureans and Cartesians is primarily apologetic,
rather than methodological, in its aim. His examples are designed to 'strengthen and
enlarge' an argument which has shown itself to be the most effective in inspiring belief
in, and devotion to, the Deity. Viewing the *Disquisition* as a whole, the 'scientific'
usefulness of considering final causes is for Boyle an issue of decidedly secondary
importance.

Indeed, he emphasises that, although the search for final causes is a worthwhile and
indeed necessary endeavour, it is *not* the task of the naturalist *qua* naturalist to search
for final causes. Instead, he believes that,

> to inquire to what purpose nature would have such and such effects produced, is
> a curiosity worthy of a rational creature, upon the score of his being so; but this
> is not the proper task of a naturalist, whose work, as he is such, is not so much to
> discover, why, as how, particular effects are produced. (443)

Considerations of final causes have their uses in natural philosophy, but they are
secondary to the usefulness of efficient causes. The issue here is not whether final
causes are important, but whether it is the task of the *naturalist* to be concerned with

them. In Boyle's view, 'a naturalist, who would deserve that name, must not let the search or knowledge of final causes make him neglect the industrious indagation of efficients' (444).

What emerges, then, is that for Boyle it is not the proper task of *the naturalist* to be concerned with final causes. Recall that his explicit aim in the *Disquisition* (as expressed in its full title) was to determine whether a *naturalist* should admit the final causes of natural things. His answer is that the naturalist *should* admit them (insofar as he is a rational being), but that it is not his *task, qua naturalist*, to be concerned with them. In the *Disquisition* he is concerned to demonstrate both the existence and possibility of knowing final causes in nature, not for the sake of scientific utility but because he believes that a detailed knowledge of final causes provides the most powerful incentive for acknowledging the Deity.

Lennox laments the fact that Marie Boas Hall, in her book on Boyle's natural philosophy, includes the *Disquisition* under the heading of 'Natural Religion'.[19] But the *Disquisition* was categorised by Boyle himself as a *theological* work, rather than as a work in natural philosophy proper.[20] In Lennox's interpretation this would seem unaccountable. But if the argument of the present paper is correct, this is *precisely* how the *Disquisition on Final Causes* should be understood.

Conclusion

As John Hedley Brooke has pointed out, it comes as something of a shock to contemporary readers to discover that the mechanical philosophy was seen by its adherents as being *more* conducive to religion and less friendly to atheism than its more metaphysically pregnant rivals.[21] But if matter is thought of as completely (or at least substantially) inert and without internal principles of activity, then it would be natural to assume that the order and complexity observed in the world could only be the product of an intelligent Designer. Seen in this light, Boyle's affirmation of final causes in nature fits in neatly with both his philosophical and theological views.

Natural philosophy would soon reach a crossroads, however, at which the study of the natural was to be sharply separated from the study of the supernatural. Ironically, the mechanical philosophy would be largely responsible for this 'secularisation of science'. Boyle's writings, and the *Disquisition* in particular, are thus especially valuable for understanding the development of natural philosophy at a crucial stage in its historical development.

Acknowledgements

I would like to thank Ernan McMullin, Rose-Mary Sargent and Michael Hunter for helpful comments on and criticisms of earlier drafts of this paper.

NOTES

1   'To know the Supreme Cause from the causes of things.' This motto appears on the frontispiece of *The Works of the Honourable Robert Boyle*, ed. Thomas Birch, 6 vols (London, 1772).

2   Marie Boas Hall, in her *Robert Boyle on Natural Philosophy: An Essay with Selections from his Writings* (Bloomington: Indiana University Press, 1965), pp. 153–4, discusses Boyle's views on teleology briefly. Harold Fisch, 'The scientist as priest: a note on Robert Boyle's natural theology', *Isis* 44 (1953): 252–65, discusses Boyle's views in passing on p. 264.

3   James G. Lennox, 'Robert Boyle's defense of teleological inference in experimental science', *Isis* 74 (1983): 38–53.

4   Lennox, 'Boyle's defense' (n. 3), p. 39 explicitly connects Boyle's *Disquisition* with George C. Williams *Adaptation and Natural Selection* (New York: Columbia University Press, 1966), and Larry Wright, *Teleological Explanation* (Berkeley, Los Angeles: University of California Press, 1975), both of which are concerned with assessing the *scientific* value of teleological explanations.

5   My efforts may be understood as a much belated response to Boyle's own request. In the preface to the *Disquisition* Boyle admitted that the work was 'overly long' and lacked 'symmetry', and confided his fear that he had at best 'thrown together materials for a just discourse on my subject' (394: here and throughout this chapter, numbers in parentheses refer to page numbers in Birch's edition of the text, *Works*, v, 392–444). He expresses his hope that 'if the materials be good and solid, they will . . . find an architect, that will dispose them in a more artful way than I was either at leisure or solicitous to do' (394). My modest goal, therefore, is to play the role of 'architect' for Boyle by setting out the logical structure of his argument and highlighting those aspects of the work that display most clearly his central aims and concerns in writing this treatise.

6   Thomas Birch, *The History of the Royal Society*, 4 vols (London, 1756–7), iv, 553. For further information on the composition of this work, see Davis's essay, above.

7   Aristotle, *Politics* 1256b 20–2. See *The Politics*, trans. T. A. Sinclair (Baltimore: Penguin, 1962), p. 40.

8   Descartes makes this claim in both *The World* and in *The Principles of Philosophy* (e.g., Principle xxxvi). See *The Philosophical Writings of Descartes*, trans. J. Cottingham, R. Stoothoff and D. Murdoch, 2 vols (Cambridge University Press, 1985), i, 96, 240, 256.

9   It is interesting to note that, although Boyle devotes a good deal of effort in the *Disquisition* to arguing that structures like the vertebrate eye clearly manifest the wisdom of God, he attaches to this treatise, by way of an Appendix, 'Some Uncommon Observations about Vitiated Sight' (*Works*, v, 445–52), in which he catalogues a number of diseases that may afflict the eyes!

10 The reference is to Cristof Scheiner, *Oculus, hoc est: Fundamentum opticum* (Innsbruck, 1619; London, 1653).

11 On Boyle's account of Harvey's discovery of the circulation of the blood, see Richard A. Hunter and Ida Macalpine, 'William Harvey and Robert Boyle', *NRRS* 13 (1958): 115–27; Jerome J. Bylebyl, 'Boyle and Harvey on the valves in the veins', *Bulletin of the History of Medicine* 56 (1982): 351–67.

12 Lennox, 'Boyle's defense' (n. 3), pp. 39–40.

13 *Ibid.* p. 40.

14 *Ibid.* p. 39.

15 *Ibid.* pp. 39–40.

16 *Ibid.* p. 40.

17 On Boyle's conception of God's activity in nature, see Timothy Shanahan, 'God and nature in the thought of Robert Boyle', *Journal of the History of Philosophy* 26 (1988): 547–69.

18 Lennox, 'Boyle's defense' (n. 3), p. 39.

19 *Ibid.*; Marie Boas Hall, *Robert Boyle* (n. 2), pp. 153–4.

20 BP 35, fol. 187; R.S. MS 185, fols 1–3.

21 John Hedley Brooke, *Science and Religion: Some Historical Perspectives* (Cambridge University Press, 1991), pp. 117–18.

Locke and Boyle on miracles and God's existence

J. J. MACINTOSH

Though twentieth-century philosophers often write as if there were no significant discussions of miracles between Aquinas and Hume, the late seventeenth and early eighteenth centuries were the home of what R. M. Burns has rightly called 'The Great Debate on Miracles'. He suggests, only slightly intemperately, that Hume's 'essay was very much a tail-end contribution to a flagging debate', a debate which both started and peaked considerably earlier.[1]

Two of the major English philosophers of the second half of the seventeenth century, Robert Boyle and John Locke, wrote interestingly and at length on the topic. Both were aware that belief is often a product of upbringing. Locke asked, 'How much more does [a child's notion of God] resemble the Opinion, and Notion of the Teacher, than represent the True God?',[2] and Boyle noted that 'usually, such as are born in such a place, espouse the opinions, true or false, that obtain there', indeed, 'the greatest number of those that pass for Christians, profess themselves such only because Christianity is the religion of their Parents, or their Country, or their Prince, or those that have been, or may be, their Benefactors; which is in effect to say, that they are Christians, but upon the same grounds that would have made them Mahometans, if they had been born and bred in Turky',[3] but both felt that *their* faith was more solidly grounded.

Neither Locke nor Boyle relied on miracles to provide evidence of God's existence. Both agreed that such evidence must come from elsewhere, and that the point of miracles is to teach us something about the world once we have established God's existence independently. Their arguments have a family resemblance, and it is hard not to suspect a certain amount of influence. Boyle's arguments, mainly unpublished and almost certainly earlier, are both subtler and more persuasive than Locke's, so I shall discuss Locke's arguments first, offering them in effect as a background to Boyle's.[4]

Locke's arguments for God's existence

'We have the Knowledge of *our own Existence* by Intuition; of the *Existence of GOD* by Demonstration; and of other Things by Sensation', said Locke,[5] but his 'Demonstration' is almost startlingly weak. He drew a conclusion considerably stronger than his premises allow, and relied at one crucial stage on a further undeveloped argument which renders his own unnecessary. His argument, strongly resembling one offered by Cudworth, begins by pointing out that at least one thing, and indeed one knowing or cogitative thing, exists, namely himself.[6] Now, we know

> by an intuitive Certainty, that bare nothing can [not] produce any real Being. If
> therefore . . . there is some real Being . . . it is . . . evident . . . that from Eternity
> there has been something; Since what was not from Eternity, had a Beginning; and
> what had a Beginning, must be produced by something else.[7]

There are a number of points involved in this highly compressed argument. Leaving aside Locke's suspect 'intuitive Certainty', the phrase 'from Eternity', taken by Locke to involve an infinite past time, is also suspect. Unlike earlier philosophers and his contemporaries Leibniz and Newton, Locke simply ignored the possibility that time and the world may have been co-created.[8]

Equally ignored was the possibility of a sempiternal universe created by an eternal deity. 'By faith alone do we hold, and by no demonstration can it be proved, that the world did not always exist', argued Aquinas,[9] but such a universe was unacceptable to both Locke and Boyle. Indeed, Boyle took the doctrine to lead to atheism:

> [T]hough I will not say, that *Aristotle* meant the mischief his doctrine did, yet I am
> apt to think, that the grand enemy of God's glory made great use of *Aristotle*'s
> authority and errors, to detract from it. *Aristotle*, by introducing the opinion of the
> eternity of the world . . . did . . . in almost all men's opinions, openly deny God the
> production of the world.[10]

Boyle was aware that Aristotle felt his 'conceipt of the Eternity (not of Matter only but) of the world' to be compatible with the existence of a God, but found that 'whatever may be the meaneing of those darke & puzzleing Expressions wherein Aristotle has deliuer'd his Doctrine about the first mouer', most people would take it to be an atheistic doctrine.[11]

Like others before and after him, Boyle found such infinity suspect since it would involve a denumerably infinite set having another as a proper subset, but 'yᵉ infinite Number of hours must vastly exceed the number of Years, wᶜʰ yet must be granted to be Infinite'. Unlike Locke, Boyle saw that this position also raises problems for a temporally infinite (as opposed to an eternal) deity, which both accepted.[12] Boyle noted that we must be careful about assuming that 'the common rules, whereby we judge of other things, [are] applicable to infinites' ('by which the intellect is non-plus'd'), but he often found it difficult to apply his own good advice.[13]

Given Locke's premises, then, we cannot correctly conclude either that there was a *single* thing existing 'from Eternity' nor even that there was a single *kind* of thing, some member of which has always existed. However, assuming such a being, it was, Locke argued, either cogitative or incogitative. It cannot have been the latter because then a 'knowing Being" would have been produced by 'Things wholly void of Knowledge, and operating blindly, and without any Perception'. But: 'it is . . . repugnant to the *Idea* of senseless Matter that it should put into it self Sense, Perception, and Knowledge'.[14]

This looks like Zeno of Citium's 'Nothing lacking consciousness and reason can produce out of itself beings with consciousness and reason',[15] but Locke's claim is slightly different: it is conceptually impossible for *matter* to give rise to sentience. This does not directly entail that *nothing* non-sentient could give rise to sentience, but since Locke limited the possibilities to 'two sorts of Beings in the World, that Man knows or conceives', the 'purely material' and the 'Sensible, thinking perceiving Beings, such as we find our selves to be', he took his result to follow, though allowing it requires that our epistemological limits exhaust the ontological possibilities, an assumption Locke elsewhere explicitly denies. Additionally, not only can matter not give rise to *sentience*, but 'Matter has not the power to produce Motion in it self'[16] so that a problem would remain, even if we took thought to be a brain process, a possibility which Locke was explicitly unwilling to rule out. It is

> impossible for us . . . without revelation, to discover, whether Omnipotency has not given to some Systems of Matter fitly disposed, a power to perceive and think . . . : It being . . . not much more remote from our Comprehension to conceive, that GOD can, if he pleases, superadd to Matter a Faculty of Thinking, than that he should superadd to it another Substance, with a Faculty of Thinking; since we know not wherein Thinking consists, nor to what sort of Substances the Almighty has been pleased to give that Power.[17]

What is 'repugnant to the *Idea* of senseless Matter' is that 'it should put into *it self* Sense, Perception, and Knowledge'. It is not demonstrably impossible for matter to *have* such faculties, but Hume's 'little agitation of the brain which we call *thought*' requires divine assistance to be *thought*.[18] Leibniz felt that Locke was confused here; even God cannot give matter the power to think: 'To speak of sheerly "giving" or "granting" powers is to return to the bare faculties of the Scholastics . . . [I]t is not within the power of a bare machine to give rise to perception, sensation, reason.'[19] God could no more add thought to matter than he could add weight to a number, but Locke held that our inability to conceive thinking matter may involve a 'weakness of our apprehensions [which] reaches not the power of God'.[20] Many contemporaries of Locke were content simply to let observation override theory:

> I see that [non-human animals] have feeling and consciousness, although the manner in which they are conscious is unknown to me. Thus, although there is nothing more shocking . . . than granting some consciousness tö matter, it nevertheless seems to me that we should first . . . see whether animals who have every appearance of being nothing but matter do not have some consciousness, rather than be dismayed by the absurdities, either real or apparent, which occur in this matter.[21]

For Locke, we are either bits of thinking matter, or we are incorporeal souls joined to corpuscularian bodies. But neither of these is naturally (i.e. mechanically) explicable. Therefore, in either case, we need to invoke a supernatural agency. Much later Darwin asked, 'Why is thought being a secretion of brain, more wonderful than gravity a property of matter?'[22] Locke would have been happy with such a question and such a comparison:

> [I]t is evident, that by mere Matter and Motion, none of the great Phænomena of Nature can be resolved, to instance but in that common one of Gravity, which I think impossible to be explained by any natural Operation of Matter, or any other Law of Motion, but the positive Will of a Superiour Being, so ordering it.[23]

Locke's argument now moves by way of a fairly straightforward scope fallacy, from the 'evident' truth that there was never a time when something (i.e. something or other) did not exist, to the result that (therefore) some (one) thing, or at any rate some one *kind* of thing, has existed at all (past) times.[24] Considering, and rejecting, the possibility that this 'Being' might be 'matter', he noted that though we

> speak of [matter] as one thing, yet really all Matter is not one individual thing. And therefore if Matter were the eternal first cogitative Being, there would not be one eternal infinite cogitative Being, but an infinite number of eternal finite cogitative Beings, independent one of another, of limited force, and distinct thoughts, which could never produce that order, harmony, and beauty which is to be found in Nature.[25]

Thus, he suggests, somewhat off-handedly, nothing short of God will do to explain the order, harmony and beauty which is to be found in nature. The argument is a strange one, for it fails entirely unless this smuggled in final move works. But if *it* works, Locke's lengthy preamble is unnecessary: we can go straight from design ('order, harmony, and beauty') to God in the standard way. Locke's proof only works if the standard design argument works. But if that works, Locke's argument is unnecessary. One can see why he offered it, however, for notoriously design arguments convince only the already committed.[26] Paradoxically perhaps, the more 'design' we find, the less need we find for a designer.

Locke concludes with a further, modal, fallacy, moving from

(1) if there is a cogitative being now, then there has always been a cogitative being; to

(2) necessarily, if something cogitative exists now, then there has been from eternity some cogitative thing; to

(3) if something cogitative exists now, then 'whatsoever is the first eternal *Being* must necessarily be cogitative'.

Given that something cogitative *does* exist now, Locke proceeds to detach the consequent to get his

(4) 'discovery of the *necessary Existence of an eternal Mind*'.[27]

It is obvious that, though the second will follow from the first if we allow the first something akin to theoremhood (since Locke's proof is intended to be *a priori*), and the fourth from the third (given the truistic antecedent of the conditional), the third (and consequently the fourth) does not follow from the second (nor from the first). This move from $\Box(p \rightarrow q)$ to $(p \rightarrow \Box q)$ provides us with a second scope fallacy. And indeed, the first scope fallacy has reappeared in the move from (2) to (3).

Locke's 'necessity' claim may be seen to fail by considering an Epicurean world without sentience, which is not particularly 'harmonious' or 'beautiful', for the argument will not prove a God in such a world. This is not a merely fanciful counterexample, for both Boyle and Hooke, among others, believed the actual world to have been like this for a time: order and beauty were *imposed* upon an initial chaos. Consequently, since there is a possible world for which there is no (Lockean) reason to predicate a God, Locke's argument does not yield God's *de re* necessary existence. Alternatively, if God *is* to have necessary existence, something more than Locke's version of the design argument is necessary, as Leibniz pointed out.[28] Indeed, as Kant was to argue, something very like the ontological argument is required for Locke's strong modal conclusion.[29]

Locke had two background arguments for God's existence. One, barely alluded to, but essential for the success of what he believed to be his main argument, was the standard argument from design. The other, also brought in in a somewhat subterranean way, relied on the need to explain the existence of thinking things, not as in the main argument, in view of their origin, but in view of the necessity to have a non-mechanical explanation of their abilities, since both alternatives – that matter thinks, or that something immaterial interacts causally with something material – require a supernatural power.

Boyle's arguments for God's existence

Boyle's positive arguments for God's existence fall into three main groups. One group has to do with the need for God if we are to have *any* understandable account of the world. Another, overlapping group, has to do with the standard design moves, but with an emphasis, as Boyle put it, on God's watches, not His clocks.[30] Boyle thought we have an argument schema, not a single argument here, since the considering naturalist will find a host of examples that can be slotted into the argument:

> The proofe drawn for the existance of a Diety from the fabrick of the world is not so much a single *argument*, as a *Topick* . . . whence may be derived as many particular arguments as there are . . . Creatures curiously framed.[31]

I shall not dwell on Boyle's versions of the argument from design for, although they have definite affinities with the first group, they are clearly distinct, and the first group is uppermost in Boyle's writing and in his concern. Thus I disagree with at least the emphasis of James Force's suggestion that 'Boyle actively espouses the design argument to emphasise the role of God as the generally provident creator of order and regularity as observed in the operations of natural law'.[32] A better way of emphasising this role of God's is to emphasise it directly – and this Boyle did, again and again and again. It was not necessary for him to go through the design argument in order to emphasise what he perceived as the scientific need for a deity, and in general he did not wander from the direct path, though he often *gestured* at the design argument.

The third argument has to do (as in Locke's case) with the existence of rational entities and the belief that their abilities need *special* explanations. All these arguments – or considerations, for in what was a common move at the time, Boyle offered them more as props for the weakly committed, rather than as conversion factors for the strongly opposed[33] – are intended to show at least the *possibility* of God's existence, which in his more Pascalian moods Boyle felt would be enough to motivate belief: 'if we can once evince . . . that 'tis not impossible there should be a Deity, nor absurd to believe there is one, we have not gain'd a little for Religion', Boyle thought, and 'if Physicks do but stand neuter', we may be led via miracles to the Christian religion.

Clearly aware of the distinction between the two questions: 'Is there a God?' and 'Is it rational to believe that there is a God?', Boyle sometimes wrote as if his arguments could answer the first as well as the second question, but in general his caution continually tempted him to the less committal conclusion. His confidence in his arguments varied. Like many of his contemporaries he sometimes felt that only a vicious character could ground atheism,[34] but at other times he was clearly aware of the force of the hard-headed non-believer's position. Despite the fact that 'God was pleas'd to giue the world a contrivance so conspicuously admirable that the proofe of his existence should be as obvious, as the objections that Atheists make against it',[35] refuting Epicurean atheism is a 'Taske much harder to be well performd than perhaps any man that has not tryed will imagine', and so Boyle 'never pretended to convert resolved Atheists'.[36]

God as a theoretical entity

Locke's interest in God was a philosopher's interest; but Boyle's interest was the interest not only of a committed believer, but of a scientist who was convinced that his science would not be a possible endeavour without belief in God. For Boyle God was necessary to explain the fact that we have laws of nature, that is, predictable and retrodictable regularities that are worth our while discovering. Further, Boyle was well aware that 'laws' of nature do not, and cannot, *govern* anything at all. Moreover, even if these laws did, in some sense 'govern' what occurred, that would be a further (and unnecessarily complicating) hypothesis that would itself need explication and explanation. If we seek explanations for the lawlike behaviour of physical objects, we must turn to something *extra*-systemic.[37] The two main candidates canvassed in Boyle's time were God and a '*Plastic Nature* under [God], which as an Inferior and subordinate Instrument, doth Drudgingly Execute that Part of his Providence, which consists in the Regular and Orderly Motion of Matter', a 'nature' which interposed itself between God and the universe God had created.[38]

A third alternative, that something called 'chance' was the operative principle, was rejected by Boyle on the grounds that either it is a pure extravagance – the probabilities are ridiculously high against such 'chance' – or it involves a conceptual confusion: the belief that something called *chance* can operate as a causal explanatory factor is not a notion that will bear inspection. Such a confusion might be natural when *fate* had been believed to be a divinity, but by the late seventeenth century it had no part to play.[39]

The notion of 'chance' was undergoing a sea change during Boyle's lifetime,[40] but it is clear that Boyle, like Cudworth on the opposite side of the fence, was right to reject the quasi-causal principle that was in effect on offer at the time. In the first part of *The Usefulness of Natural Philosophy*, Boyle refers us to 'that witty father *Lactantius*' for a refutation of the Epicurean doctrine of the origin of animals on the basis of its improbability.[41] It is massively improbable that Epicurean atoms would come together in just the right combination to produce an animal. 'So ridiculous a sophism', said Henry More, 'is more fit to impose on the inconsiderate Souls of Fools and Children, then upon men of mature Reason . . . well exercised in Philosophy'.[42] Even allowing that atoms could coalesce by chance into an animal, the improbability increases alarmingly when we consider (to list only some of the more obvious problems) (a) the need for the animal to have working generative organs, (b) the need for there to be *another* animal (c) with working generative organs but (d) of the opposite sex and (e) produced in the same spatio-temporal neighbourhood as the first. Boyle and Lactantius were clearly right to be sceptical of the *chance* hypothesis, but we need not fly so quickly to the Almighty:

> Some scientists have argued that, given enough time, even apparently miraculous events become possible – such as the spontaneous emergence of a single-cell organism from the random couplings of chemicals. Yet Fred Hoyle, the iconoclastic British astronomer, has said such an occurrence is about as likely as the assemblage of a 747 by a tornado whirling through a junkyard.
>
> Most researchers agree with Hoyle on this point (although on little else). The one belief almost everyone shares is that matter quickened through a succession of steps, none of which is wildly improbable.[43]

Nowadays Boyle's paradigm, biology, no longer provides material for design arguments among serious thinkers: the last hope is pinned not merely on cosmology, but on its account of the world's first few milliseconds. However, formally the arguments used are remarkably similar to those Boyle plucked from genetics: any theory that does not incorporate a designer leaves us with massive improbabilities. Alas, just as the rise of modern genetics and contemporary evolutionary theory gives us a designerless explanation for the existence of animals and their kinds, so the rise of quantum cosmology bids fair to disinherit the anthropic principle.[44]

What of the plastic nature? Cudworth (enthusiastically endorsed by Ray)[45] argued that its absence would mean that God had to work immediately in nature, which 'would render the Divine Providence Operose, Solicitous and Distractious: and thereby make the Belief of it entertained with greater difficulty, and give advantage to Atheists'. Moreover, 'It is not so decorous in respect of God . . . that he should . . . set his own hand . . . to every work, and immediately do all the meanest and triflingest things himself drudgingly.'[46]

Again, if God were to do things immediately, 'The Slow and Gradual Process that is in the Generation of things . . . would seem to be a Vain and Idle Pomp, or a Trifling Formality'. Finally, the plastic nature allows for 'Errors and Bungles . . . Whereas an *Omnipotent Agent* . . . would always do [its Work] *Infallibly* and *Irresistibly*'.[47] Cudworth is on to an important point. There *are* what would be bungles in nature, if nature were designed, and it is the part of the good scientist to note them: 'Natural selection, as a historical process, can only work with material available . . . The resulting imperfections and odd solutions . . . record a process that unfolds in time from unsuited antecedents, not the work of a perfect architect creating *ab nihilo*.'[48]

In fact, Boyle *did* notice them, but thought that either they were explicable in the grand scheme of things, though not necessarily by us,[49] or they were to be explained by the consideration that, given the complexity involved, God was really doing rather well. Boyle had an acute sense of the difficulties that even apparently simple experiments could give rise to, and perhaps in consequence had more sympathy than stricter theologians for the troubles that could beset the divine combination of omnipotence and omniscience. Boyle usually speaks of God as being 'infinitely perfect', but sometimes adds a rider: 'God [is] a Being infinitely Perfect or at least Absolutely Perfect, that is, as perfect as is possible for us to conceiue.' Elsewhere, after offering the same qualification, he adds the modest claim that God is 'intelligent and highly potent'.[50]

Boyle believed in, but often underestimated, God's abilities. While admonishing us not to think that God must have made the universe for us he apparently forgot that God's foreknowledge would allow him to look ahead to human needs and capacities:

> [T]he world itself was first made before . . . man: whence we may learn, that the author of nature consulted not, in the production of things, with human capacities; but first made things in such a manner as he was pleased to think fit, and afterwards left human understandings to speculate as well as they could upon those . . . things.[51]

Boyle recognised that many apparent deviations from (accepted) laws of nature may be instances of long term regularities. Although

> [T]he grand rules whereby things corporeal are transacted, and which suppose the constancy of the present fabrick of the world, and course of things, are not altogether so uniformly complied with, as we are wont to presume . . . [nonetheless] there are cases, wherein . . . we may sometimes take such things for deviations and

exorbitancies from the settled course of nature, as, if long and attentively enough considered, may be found to be but periodical phaenomena, that have long intervals between them.[52]

That is, there *are* exceptions, but not all apparent exceptions are so. In the next century, Hume's Philo was to take a sterner view of 'the inaccurate workmanship of all the springs and principles of the great machine of nature. One would imagine, that this grand production had not received the last hand of the maker; so little finished is every part, and so coarse are the strokes, with which it is executed.'[53]

Rejecting both the plastic principle and 'chance' left Boyle with the need for *some* explanatory principle, given his views about causality and creation. Rejecting the view that the world is infinite in past time, he rejected too the option of a demiurge, a designer who creates as we do by manipulating already existent matter, opting instead for a God who created matter out of nothing. *How* this could be was an impenetrable mystery for Boyle – 'The Worlds creation by God is as unaccountable a thing to meer Reason, as what the Scripture teaches us of a Trinity' – but *that* it was so he had no doubt.[54]

Creating matter left God with a number of further tasks. Matter *in itself* lacks many of the properties we observe in the world around us. Matter is 'dead', 'stupid', 'lazy', 'inert', 'brute'. Thus we need explanations for (at least) the following observable properties of things in the world:

   (i)   some of them move; indeed
  (ii)   some of them move spontaneously: they initiate motion;
 (iii)   they appear to obey certain laws of motion;[55]
  (iv)   they have an apparent gravitational attractive power;
   (v)   they are moving in just the directions and with just the velocities that they are, and have just the sizes they do;
  (vi)   they have a principle of coherence: they do not simply fall apart, some binding force produces individual items;
 (vii)   some of them attract others electrically;
(viii)   some of them attract others magnetically;
  (ix)   some of them have the power to produce offspring of the same species;
   (x)   some of them can think.

It was the hope of the time that many of these could be explained mechanically, but it seemed merely fanciful (to both Boyle and Locke) to hope that all of them could. Boyle's picture of creation looks something like this. At a particular point in absolute time, God made matter. He could have started things off earlier or later, but He did not.[56] Having created matter, He broke it up and started it moving. *Then* He gave it laws, since the 'casual justlings of atoms' would not have given rise to *this* world.[57] For Boyle, a world without the laws of nature was not only possible, our world was such a world for a time. After having made matter, broken it up, started it moving in the right directions, with the right velocities, and given it laws, God then added some 'seminal' principles.

Boyle now runs two arguments in tandem. One is to the effect that to think our universe could have arisen out of an initial chaos by 'chance' is to strain the notion of probability well past its legitimate limits. Even allowing for chance, some of these things cannot be explained mechanically. In particular, the fact that physical objects appear to be law governed cannot be; nor can the fact that they *continue* in existence; nor can the fact that some of the entities we encounter in our daily life can think.

Thus we find a theoretical need for God both initially and constantly. Initially to create and design the universe; constantly to keep things in existence and to sustain the law-like behaviour of God's inanimate creatures, as well as the law-like behaviour of much of the animate portion of creation. Unlike Descartes, Boyle does not use this necessity for a sustaining power as a proof of God's existence.[58] Nor, though he often refers to it, does he ever attempt to utilise Descartes' argument from the supposedly innate idea of God. In common with Locke and Bentley, Boyle's version of Descartes' argument is highly simplified, perhaps owing more to More than to Descartes. Claiming that our (innate) idea of God is 'no arbitrarious figment taken up at pleasure', More moved, fallaciously but speedily, to the conclusion that God must *necessarily* exist. Descartes' argument, based on the claim that our idea of God has too much objective reality to have been produced by anything with less formal reality than God, is at least formally valid, even though its scholastic premises need more justification than they receive.[59]

The argument from the existence of rational entities

Following a long Aristotelian tradition, codified in the early Middle Ages, practically all seventeenth-century thinkers, including Locke and Boyle, were agreed in placing many mental acts in the brain. In particular, Boyle included the 'functions that belong to the soul of a man, as he is an animal, as imagination and memory, distinct from intellectual operations'.[60] There was also general agreement on the further Aristotelian point that various mental activities could not be accomplished neurophysiologically. In particular, various mathematical results, such as Boyle's perennial favourite, the proof of the incommensurability of the side and diagonal of a square, could not have been achieved mechanically.[61]

If there are mental abilities which are demonstrably *not* available to matter, human beings must have incorporeal souls, and two consequences follow: that human immortality is at least *possible*, and that there is a clear need for something to effect a connection between the incorporeal and the corporeal. Empirical observation reveals that there is a consistency between our interaction with nature and our conscious sensations, yielding a further area of law-like behaviour which cannot be explained within a mechanical framework. Parsimony rules out a *number* of entities working in harmony to accomplish this, and clearheadedness rules out chance. What is left is one benevolent, wise and powerful agent. Boyle did not have to seek far to put a name to this agent.

Locke on miracles

In mid-century Hobbes had set out a fairly standard view: miracles must be 'strange', and such that we cannot conceive natural causes for them. Thus the generation of 'some new shape of a living creature' could be a miracle, but 'when a man, or other animal, engenders his like, though we know no more how this is done, than the other; yet because it is usual, it is no miracle'. Hobbes continued, 'Again, it belongeth to the nature of a miracle, that it be wrought for the procuring of credit to God's messengers, ministers, and prophets, that thereby men may know, they are called, sent, and employed by God, and thereby be the better inclined to obey them.' Hence creation is not called a miracle, nor is the flood. Hobbes concluded: 'a miracle is a work of God, (besides his operation by the way of nature, ordained in the creation) done, for the making manifest to his elect, the mission of an extraordinary minister for their salvation'.[62]

Locke agreed: the function of a *divine* miracle is to validate a revelation by showing that the messenger comes from God. However, for Locke this function is not a *necessary* condition of some event being miraculous. He allowed that miracles could be performed by entities who not only did not have God's sanction, but who were indeed actively opposed to God. To some extent this issue is merely terminological. Should we say 'only God can perform miracles' and speak of miracles and pseudo-miracles, or should we think in terms of a single group of miracles, some but not all of which are divinely instigated? In either case, Locke thinks, miracles are 'fewer than perhaps is imagined'.

For Locke the following are necessary conditions of some event being miraculous. It must

(1) be a 'sensible operation',
(2) be 'above the comprehension of the spectator',
(3) be in the observer's 'opinion contrary to the established course of nature', and
(4) be taken by the observer to be divine.[63]

Locke noted but discounted two objections to this observer-oriented view: first, that the doctrine *is* relativistic, so that what is unmiraculous for the knowledgeable goose may appear (and therefore *be*) miraculous to the credulous gander; secondly, that things will be (correctly) called miraculous which are entirely within the compass of nature, which have 'nothing extraordinary or supernatural in them'. He seems not to have noticed the natural consequence of point (4): that no miracles could occur in a society of atheists. Credulity is a *sine qua non* for Lockean miracles.

Only three religions claim to be grounded in revelation, Locke believed, namely those of Moses, Jesus and Mohammed. But 'Mahomet having none to produce, pretends to no miracles for the vouching of his mission; so that the only revelations that come attested by miracles, being those of Moses and Christ, and they confirming each other', Locke concludes, somewhat airily, 'the business of miracles . . . has no manner of difficulty in it'.[64] Locke did not consider the possibility that the *Koran* itself is miraculous (in view of the disparity between it and what might reasonably be expected of its author in the absence of divine inspiration), but Boyle had already taken explicit notice of this point:

> [T]he Saracens, depressed with their religion's being destitute of attesting miracles, . . . reply, that though there were no other miracle to manifest the excellency of their religion . . . , yet the Alkoran it self were sufficient, as being a lasting miracle that transcends all other miracles. How charming its eloquence may be in its original, I confess my self too unskilful in the Arabick tongue, to be a competent judge . . . but the recent translations I have seen of it in French, and . . . Latin, elaborated by great scholars, and accurate Arabicians, by making it very conformable to its eastern original, have not so rendered it, but that persons, that judge of rhetorick by the rules of it current in these western parts of the world, would, instead of extolling it for the superlative, not allow it the positive degree of eloquence; [and] would think the style as destitute of graces, as the theology of truth.[65]

So far we have spoken only of necessary conditions (and Locke will add more). But 'what shall be a sufficient inducement to take any extraordinary operation to be a miracle, i.e. wrought by God himself for the attestation of a revelation from him? . . . to this I answer, the carrying with it the marks of a greater power than appears in opposition to it.'[66]

Given God's power, God *need* not allow any opposition to Him to succeed. Given God's goodness, God *will* not. God will not 'suffer his messenger and his truth to be borne down by the appearance of a greater power on the side of an impostor, and in favour of a lie'.[67] However, even allowing that – though it raises problems for Scriptural predictions of misleading miracles – there are still problems, for two fairly obvious further cases present themselves at once. The first is that in which, though one side is clearly more powerful, *neither* side represents the deity: after all, the most that Locke's argument to this point can show is that the loser must be an impostor: it cannot, despite his claim, show us that the winner is not. The second is the case in which there is *no* opposition, in which there is a purported miracle, without any competing miracle.

Locke explicitly allowed non-divine miracles. He remarked that God has taken care 'that no pretended revelation should stand in competition with what is truly divine', but did not notice how very far this was from licensing his further comment, 'we need but open our eyes to see and be sure which came from him'.

Even if we were to allow Locke his claim that where we have two or more competing miracles, the winner is from God, why should we allow his further claim: 'upon the same ground of superiority of power, uncontested revelation will stand too'? To argue for this, Locke introduced three new necessary conditions as premises.

First, no putative revelation should 'derogat[e] from the honour of . . . God, or [be] inconsistent with natural religion and the rules of morality'.[68] If, for example, someone purporting to be bearing a message from God were to say: 'Now go and smite Amalek, and utterly destroy all that they have, and spare them not; but slay both man and woman, infant and suckling, ox and sheep, camel and ass', we would know at once that s/he was either an impostor or badly mistaken.[69]

Secondly, the message should not be 'of things indifferent, and of small moment, or that are knowable by the use of . . . natural faculties'. Here Locke must expect more serious opposition, at least regarding the third restriction. It is not God's intention, St Thomas argued persuasively,[70] to bar the gates of Heaven to those who are too stupid, or too busy, or just plain too lazy to go in for hard thinking, but we may perhaps take Locke's general point that the putative revelation should be *weighty*.

Thus, argued Locke, making his third preliminary point, the matter revealed 'must be the revelation of some supernatural truths relating to the glory of God, and some great concern of men'. Having given us his preliminary premises, how does Locke achieve his conclusion?

> Supernatural operations attesting such a revelation may with reason be taken to be miracles, as carrying the marks of a superior and over-ruling power, as long as no revelation accompanied with marks of a greater power appears against it. For these supernatural signs being the only means God is conceived to have to satisfy men, as rational creatures, of the certainty of any thing he would reveal, as coming from himself, can never consent that it should be wrested out of his hands, to serve the ends and establish the authority of an inferior agent that rivals him.[71]

Thus, for Locke, miracles are (a) observer relative and (b) subject to their weightiness and lack of positive immorality, self-substantiating: this is the one case, Locke claimed, where implausibility entails credibility.[72] Locke's position depends heavily on a prior proof of God's existence, and *should* also require a prior proof of God's omnipotence, benevolence and jealousy. God must *care* about us not being mistaken about God. But it is far from clear how Locke would establish this. Not, clearly, from the Judaeo–Christian revelation: for that is what miracles are supposed to lead us to. Not from empirical evidence, for that is at best balanced. Locke *says* God will not allow what Boyle calls 'Lying Miracles',[73] but his argument for this claim is very weak. Boyle, by contrast, makes more modest claims, but argues for them more cogently even though his position, too, is not without serious problems.

Boyle on miracles

Locke was clear that miracles have a function, and it is implicit in his view that miracles were a historical phenomenon. He did not seriously consider contemporary reports of miracles, even if they were uncontested.[74] For Boyle the line was less sharp, and he shared the worries of earlier writers that miracles might deceive us.[75] He wrote:

> I first assent to a Natural Religion upon the score of Natural Reason antecedently to any particular Revelation. And *then*; if a Miracle be wrought to attest a particular Doctrine . . . , I endeavour according to ye Principles of Natural Religion & right Reason, to discover whether or no this propos'd Doctrine be such, yt I ought to looke upon a Miracle yt is vouch'd for it, as comeing from God or not. And *lastly* if I find by ye Agreeableness of it to ye best notions that natural Theology giues us

of God & his Attributes, that this Religion cannot in reason be doubted to come from him; I then judge ye Body of ye Religion to be true.

But this is to be understood of the miracles yt are wrought in favor of a Religion when 'tis *first proposed* to be embrac'd. For when a Religion has been once solidly establish'd by a competent number of true & uncontroald Miracles, then I haue a new Criterion to judge of a subsequent miracle by; for ye Veracity of God being founded upon his Justice & Goodness (Three Attributes knowable to us by ye Light of Natural Reason) it is not to be supposed that he would contradict himselfe . . . : And therefore also, 'tis not to be suppos'd, that ye subsequent miracle proceeds from him unless it be control'd by a greater miracle: as in ye case of the Egyptian Sorcerers and Moses;[76] Simon Magus and ye two Apostles, Peter & John;[77] or unless God permit it to try men, wch may be reasonably presum'd in case some express prediction were made that such Lying Miracles should be sufferd to be wrought for that end. And upon yt ground it might reasonably be said by Paul (the Apostle of him yt had foretold, yt false Christs & other Seducers should arise with *lying wonders*[78] to seduce if it were possible the *choicest Christians*) to the Galatians, that if either he or an Angell from heaven should teach them another Gospell,[79] . . . they should reject him, knowing that . . . even Satan may, by God's permission, *transform himselfe into an Angel of Light.*[80]

Boyle's focus was importantly different from Locke's. For him there was the realm of things to be explained scientifically, 'the world . . . as the great system of things corporeal, as it once really was towards the close of the sixth day of creation, when God had finished all his material works, but had not yet created man',[81] in which all explanations were to be intelligible,[82] but there were also many things which required supernatural intervention, including God's sustaining His creatures in existence, sustaining the system as an on-going law-like entity or *automaton*, granting incorporeal souls to corporeal animals ('physical miracles' which occur hundreds of times each day),[83] and ensuring that there was a law-like connection between the sensory input of the animals and the intellectual abstractions they were able to perform. Also in need of explanation was the mechanically inexplicable ability of people, i.e. incorporeal souls, to move matter. Additionally, people sometimes seemed able to acquire knowledge beyond their ordinary ken, providing a *prima facie* case for angelic intervention.[84] There were apparent miracles of healing and apparent cases of diabolic communication. There were cases of things which could in some sense be 'explained' naturally, but which seemed to Boyle much happier wearing supernatural explanations than natural ones, the most interesting case being the spread and survival of the Christian religion – a 'permanent' as opposed to other, 'transient', miracles.[85] There were cases of created intellects apparently being able to foretell the future. And then there were cases of supernatural intervention with an apparent intent to validate a particular instituted religion, or one in the process of becoming instituted.

For Boyle the world split up into events which have a mechanical explanation and those which do not. The ones which have a mechanical explanation are thereby law-like. Of the ones which do not, some are law-like in their regularity and some are not, but it is clear that supernatural intervention in Boyle's system is pretty much a commonplace.[86] We might decide, with Hobbes and Locke, to call only those which bore a validating message 'miracles', but with Boyle we are a long way from the notion that miraculous (or at least supernatural) intervention is extraordinary. Often it is law-like, as in the case of the mechanically inexplicable interaction between the human soul and the human body. We are also a long way from Locke's notion that miraculous intervention is agent relative.

Prescience was both a miracle and a problem for Boyle. Following Aristotle, St Thomas held that future events fall into three distinct classes: those which are already present in their causes; those which the present situation makes probable; and those which depend on random or casual factors, including free will. Members of the first two groups are predictable in principle by created intellects; members of the third group are known 'in themselves, and such knowledge of the future is proper to God alone'.[87]

Future contingents, then, must not be predictable in the usual way, for if they were they could be known *now*, and hence it would be true, for us, here and now, that they would happen. But then the necessity that the present has would infect the future. Moreover, they must not be unpredictable merely because of the weakness of our intellect, for then they would still be *true* (*false*) now whether or not we knew it. They must be unpredictable *in principle* to any created intellect. Divination is an attempt to pre-empt a divine prerogative. Since that is not possible, prediction, genuine, trustworthy prediction, can come only from God. But then it must involve a communication, i.e. an intervention in the natural order. And since there must be a weighty *point* to such intervention (in view of what is presumed to be known about God's character) such intervention is miraculous.

Thus, if someone did *predict* future contingents accurately (as opposed to making a lucky guess), that person must have been divinely inspired, for *only* God has the requisite foreknowledge.[88] Boyle points out that the second coming of Christ is not known to humans, 'the angels in heaven, nor . . . the Messiah himself, but the Father only'.[89] All this involves accepting bivalence, the doctrine that every (indicative) sentence is either true or false, but, unlike the great philosophers of the thirteenth and fourteenth centuries, Locke and Boyle seem not to be aware of even the possibility of rejecting it.[90] Accepting bivalence led Locke to write: 'I cannot make freedom in man consistent with omnipotence and omniscience in God, though I am . . . fully perswaded of both. And therefore I have long since given off the consideration of that question, resolving . . . that if it be possible for God to make a free agent, then man is free, though I see not the way of it',[91] and led to Boyle's perplexity:

> The Generality of our Philosophers as well as Divines believe that God has a foreknowledge of <all> future contingencys: <& yet how> a Certaine Prescience <can consist with> the free will of <Man> . . . is soe difficult to discerne that y[e] Socinians are wont to deny such things as depend upon y[e] Will of Free Agents to be the proper objects of Omniscience, & the Head of y[e] Remonstrants . . . confesses he knows not how <clearly> to make out y[e] Consistency of <Gods> Prescience & mans freedome. And the reconsilement of these Truths is not a difficulty peculiar to y[e] X[tian] Religion, but concernes speculative men in all Religions, who

acknowledge yᵉ Diety to be infinitely perfect, & allow man, as they doe, to be a Free Agent.⁹²

Leibniz's Ockhamist attempt to clear the matter up: 'It is universally agreed that future contingents are certain, since God foresees them, but this does not make us say that they are necessary',⁹³ would not have commended itself to either Locke or Boyle.

Given the points we have already noticed, there is a Leibnizian problem for Boyle concerning miracles. The important features are these:

(1) laws depend solely on the free will of God;
(2) God has foreknowledge of the ways of the universe;⁹⁴
(3) God sustains us in existence;
(4) God sustains law-like regularities in nature;
(5) 'law' is a metaphorical term: what we have, simply, is the ordering that God has designed for the universe. God's laws need not be intelligible to us: they may be the regularities of a universe designed for the angels, or indeed, designed by God for God: God doesn't need *us*, and we have no reason apart from arrogance to think that the universe was intended exclusively for us.

Now, suppose we jot down a number of points at random. How many curves are there which pass through the points we jotted down in just the order in which we put them down? An infinite number. How many sets of laws are there which will encompass the events between creation and the last trump, all of them foreseen by God? Again, an infinite number. More important than the infinity of encompassing theories is the fact that there is at least *one* such encompassing theory: *every* event must be part of *some* law-like sequence.⁹⁵

Boyle's response was that the sorts of regularity that the naturalist *qua* naturalist is interested in are regularities within the realm of the understandable:

If it bee said, yᵗ God hath been pleas'd to make an union of [body and soul] according to certaine Laws, by vertue of wᶜʰ, they are inabled to act upon one another; I readily grant it: but . . . this is not to giue a Mechanical or Physical account . . . , but to haue recourse to yᵉ arbitrary & iṁediate agency of yᵉ first cause.⁹⁶

The Christian Virtuoso will be aware that there are both regularities and singularities that do *not* fall within the realm of the understandable (i.e. the mechanically explicable), but s/he will not *qua* naturalist, be concerned with them: 'instead of dividing the operations of God, here below, into two sorts only, natural and supernatural; I think we may take in a third sort, and divide the same operations into supernatural, natural in a stricter sense, that is mechanical, and natural in a larger sense, which I call supra-mechanical.'⁹⁷

These naturalistically inexplicable features of the world contain the miracles as a subset. Boyle is less interested in them *as* a subset than (for example) Locke is; his real interest lies in *all* the interventions which can, and (he thinks) should, lead us to belief in God. As with Locke, there is an epistemological problem, but here it is a problem that was fully recognised. Holding that 'divers . . . supernatural things that are wont to be accounted miracles, may proceed from two extreamly differing Causes, namely God & the Divel', Boyle realised that 'it is necessary to consider the nature of the Doctrines for wch Miracles are vouch'd, that we may not perniciously mistake Diabolical workes for Divine Mir[a]cles'. First, Boyle suggested, if the putative miracles are 'substantially attested' we should simply accept the 'historical part'. Then we should decide whether they 'can be made out to haue God for their author'.

The decision that a particular miracle is 'Divine and not Diabolical' is made, *not* on grounds of doctrine, but on 'ye *General Principles of Natural Reason & Religion*, wch teach . . . antecedently to all particular Revelations, That *there is a G[od;]* That *he is*, & can be, *but One*, That [*he*] *is Just, Wise, Good, Gracious* &c.; and That *he has the Care and Governm^t of Humane Affairs*'. Then 'if ye Revelation backd by a Miracle proposes nothing that contradicts any of these Truths . . . & much more if it proposes a Religion that illustrates and confirms them; I then thinke my self oblig'd to admit both ye Miracle, and the Religion it attests'.[98]

Ultimately, then, what we have is a probabilistic account: we can have moral certainty that this or that miracle proceeds from God, provided that (a) we have independent evidence of God's existence or, at least, of the *possibility* of God's existence, and (b) the miracle fits the imposed necessary conditions. Even then, said the cautious Boyle, we should hold to our opinion loosely: in particular we should not let our opinion lead us to be intemperate in our dealings with our fellow Christians.[99] Chillingworth remarked that God would not damn men for misinterpreting dark sayings, and it is clear that Boyle will not do so either.

Acknowledgements

Like every worker on the Boyle manuscripts, I owe a debt of gratitude to the Librarian and staff of the Royal Society Library. They have provided me with a great deal of unobtrusive help over the years. I am also grateful to Michael Hunter, M. A. Stewart and Margaret J. Osler for a number of helpful suggestions.

NOTES

1 R. M. Burns, *The Great Debate on Miracles: From Joseph Glanvill to David Hume* (Lewisburg: Bucknell University Press, 1981), p. 10. For an excellent account of the historical background to Boyle's contribution, see Daniel Beck, 'Miracle and the mechanical philosophy: the theology of Robert Boyle in its historical context', unpublished Ph.D. thesis, University of Notre Dame, 1986.

2 John Locke, *An Essay Concerning Human Understanding*, ed. P. H. Nidditch (Oxford: Clarendon Press, 1975), 1.4.13. (References are by book, chapter and section.)

3 *Christian Virtuoso I, Appendix, Works*, vi, 712; BP 7, fol. 233.

4 For a kinder view of Locke's comparative worth, see Burns, *Great Debate* (n. 1), p. 57.

5 Locke, *Essay*, 4.9.2.

6 *Ibid.* 4.10.2. For details of the relation to Cudworth's argument, see M. R. Ayers, 'Mechanism, superaddition, and the proof of God's existence in Locke's *Essay*', *The Philosophical Review* 90 (1981): 210–51.

7 Locke, *Essay*, 4.10.3.

8 The young Newton considered possible biblical evidence for the creation of time: 'Whither Moses his saying . . . y^t ye evening & ye morning were ye first day &c do prove y^t God created time. Coll 1.16 or heb 1 ch 2v τὸς αἰῶνας ἐποίησεν expoūded, he made ye worlds. prove y^t God created time.' *Certain Philosophical Questions*, eds J. E. McGuire and M. Tamny (Cambridge University Press, 1983), p. 448.

9 Thomas Aquinas, *Summa Theologiae*, 1^a 46.2 resp.

10 *Notion of Nature, Works*, v, 163.

11 BP 2, fol. 2. For Aristotle's argument see *de Caelo*, 270^a 13–27.

12 BP 1, fol. 51, and cf. *Things above Reason, Works*, iv, 415–16.

13 *Ibid.* pp. 409, 416. Mathematically more powerful minds also found the issue confusing, as Newton's second letter to Bentley reveals. *The Works of Richard Bentley, D.D.*, ed. A. Dyce, 3 vols (London, 1838; reprinted New York: AMS Press, 1966), iii, 207–10 in reprinted edition.

14 Locke, *Essay*, 4.10.5.

15 'Nihil quod animi quodque rationis est expers, id generare ex se potest animantem conpotemque rationis', quoted by Cicero, *De natura deorum* ii 22.

16 Locke, *Essay*, 4.10.9. Locke took this to be self-evident, but Boyle produced an argument from parsimony and explanatory clarity for it. See J. J. MacIntosh, 'Robert Boyle on Epicurean atheism and atomism', in *Atoms, Pneuma, and Tranquillity: Epicurean and Stoic Themes in European Thought*, ed. M. J. Osler (Cambridge University Press, 1991), pp. 197–219.

17 Locke, *Essay*, 4.3.6.

18 David Hume, *Dialogues Concerning Natural Religion*, Humetext 1.0, 1990, eds T. L. Beauchamp, D. F. Norton and M. A. Stewart, p. 148.

19 *New Essays on Human Understanding*, trans. and eds Peter Remnant and Jonathan Bennett (Cambridge University Press, 1981), p. 379 (pagination as in *Nouveaux essais sur l'entendement humain*, eds André Robinet and Heinrich Schepers, Berlin: Academie-Verlag, 1962), and see further *Monadology* § 17. Bentley, in his second Boyle lecture, was equally adamant: 'Omnipotence it self cannot create cogitative Body.' *Eight Boyle Lectures on Atheism* (London, 1692–3; reprinted New York: Garland, 1976), ii, 29 in reprinted edition.

20 *The works of John Locke*, 10 vols (London, 1823; reprinted Aalen: Scientia Verlag, 1963), iv, 468 in reprinted edition.

21 Jean Baptiste Du Hamel, to Oldenburg, 16 October 1673, *Oldenburg*, x, 298–300.

22 Quoted in Stephen Jay Gould, *Ever Since Darwin* (New York: W. W. Norton, 1979), p. 25.

23 *Some Thoughts Concerning Education*, eds J. W. and J. S. Yolton (Oxford: Clarendon Press, 1989), sect. 192, p. 246; Ayers, 'Mechanism' (n. 6), p. 216 & note; Ayers notes that Locke's phraseology here strongly suggests Boyle's influence.

24 Locke, *Essay*, 4.10.8. Leibniz, in the *New Essays*, noted the fallacy, but other contemporaries missed it completely. Henry Lee's *Anti-Scepticism: or, Notes Upon each Chapter of Mr. Lock's Essay* (London, 1702) provides a striking example.

25 Locke, *Essay*, 4.10.10.

26 Wrapping a serious point in a joke, Pascal noted that even though 'la nature [a] gravé son image et celle de son auteur dans toutes choses', the canonical authors never used arguments of the form '"Il n'y a point de vide, donc il y a un Dieu."' *Pensées*, ed. L. Lafuma (Paris: Éditions du Luxembourg, 1952), p. 137 (Fr 199–390; Br 72), p. 289 (Fr 463–19; Br 243).

27 Locke, *Essay*, 4.10.11–12.

28 Leibniz, *New Essays* (n. 19), pp. 435–8 (4.10.6–7).

29 *Critique of Pure Reason*, trans. Norman Kemp Smith (London: Macmillan, 1958), A629 = B657.

30 *Usefulness, Works*, ii, 22. Fastening on insects for design arguments was common at the time. See David Berman, *A History of Atheism in Britain* (London: Croom Helm, 1988), pp. 3–7.

31 BP 7, fols. 164v–165r.

32 James E. Force, 'The breakdown of the Newtonian synthesis of science and religion', in James E. Force and Richard H. Popkin, *Essays on the Context, Nature, and Influence of Isaac Newton's Theology* (Dordrecht: Kluwer Academic Publishers, 1990), pp. 143–63, on pp. 145–6.

33 Similarly, Cudworth offered his 'Labour . . . for the Confirmation of *Weak, Staggering*, and *Sceptical Theists*.' *The True Intellectual System of the Universe* (London, 1678; reprinted Stuttgart-Bad Cannstatt: Friedrich Frommann, 1964), preface.

34 'The sinful Lusts, unruly Passions, and corrupt Interests of such vitious Persons as Atheists are
    commonly obserued to be', said Boyle, 'cannot but haue a great stroke in . . . the Judgements they pass
    . . . and the force of Arguments that are employd to proue them' (BP 6, fol. 301), and Richard
    Allestree thought that fear of a future life and the consequent 'damps and shiverings' of atheists
    provided 'the original and first rise of this impiety, it being impossible for any man that sees the whole,
    nay but the smallest part of the Universe, to doubt of a first and supreme Being'. *The Government of
    the Tongue* (Oxford, 1674), pp. 15, 18. That Bentley's 'delightfull and ravishing *Hypothesis* of
    Religion', *Eight Boyle Lectures* (n. 19), ii.12, might also have a causal influence on belief was deemed
    irrelevant, since that hypothesis was accepted as the true one.
35 BP 2, fol. 144v.
36 *Ibid.* fols. 63–4.
37 Boyle was well aware that all we can *observe* are regularities. To obtain and account for what
    contemporary philosophers sometimes call *strong* laws of nature, some external feature must be
    invoked. For an ingenious contemporary suggestion, see D. M. Armstrong, 'What makes induction
    rational?', *Dialogue* 30 (1991): 503–11.
38 Cudworth, *True Intellectual System* (n. 33), p. 150.
39 For an important discussion of the notion earlier in the century, see Margaret J. Osler, 'Fortune, fate
    and divination: Gassendi's voluntarist theology and the baptism of Epicureanism', in Osler, *Atoms,
    Pneuma, and Tranquillity* (n. 16).
40 See Ian Hacking, *The Emergence of Probability* (Cambridge University Press, 1975), and Ivo Schneider,
    'Why do we find the origin of a calculus of probabilities in the seventeenth century?', *Probabilistic
    Thinking, Thermodynamics and the Interaction of the History and Philosophy of Science*, eds. J. Hintikka,
    D. Gruender and E. Agazzi, 2 vols (Dordrecht: D. Reidel, 1980), ii, 3–24.
41 *Works*, ii, 44.
42 Henry More, *An Antidote Against Atheism*, in *A Collection of Several Philosophical Writings*, 2 vols
    (London, 1662; reprinted New York: Garland, 1978), i, 52.
43 John Horgan, 'In the beginning . . . ', *Scientific American*, 264.2 (1991): 116–25, on p. 118.
44 See Jonathan J. Halliwell, 'Quantum cosmology and the creation of the universe', *Scientific American*,
    265.6 (1991): 76–85.
45 John Ray, *The Wisdom of God Manifested in the Works of the Creation* (London, 1691), pp. 32–5.
46 Cudworth attributes this point to Thomas White. More mathematically inclined writers such as
    Descartes and Leibniz saw clearly that all such tasks were of equal, because no, difficulty to the
    Almighty.
47 Cudworth, *True Intellectual System* (n. 33), pp. 149–50 (1.3.4).
48 Stephen Jay Gould, *The Flamingo's Smile* (New York: Norton, 1987), p. 35.
49 The young Boyle emphasised that not everything was designed for us. God explicitly commanded
    Noah to save *all* the animals, noxious as well as beneficial, 'when there needed nothing to destroy them
    but not to saue them', which suggests that they were made 'for other Ends, besides Man's Seruice and
    Aduantages', BP 37, fol. 184r. Later he asked, why should God 'in his creating of things, have respect
    to the measure and ease of human understandings; and not, rather, if of any, of angelical intellects?'
    Usefulness, I, Works, ii, 46.
50 BP 2, fol. 89; BP 6, fol. 326.
51 *Christian Virtuoso I, Appendix, Works*, vi, 694.
52 *Cosmical Qualities, Works*, iii, 322–3. Boyle gives as an example periodic stars, especially those studied
    by 'the justly famous Bullialdus', *ibid.* p. 324. Bullialdus (Ismael Boulliau, 1605–95) was the first to
    establish the periodicity of a variable star: in *Ad astronomos monita duo* (1667) he put the period of Mira
    Ceti (discovered by Fabricius in 1596) as 333 days, an overestimate of less than two days. Boyle clearly
    kept his reading up to date. There is a reference to the same phenomenon four years later in *Reason
    and Religion, Works*, iv, 180. Discussing such seeming irregularities late in life, he still held that, given
    'an artificer of so vast a comprehension . . . as is the Maker of the world . . . things [which] may appear
    to . . . break out abruptly and unexpectedly . . . really have [an explanation] that would, if we discerned
    it, keep us from imputing [them] either to chance or to nature's aberrations', *Notion of Nature, Works*,
    v, 216.
53 David Hume, *Dialogues Concerning Natural Religion* (n. 18), pp. 209–10.
54 BP 1, fol. 62.

55 Leibniz held that God would never allow 'matter . . . of itself [to] move in a curved path, because it is
    impossible to conceive how this could happen – that is, to explain it mechanically', *New Essays* (n. 19),
    preface, p. 66, but Boyle held it to be logically possible that other parts of the universe might have
    different natural laws, *High Veneration, Works*, v, 139; cf. BP 9, fol. 60r.
56 *Seraphic Love, Works*, i, 270.
57 *Usefulness I, Works*, ii, 48; *Forms and Qualities, Works*, iii, 48. Hooke argued for the same ordering: 'Of
    Comets & Gravity', *Posthumous Works*, ed. Richard Waller (London, 1705), pp. 174–5.
58 The vexed question of *how* God sustains both creatures and the laws of the corporeal universe is
    discussed in Timothy Shanahan's 'God and nature in the thought of Robert Boyle', *Journal of the
    History of Philosophy* 26 (1988): 547–69.
59 More, *Antidote* (n. 42), i, 22. For Descartes' argument see *Meditation* III.
60 *Christian Virtuoso, II, Works*, vi, 748.
61 Some contemporary philosophers make similar points on Gödelian grounds. See for example Gödel's
    own 'Some basic theorems on the foundations of mathematics and their implications', forthcoming in
    vol. 3 of his *Collected Works* (New York: Oxford University Press, forthcoming).
62 Hobbes, *Leviathan* (1651), ed. Michael Oakeshott (Oxford: Basil Blackwell, nd), ch. 37, pp. 286, 288.
63 Locke, *A Discourse of Miracles* (1706, written 1702), in Locke, *Works* (n. 20), ix, 256.
64 *Ibid.* p. 258.
65 *Style of the Scriptures, Works*, ii, 298. Boyle does not deny that the style could have been miraculous,
    for he runs a similar argument concerning the Apostles and the miracle of Pentecost: the Apostles'
    'Hearers . . . knew it was <not> naturally possible, that uninspir'd Persons, and especially illiterate
    Fishermen, should <grow> able, in a trice, to make <weighty> discourses to many differing Nations,
    in their respective Languages': BP 7, fol. 99, Boyle's deletions omitted. Boyle accepts, and indeed uses,
    the form of the Koran argument: it is the premise he disputes.
66 Locke, *Works* (n. 20), ix, 259. However, 'we have . . . a very imperfect, obscure idea of active power'
    (Locke, *Essay*, 2.21.4).
67 Locke, *Works* (n. 20), ix, 260.
68 *Ibid.* p. 261.
69 I Samuel xv, 3.
70 *Summa Contra Gentiles*, book I, ch. 4.
71 Locke, *Works* (n. 20), ix, 262.
72 Locke, *Essay*, 4.16.13. Locke does, however, point out that the number and variety of Christ's miracles
    increases the probability that the individual miracles were correctly so called. *Discourse on Miracles*, in
    Locke, *Works* (n. 20), ix, 259.
73 BP 7, fol. 123.
74 In 'Miracles, experiments and the ordinary course of nature', *Isis* 81 (1990): 663–83, Peter Dear argues
    that this was the case generally for the English in the seventeenth century. Contemporary miracles
    were typically disallowed by Protestants. Early in the century, writing to show that 'Popish Miracles
    are either counterfeit or diuellish', the Spaniard Fernando Tejeda urged an Augustinian distinction
    between 'Christian Faith when it was to be propagated', and the present time when 'Faith is already
    greatly defused throughout the world, and sufficiently fortified by the miracles of the first ages, so that
    now it is not safe, but rather very dangerous to iudge of Religion meerly by Miracles'. *Miracles
    Vnmasked* (London, 1625), p. 4. See also, for example, Richard Sheldon's *The Svrvey of the miracles of
    the Chvrch of Rome, prouing them to be antichristian* (London, 1616), pp. 32–6. In the next century,
    Hume simply assumed the absence of contemporary miracles: '*It is strange*, a judicious reader is apt to
    say, upon the perusal of these wonderful historians, *that such prodigious events never happen in our days.*
    But it is nothing strange, I hope, that men should lie in all ages', *An Enquiry Concerning Human
    Understanding* (n. 18), pp. 119–20, but Boyle, as always, was more cautious. Writing to Stubbe on
    9 March 1666, he was sharply dismissive of the suggestion that '*Greatraks*'s stupendous performances'
    were miracles, but noted that he had never 'met with . . . any cogent proof, that miracles were to cease
    with the age of the apostles', *Works*, i, lxvi.
75 Even deceptive miracles should strengthen belief, for 'though all *Miracles* . . . do not immediatly prove
    the *Existence of a God*, nor confirm *a Prophet*, or whatsoever *Doctrine*; yet do they all of them evince,
    that there is a Rank of *Invisible Understanding Beings*, Superiour *to men*, which the Atheists commonly
    deny'. Cudworth, *True Intellectual System* (n. 33), p. 708. Combating such atheists was part of the

reason, Glanvill tells us in the preface to *Saducimus Triumphatus*, 3rd edn (London, 1689), for his offering a work containing 'Full and Plain Evidence Concerning Witches and Apparitions'. Boyle agreed that there were such beings, though cautioning that 'most [accounts of them] are false, and occasioned by the credulity or imposture of men.' *Reason and Religion, Works*, iv, 170.

76 Exodus vii–xi.

77 Acts viii, 9–24.

78 II Thessalonians ii, 9. When offering Pentecost as the most convincing of miracles, Boyle thought it necessary to point out the unlikelihood of its being a 'false Miracle': 'there was no *place* on Earth, nor no *time* of the year, where 'twas less likely that the Infinitely good and wise God, would have suffer'd a *false Miracle*, and which pretended to establish a new Doctrine, to be wrought in his name'. BP 7, fol. 97r.

79 Galatians i, 8.

80 BP 7, fols 122–3. The Angel of Light reference is to II Corinthians xi, 14. Compare William Chillingworth, *The Religion of Protestants* (Oxford, 1638; reprinted Menston: Scholar Press, 1972), The Preface to the Author of Charity Maintained, sect. 43:

> this book, . . . confirm'd by innumerable Miracles, fortels me plainly, that in after ages great signes and wonders shall be wrought in confirmation of false doctrine, and that I am not to believe any doctrine which seems to my understanding repugnant to the first, though an Angell from Heaven should teach it . . . : But that true doctrine should in all ages have the testimony of Miracles, that I am no where taught; So that I have more reason to suspect and be afraid of pretended Miracles, as signes of false doctrine, then much to regard them as certain arguments of the truth.

81 *Notion of Nature, Works*, v, 166–7.

82 'Mr. *Boyle* made it his chief Business to inculcate, that every Thing was done *mechanically* in natural Philosophy.' Leibniz to Clarke, 5.114, *The Works of Samuel Clarke, D.D.*, 4 vols (London, 1738; reprinted New York: Garland, 1978), iv, 667. Boyle makes the point that intelligible and mechanical explanations are coextensive in a number of places. See, for example, *The Excellency and Grounds of the Mechanical Hypothesis, Works*, iv, 73, and *Forms and Qualities, Works*, iii, 48. Leibniz agreed for all practical purposes: '[H]owever much I agree with the Scholastics in this general and, so to speak, metaphysical explanation of the principles of bodies, I am as corpuscular as one can be in the explanation of particular phenomena', he wrote to Arnauld in 1686, and later (29 December 1691), writing to Huygens about Boyle he remarked, 'we all know [that] everything happens mechanically'. C. I. Gerhardt (ed.), *Die Philosophische Schriften von G. W. Leibniz*, 7 vols (Berlin, 1875–90), ii, 58; *Œuvres Complètes de Christiaan Huygens*, 23 vols (The Hague: Martinus Nijhoff, 1888–1950), x, 228.

83 BP 2, fol. 62. For Boyle, the soul is an important factor in personal identity. By way of contrast, the young Newton, referring to Ecclesiasticus xxxiii, 10 ('All men alike come from the ground; Adam was created out of earth') considered seriously the possibility that the soul is *simply* the vivifying factor, while both individual and species differences result from 'yᶜ different tempers & modes' of the bodies in question. *Certain Philosophical Questions* (n. 8), p. 448.

84 ' . . . though I dare not affirm . . . that God discloses to men the great mystery of chymistry by good angels, or by nocturnal visions . . . yet persuaded I am, that the favour of God (does much more than most men are aware of) vouchsafe to promote some men's proficiency in the study of nature', *Usefulness, I, Works*, ii, 61. Michael Hunter notes the same suggestion coming from Meric Casaubon: 'it is not improbable that divers secrets of [chemistry] came to the knowledge of man by the Revelation of Spirits', in 'Alchemy, magic and moralism in the thought of Robert Boyle', *BJHS* 23 (1990): 387–410, on pp. 398–9.

85 *Christian Virtuoso, I, Works*, v, 535–6.

86 In *Science and Religion in Seventeenth-Century England* (New Haven: Yale University Press, 1958), R. S. Westfall argued for a different conclusion: 'Boyle's opinion about miracles stood in absolute contradiction to the rest of his thought' (p. 89), but I think Boyle's position was quite consistent here.

87 Aquinas, *Summa Theologiae*, 1ᵃ 57.3 resp., 1ᵃ 86.4 resp. For Aristotle's classification see *Physics*, 196ᵇ10–197ᵃ36. Subsequently Ockham restricted members of the third group to events depending on a free agent: *Expositio in Librum Perihermeneias Aristotelis*, in *Opera Philosophica*, eds A. Gambatese and S. Brown, 7 vols (New York: St Bonaventure, 1978), ii, 422 (I.6.15).

88 Following Boethius, philosophers such as St Thomas made much of God being an eternal, rather than a sempiternal, deity, but the God of Locke and Boyle seems to have been very much a temporal deity.

89 *Christian Virtuoso, I, Appendix, Works*, vii, 697. See Mark xiii, 32, Matthew xxiv, 36.

90 Even when they were aware of the problem future contingents raise, seventeenth-century thinkers usually accepted bivalence without further discussion. Thus Gassendi, though aware of the difficulty, offers no argument for his Canon XII: '*Of two opposite contingent propositions one is true, the other false, whether they refer to present, past or future time*', *Institutio Logica, Opera Omnia*, 6 vols (Lyon, 1658), i, 29; trans. Howard Jones (Assen: van Gorcum, 1981), p. 111. Bivalence should be distinguished from the law of excluded middle. Bivalence tells us that precisely one of the sentences

> There will be a sea battle tomorrow.
>
> There will not be a sea battle tomorrow.

is (here and now) true. Excluded middle tells us that the compound sentence

> Either there will be a sea battle tomorrow or there will not be.

is true, but does not thereby decide on whether or not the *contained* sentences have truth values. Excluded middle need raise no problems for free will.

91 Locke to William Molyneux, 20 January 1693, *The Correspondence of John Locke*, ed. E. S. De Beer, 8 vols (Oxford: Clarendon Press, 1979), iv, 625–6.

92 BP 2, fols 49–50. Boyle's deletions omitted.

93 G. W. Leibniz, *Discourse on Metaphysics*, eds and trans. P. G. Lucas and L. Grint (Manchester University Press, 1953), sect. 13.

94 Boyle goes very far in this direction: 'God loved us even before we had a being; and our felicity in his decrees preceded our existence in this world. God loved you numerous ages before you were; and his goodness is so entirely its own motive, that even your creation . . . is the effect of it.' *Seraphic Love, Works*, i, 266.

95 Leibniz, *Discourse on Metaphysics* (n. 93), sects 6, 7.

96 BP 3, fol. 105.

97 *Christian Virtuoso, II, Works*, vi, 754. Others were even more generous in allowing non-mechanical *laws*. In his *Dialogues on Metaphysics*, Malebranche offered five distinct types: laws of communication of motion, laws of the union of mind and body, laws concerning the union of our minds with 'Universal Reason', laws pertaining to angelic action and laws concerning the distribution of 'internal grace'. See further R. C. Sleigh, *Leibniz and Arnauld* (New Haven: Yale University Press, 1990), pp. 158ff.

98 BP 7, fols 120–1. Brackets enclose conjectural readings: the ms is damaged.

99 See, for example, *CPE, Works*, i, 302–3.

Bibliography of writings on Boyle published since 1940

AGASSI, JOSEPH, 'Robert Boyle's anonymous writings', *Isis* 68 (1977): 284–7

AGASSI, JOSEPH, 'Who discovered Boyle's law?', *Studies in History and Philosophy of Science* 8 (1977): 189–250

ALEXANDER, PETER, 'Boyle and Locke on primary and secondary qualities', *Ratio* 16 (1974): 51–67

ALEXANDER, PETER, 'Curley on Locke and Boyle', *Philosophical Review* 83 (1974): 229–37

ALEXANDER, PETER, 'The names of secondary qualities', *Proceedings of the Aristotelian Society* 72 (1977): 203–20

ALEXANDER, PETER, *Ideas, Qualities and Corpuscles: Locke and Boyle on the External World* (Cambridge University Press, 1985)

APPLEBAUM, W., 'Boyle and Hobbes: a reconsideration', *JHI* 25 (1964): 117–19

BECK, D. A., 'Miracle and the mechanical philosophy: the theology of Robert Boyle in its historical context', University of Notre Dame Ph.D. thesis, 1986

BIRLEY, ROBERT, 'Robert Boyle's headmaster at Eton', *NRRS* 13 (1958): 104–14

BIRLEY, ROBERT, 'Robert Boyle at Eton', *NRRS* 14 (1960): 191

BOAS, MARIE: *see* [Hall], Marie Boas

BOWEN, M. E. C., '"This great automaton, the world": the mechanical philosophy of Robert Boyle, F.R.S.', Columbia University Ph.D. thesis, 1976

BROWN, T.J., 'British scientific autographs, 2: Robert Boyle, 1627–91', *The Book Collector* 13 (1964): 487

BUCHAHAN, P. D., GIBSON, J. F., and HALL, M. B., 'Experimental history of science: Boyle's colour changes', *Ambix* 25 (1978): 208–10

BUEHLER, C. F., 'A projected but unpublished edition of the "Life and works" of Robert Boyle', *Chymia* 4 (1953): 79–83

BURNS, D. T., 'Robert Boyle (1627–91): a foundation stone of analytical chemistry in the British Isles', *Analytical Proceedings* 19 (1982): 224–33, 288–95; 22 (1985): 253–6; 23 (1986): 75–7, 349–51; 28 (1991): 362–4

BURNS, D. T., '"A man of God and gases". D. Thorburn Burns on the life, work and risks of working with Robert Boyle', *Chemistry and Industry*, 16 Dec. 1991: 921

BURNS, R. M., *The Great Debate on Miracles: From Joseph Glanvill to David Hume* (Lewisburg: Bucknell University Press, 1981)

BYLEBYL, J. J., 'Boyle and Harvey on the valves in the veins', *Bulletin of the History of Medicine* 56 (1982): 351–67

CAMPODONICO, ANGELO, *Filosofia dell'Esperienza ed Epistemologia della Fede in Robert Boyle* (Florence: Felice le Monnier, 1978)

CANNY, NICHOLAS, *The Upstart Earl: A Study of the Social and Mental World of Richard Boyle, First Earl of Cork, 1566–1643* (Cambridge University Press, 1982)

CARRÉ, M. H., 'Robert Boyle and English thought', *History Today* 7 (1957): 322–7

CLERICUZIO, ANTONIO, 'Le transmutazioni in Bacon e Boyle', in *Francis Bacon. Terminologia e Fortuna nel XVII Secolo*, ed. M. Fattori (Rome: Edizioni dell'Ateneo, 1985), pp. 29–42

CLERICUZIO, ANTONIO, '*Spiritus vitalis*: studio sulle teoria fisiologicha da Fernel a Boyle', *Nouvelles de la République des Lettres* 2 (1988): 33–84

CLERICUZIO, ANTONIO, 'Robert Boyle and the English Helmontians', in *Alchemy Revisited*, ed. Z. R. W. M. von Martels (Leiden: E. J. Brill, 1990), pp. 192–9

CLERICUZIO, ANTONIO, 'A redefinition of Boyle's chemistry and corpuscular philosophy', *Ann. Sci.* 47 (1990): 561–89

CLERICUZIO, ANTONIO, 'From van Helmont to Boyle: a study of the transmission of Helmontian chemical and medical theories in seventeenth-century England', *BJHS* 26 (1993): 303–34

COHEN, I.B., 'Franklin, Boerhaave, Newton, Boyle and the absorption of heat in relation to color', *Isis* 46 (1955): 99–104

COHEN, I. B., 'Newton, Hooke and "Boyle's law" (discovered by Power and Towneley)', *Nature* 204 (1964): 618–21

COHEN, L. A., 'An evaluation of the classic candle-mouse experiment', *Journal of the History of Medicine* 11 (1956): 127–32

COLIE, R. L., 'Spinoza in England 1665–1730', *Proceedings of the American Philosophical Society* 107 (1963): 183–219

CONANT, J. B., ed., 'Robert Boyle's experiments in pneumatics', in *Harvard Case Histories in Experimental Science*, 2 vols (Cambridge, Mass.: Harvard University Press, 1948; new edn, 1957), i, 1–63

CONRY, YVES, 'Robert Boyle et la doctrine cartesienne des animaux-machines', *Revue d'Histoire des Sciences* 33 (1980): 69–74

COPE, J. I., 'Evelyn, Boyle and Dr Wilkinson's "Mathematico-Chymico-Mechanical School"', *Isis* 50 (1959): 30–2

CRINÓ, A. M., 'Robert Boyle visto da due contemporanei', *Physis* 2 (1960): 318–20

CRINÓ, A. M., 'An unpublished letter on the theme of religion from Count Lorenzo Magalotti to the Honourable Robert Boyle in 1672', *Journal of the Warburg and Courtauld Institutes* 45 (1982): 271–8

CROWLEY, M. E., 'The notion of nature in the corpuscular philosophy of Robert Boyle', Marquette University Ph.D. thesis, 1970

CROWTHER, J. G., *Founders of British Science* (London: Cresset Press, 1960), pp. 51–93

CURLEY, E. M., 'Locke, Boyle, and the distinction between primary and secondary qualities', *Philosophical Review* 81 (1972): 438–64

DAUDIN, H., 'Spinoza et la science experimentale: sa discussion de l'experience de Boyle', *Revue d'Histoire des Sciences* 2 (1948): 179–90

DAVIS, E. B., 'Creation, contingency and early modern science: the impact of voluntaristic theology on seventeenth-century natural philosophy', Indiana University Ph.D. thesis, 1984, ch. 4

DEACON, MARGARET, *Scientists and the Sea 1650–1900* (London: Academic Press, 1971), ch. 6

DEAR, PETER, 'Miracles, experiments and the ordinary course of nature', *Isis* 81 (1990): 663–83

DEBUS, A. G., 'Robert Boyle and his *Sceptical Chymist*', *Indiana Quarterly for Bookmen* 5 (1949): 39–47

DEBUS, A. G., 'Solution analyses prior to Robert Boyle', *Chymia* 8 (1962): 141–61; reprinted in A. G. Debus, *Chemistry, Alchemy and the New Philosophy 1550–1700* (London: Variorum Reprints, 1987), ch. 8

DEBUS, A. G., 'Fire analysis and the elements in the sixteenth and the seventeenth centuries', *Ann. Sci.* 23 (1967): 127–47; reprinted in A. G. Debus, *Chemistry, Alchemy and the New Philosophy 1550–1700* (London: Variorum Reprints, 1987), ch. 7

DEWHURST, KENNETH, 'Locke's contribution to Boyle's researches on the air and on human blood', *NRRS* 17 (1962): 198–206

DIBNER, B., 'Early electrical machines', *Electrical Engineering* 76 (1957): 367–9

DIJKSTERHUIS, E. J., *The Mechanisation of the World Picture*, Eng. trans. (Oxford: Clarendon Press, 1961), part iv

DOBBS, B. J. T., *The Foundations of Newton's Alchemy* (Cambridge University Press, 1975), ch. 6

DRUMIN, W. A., 'The corpuscular philosophy of Robert Boyle: its establishment and verification', Columbia University Ph.D. thesis, 1973

DUFFY, EAMON, 'Valentine Greatrakes, the Irish stroker: miracle, science and orthodoxy in Restoration England', *Studies in Church History* 17 (1981): 251–73

EAMON, WILLIAM, 'New light on Robert Boyle and the discovery of colour indicators', *Ambix* 27 (1980): 204–9

'The effigy of Robert Boyle', *Nature* 168 (1951): 638

EMERTON, NORMA, *The Scientific Reinterpretation of Form* (Ithaca: Cornell University Press, 1984)

FEISENBERGER, H. A., 'The libraries of Newton, Hooke and Boyle', *NRRS* 21 (1965): 42–55

FIRTH, D. C., 'Robert Boyle, 1627–91', in *Late Seventeenth-Century Scientists*, ed. D. Hutchings (Oxford: Pergamon Press, 1969), pp. 1–32

FISCH, HAROLD, 'The scientist as priest: a note on Robert Boyle's natural theology', *Isis* 44 (1953): 252–65

FISCH, HAROLD, *Jerusalem and Albion: The Hebraic Factor in Seventeenth-Century Literature* (London: Routledge & Kegan Paul, 1964), ch. 13.

FISHER, M. S., *Robert Boyle, Devout Naturalist* (Philadelphia: Oshiver Studio Press, 1945)

FRANK, R. G., *Harvey and the Oxford Physiologists: a Study of Scientific Ideas and Social Interaction* (Berkeley and Los Angeles: University of California Press, 1980)

FULTON, J. F. 'Robert Boyle: a reappraisal', in *Aviation Medicine in its Preventive Aspects* (University of London, 1948), pp. 4–22.

FULTON, J. F., 'A book from Robert Boyle's library', *Journal of the History of Medicine* 11 (1956): 103–4

FULTON, J. F., 'Boyle and Sydenham', *Journal of the History of Medicine* 11 (1956): 351–2

FULTON, J. F., 'The Honourable Robert Boyle, F.R.S. (1627–92)', *NRRS* 15 (1960): 119–35; also in *The Royal Society: its Origins and Founders*, ed. H. Hartley (London: Royal Society, 1960)

FULTON, J. F., *A Bibliography of the Honourable Robert Boyle*, 2nd edn (Oxford: Clarendon Press, 1961); originally published in the *Proceedings* of the Oxford Bibliographical Society in 1932

GENERALES, C. D. J., 'Notes on Robert Boyle contributory to space medicine', *New York State Journal of Medicine* 67 (1967): 1193–204

GENERALES, C. D. J., 'Robert Boyle, progenitor of space medicine', in *Proceedings of the XXIII International Congress of the History of Medicine, London 2–9 Sept. 1972* (London, 1974), ii, 1140–8

GILLESPIE, N. C., 'Natural history, natural theology and social order: John Ray and the "Newtonian ideology"', *Journal of the History of Biology* 20 (1987): 1–49

GIUA, MICHELE, *Storia della Scienza ed Epistemologia. Galilei–Boyle–Planck* (Torino: Chiantore, 1945)

GOLINSKI, JAN V., 'Robert Boyle: scepticism and authority in seventeenth-century chemical discourse', in *The Figural and the Literal: Problems of Language in the History of Science and Philosophy, 1630–1800*, eds Andrew E. Benjamin, Geoffrey N. Cantor and John R. R. Christie (Manchester University Press, 1987), pp. 58–82

GOLINSKI, JAN V., 'Chemistry in the Scientific Revolution: problems of language and communication', in *Reappraisals of the Scientific Revolution*, eds D. C. Lindberg and R. S. Westman (Cambridge University Press, 1990), pp. 367–96

GREEN, W. J., 'Models and metaphysics in the chemical theories of Boyle and Newton', *Journal of Chemical Education* 55 (1978): 434–6

GREENE, R. A., 'Henry More and Robert Boyle on the spirit of nature', *JHI* 23 (1962): 451–74

GRILLI, M., 'Robert Boyle: contributi alla fisica del calore', *Physis* 24 (1982): 489–517

HACKETT, E., 'Robert Boyle and the human blood', *Irish Journal of Medical Science* 6 (1950): 528–33

HAGNER, A. F., 'Introduction' to facsimile edition of Boyle's *Essay about the Origine and Virtues of Gems*, Contributions to the History of Geology, no. 7 (New York: Hafner Publishing Co., 1972)

HALL, A. R. and HALL, M. B., 'Philosophy and natural philosophy: Boyle and Spinoza', in *Melanges Alexandre Koyré*, 2 vols (Paris: Hermann, 1964), ii, 241–56

HALL, A. R. and HALL, M. B, eds, *The Correspondence of Henry Oldenburg*, 13 vols (Madison, Milwaukee and London: University of Wisconsin Press; Mansell; Taylor & Francis, 1965–86)

[HALL], MARIE BOAS, 'Boyle as a theoretical scientist', *Isis* 41 (1950): 261–8

[HALL], MARIE BOAS, 'The establishment of the mechanical philosophy', *Osiris* 10 (1952), 412–541

[HALL], MARIE BOAS, 'An early version of Boyle's *Sceptical Chymist*', *Isis* 45 (1954): 153–68

[HALL], MARIE BOAS, 'Acid and alkali in seventeenth-century chemistry', *Archives Internationales d'Histoire des Sciences*, 34 (1956): 13–28

[HALL], MARIE BOAS, 'La méthode scientifique de Robert Boyle', *Revue d'Histoire des Sciences* 9 (1956): 105–25

[HALL], MARIE BOAS, *Robert Boyle and Seventeenth-Century Chemistry* (Cambridge University Press, 1958)

HALL, M. B., 'What happened to the Latin edition of Boyle's *History of Cold*?', *NRRS* 17 (1962): 32–5

HALL, M. B., 'Henry Miles, F.R.S. (1698–1763) and Thomas Birch, F.R.S. (1705–66)', *NRRS* 18 (1963): 39–44

HALL, M. B., 'Introduction' to facsimile edition of Boyle's *Experiments and Considerations touching Colours* (New York: Johnson Reprint Corporation, 1964)

HALL, M. B., *Robert Boyle on Natural Philosophy: An Essay with Selections from his Writings* (Bloomington: Indiana University Press, 1965)

HALL, M. B., 'Robert Boyle', *Scientific American* 217 (1967): 96–102

HALL, M. B., 'Boyle, Robert', in *Dictionary of Scientific Biography*, ed. C. C. Gillispie, 16 vols (New York: Scribners, 1970–80), ii, 377–82

HALL, M. B., 'Boyle's method of work: promoting his corpuscular philosophy', *NRRS* 41 (1987): 111–43

HALL, M. B., 'Frederic Slare, F.R.S. (1648–1727)', *NRRS* 46 (1992): 23–41

HARRÉ, ROM, 'Robert Boyle: The measurement of the spring of the air', in Rom Harré, *Great Scientific Experiments* (Oxford: Phaidon, 1981), pp. 83–91

HARVEY, E. N., *A History of Luminescence from the Earliest Times until 1900* (Philadelphia: American Philosophical Society, 1957)

HARWOOD, JOHN T., ed., *The Early Essays and Ethics of Robert Boyle* (Carbondale and Edwardsville: Southern Illinois University Press, 1991)

HAWTHORNE, R. M., 'Boyle's/Hooke's/Towneley and Power's/Mariotte's law', *Journal of Chemical Education* 56 (1979): 741–2

HENRY, JOHN, 'Occult qualities and the experimental philosophy: active principles in pre-Newtonian matter theory', *Hist. Sci.* 24 (1986): 335–81

HENRY, JOHN, 'Henry More versus Robert Boyle: the Spirit of Nature and the nature of providence', in *Henry More (1614–87): Tercentenary Studies*, ed. Sarah Hutton (Dordrecht: Kluwer Academic Publishers, 1990), pp. 55–76

HOFF, H. E., 'Nicolaus of Cusa, van Helmont and Boyle: the first experiment of the renaissance in quantitative biology and medicine', *Journal of the History of Medicine* 19 (1964): 99–117

HOOYKAAS, REIJER, *Robert Boyle: Een Studie over Natuurwetenschap en Christendom* (Loosduinen: Electr. Drukkerij Kleijuegt, 1943)

HOOYKAAS, REIJER, 'The experimental origin of the chemical atomic and molecular theory before Boyle', *Chymia* 2 (1949): 65–80

HUNT, R. M., *The Place of Religion in the Science of Robert Boyle* (University of Pittsburgh Press, 1955)

HUNTER, J. P., 'Robert Boyle and the epistemology of the novel', *Eighteenth-century Fiction* 2 (1990): 275–81

HUNTER, MICHAEL, 'Science and heterodoxy: an early modern problem reconsidered', in *Reappraisals of the Scientific Revolution*, eds D. C. Lindberg and R. S. Westman (Cambridge University Press, 1990), pp. 437–60

HUNTER, MICHAEL, 'Alchemy, magic and moralism in the thought of Robert Boyle', *BJHS* 23 (1990): 387–410

HUNTER, MICHAEL, *Letters and Papers of Robert Boyle: A Guide to the Manuscripts and Microfilm* (Bethesda, Md.: University Publications of America, 1992)

HUNTER, MICHAEL, 'Casuistry in action: Robert Boyle's confessional interviews with Gilbert Burnet and Edward Stillingfleet, 1691', *Journal of Ecclesiastical History* 44 (1993): 80–98

HUNTER, MICHAEL, 'The conscience of Robert Boyle: functionalism, "dysfunctionalism" and the task of historical understanding', in *Renaissance and Revolution: Humanists, Scholars, Craftsmen and Natural Philosophers in Early Modern Europe*, eds J. V. Field and F. A. J. L. James (Cambridge University Press, 1993), pp. 147–59

HUNTER, MICHAEL, *Robert Boyle: the Key Biographical Texts* (London: Pickering & Chatto, forthcoming)

HUNTER, MICHAEL, 'Robert Boyle and the dilemma of biography in the age of the scientific revolution', in *Telling Lives in Science: Studies in Scientific Biography*, eds Michael Shortland and Richard Yeo (Cambridge University Press, forthcoming)

HUNTER, R. A., and MACALPINE, IDA, 'William Harvey. Two medical anecdotes, the one related by Sir Kenelm Digby, the other by the Honourable Robert Boyle', *St Bartholomew's Hospital Journal* 60 (1956): 200–8

HUNTER, R. A., and MACALPINE, IDA, 'Valentine Greatraks', *St Bartholomew's Hospital Journal* 60 (1956): 361–8

HUNTER, R. A., and MACALPINE, IDA, 'Robert Boyle – poet', *Journal of the History of Medicine* 12 (1957): 390–2

HUNTER, R. A., and MACALPINE, IDA, 'William Harvey and Robert Boyle', *NRRS* 13 (1958): 115–27

HUNTER, R. A., and MACALPINE, IDA, 'Robert Boyle's poem – an addendum', *Journal of the History of Medicine* 27 (1972): 85–8

HUNTER, R. A., and ROSE, R. A., 'Robert Boyle's "Uncommon observations about vitiated sight" (London, 1688)', *British Journal of Ophthalmology* 41 (1958): 726–31

HUTCHISON, KEITH, 'What happened to occult qualities in the Scientific Revolution?', *Isis* 73 (1982): 233–53

HUTCHISON, KEITH, 'Supernaturalism and the mechanical philosophy', *Hist. Sci.* 21 (1983): 297–333

IHDE, A. J., 'Antecedents to the Boyle concept of the element', *Journal of Chemical Education* 33 (1956): 548–51

IHDE, A. J., 'Alchemy in reverse: Robert Boyle and the degradation of gold', *Chymia* 9 (1964): 47–57

JACOB, J. R., 'The ideological origins of Robert Boyle's natural philosophy', *Journal of European Studies* 2 (1972): 1–21

JACOB, J. R., 'Robert Boyle and subversive religion in the early Restoration', *Albion* 6 (1974): 275–93

JACOB, J. R., 'Restoration, reformation and the origins of the Royal Society', *Hist. Sci.* 13 (1975): 155–76

JACOB, J. R., 'Boyle's circle in the Protectorate: revelation, politics and the millennium', *JHI* 38 (1977): 131–40

JACOB, J. R., *Robert Boyle and the English Revolution: A Study in Social and Intellectual Change* (New York: Burt Franklin, 1977)

JACOB, J. R., 'Boyle's atomism and the Restoration assault on pagan naturalism', *Social Studies of Science* 8 (1978): 211–33

JACOB, J. R., 'Restoration ideologies and the Royal Society', *Hist. Sci.* 18 (1980): 25–38

JACOB, J. R., *Henry Stubbe, Radical Protestantism and the Early Enlightenment* (Cambridge University Press, 1983)

JACOB, J. R. and JACOB, M. C., 'The Anglican origins of modern science: the metaphysical foundations of the Whig constitition', *Isis* 71 (1980): 251–67

JINGMIN, ZENG, 'Robert Boyle and China', *China Historical Materials of Science and Technology* 11, no. 3 (1990): 22–30 (in Chinese)

JONES, G. W., 'Robert Boyle as a medical man', *Bulletin of the History of Medicine* 38 (1964): 139–52

JONES, IRENE, *Robert Boyle: Lord of the Manor of Stalbridge 1643–91* (Sturminster Newton Museum, 1989)

KALUGAY, J., 'Boyle the chemist', *Koroth* 2 (1959): 129–32 (in Hebrew)

KAPLAN, B. B., 'The medical writings of Robert Boyle', University of Maryland Ph.D. thesis, 1979

KAPLAN, B. B., 'Greatrakes the stroker: the interpretations of his contemporaries', *Isis* 73 (1982): 178–85

KARGON, R. H., 'Walter Charleton, Robert Boyle and the acceptance of Epicurean atomism in England', *Isis* 55 (1964): 184–92

KARGON, R. H., *Atomism in England from Hariot to Newton* (Oxford: Clarendon Press, 1966), ch. 9

KARGON, R. H., 'The testimony of nature: Boyle, Hooke and the experimental philosophy', *Albion* 3 (1971): 72–81

KEELE, K. D., 'The Sydenham–Boyle theory of morbific particles', *Medical History* 18 (1974): 240–8

KELLAWAY, WILLIAM, 'The archives of the New England Company', *Archives* 2 (1954): 175–82

KIM, YUNG SIK, 'Another look at Robert Boyle's acceptance of the mechanical philosophy: its limits and its chemical and social contexts', *Ambix* 38 (1991): 1–10

KING, L. S., 'Robert Boyle as amateur physician', in C. W. Bodemer and L. S. King, *Medical Investigation in Seventeenth-Century England* (Los Angeles: William Andrews Clark Memorial Library, 1968), pp. 27–49

KING, L. S., *The Road to Medical Enlightenment 1650–95* (London: Macdonald, 1970)

KLAAREN, E. M., *Religious Origins of Modern Science: Belief in Creation in Seventeenth-Century Thought* (Grand Rapids, Mich.: William B. Eerdmans, 1977)

KRONEMEYER, R. J., 'Matter and meaning: dualism in the thought of Robert Boyle, Isaac Newton and John Ray', Kent State University Ph.D. thesis, 1978

KROOK, DOROTHEA, 'Two Baconians: Robert Boyle and Joseph Glanvill', *Huntington Library Quarterly* 18 (1955): 261–78

KUHN, THOMAS S., 'Robert Boyle and structural chemistry in the seventeenth century', *Isis* 43 (1952): 12–36

KULTGEN, J. H. , 'Robert Boyle's metaphysic of science', *Philosophy of Science* 23 (1956): 136–41

KUSLAN, L. I., and STONE, A. H., *Robert Boyle, the Great Experimenter* (Englewood Cliffs, NJ: Prentice-Hall, 1970)

LAUDAN, LAURENS, 'The clock metaphor and probabilism: the impact of Descartes on English methodological thought 1650–65', *Ann. Sci.* 22 (1966): 73–104

LE BRAS, J., 'Boyle et les premieres observations d'effet antioxygene', *Revue Générale du Caoutchouc* 21 (1944): 125–6

LEEUWEN, H. G. VAN, *The Problem of Certainty in English Thought 1630–90* (The Hague: Martinus Nijhoff, 1970), pp. 90–106

LEICESTER, H. M., 'Boyle, Lomonosov, Lavoisier and the corpuscular theory of matter', *Isis* 58 (1967): 240–4

LENNOX, J. G., 'Robert Boyle's defense of teleological inference in experimental science', *Isis* 74 (1983): 38–52

LIU, J., 'Robert Boyle – establisher of chemistry as a branch of science', *Ziran Zazhi* 6 (1983): 623–6 (in Chinese)

LOEMKER, L. E., 'Boyle and Leibniz', *JHI* 16 (1955): 22–43

LOEWENSON, L., 'The works of Robert Boyle and "The present state of Russia" by Samuel Collins (1671)', *Slavonic and East European Review* 33 (1955): 470–85

LUPOLI, A., 'La polemica tra Hobbes e Boyle', *Acme* [Milan] 29 (1976): 309–54

MCADOO, H. R., *The Spirit of Anglicanism: A Survey of Anglican Theological Method in the Seventeenth Century* (London: Adam & Charles Black, 1965), pp. 260–85

MCCANN, E., 'Was Boyle an occasionalist?', in *Philosophy, its History and Historiography*, ed. A. J. Holland (Dordrecht: D. Reidel, 1985), pp. 229–31

MCGUIRE, J. E., 'Boyle's conception of nature', *JHI* 33 (1972): 523–42

MACINTOSH, J. J., 'Primary and secondary qualities', *Studia Leibnitiana* 8 (1976): 88–104

MACINTOSH, J. J., 'Perception and imagination in Descartes, Boyle and Hooke', *Canadian Journal of Philosophy* 13 (1983): 327–52

MACINTOSH, J. J., 'Robert Boyle on Epicurean atheism and atomism', in *Atoms, Pneuma and Tranquillity: Epicurean and Stoic Themes in European Thought*, ed. M. J. Osler (Cambridge University Press, 1991), pp. 197–219

MACINTOSH, J. J., 'Robert Boyle's epistemology: the interaction between scientific and religious knowledge', *International Studies in the Philosophy of Science* 6 (1992): 91–121

MCKIE, DOUGLAS, 'Boyle's law', *Endeavour* 7 (1948): 148–51

MCKIE, DOUGLAS, 'Boyle's library', *Nature* 163 (1949): 627–8

MCKIE, DOUGLAS, 'Fire and the flamma vitalis: Boyle, Hooke and Mayow', in *Science, Medicine and History*, ed. E. A. Underwood, 2 vols (London: Oxford University Press, 1953), i, 469–88

MCKIE, DOUGLAS, 'Robert Boyle, F.R.S. – a "sceptical chymist"', *Pharmaceutical Journal* 187 (1961): 183–6

MCKIE, DOUGLAS, 'Introduction' to *The Works of Robert Boyle*, ed. Thomas Birch, 6 vols, reprint edn (Hildesheim: G. Olms, 1965)

MACLYSAGHT, EDWARD, ed., *Calendar of the Orrery Papers* (Dublin: Irish Manuscripts Commission, 1941)

MADDISON, R. E. W., 'Robert Boyle's library', *Nature* 165 (1950): 981

MADDISON, R. E. W., 'Studies in the life of Robert Boyle, 1: Robert Boyle and some of his foreign visitors', *NRRS* 9 (1951): 1–35

MADDISON, R. E. W., 'Studies in the life of Robert Boyle, 2: salt water freshened', *NRRS* 9 (1952): 196–216

MADDISON, R. E. W., 'Studies in the life of Robert Boyle, 3: the charitable disposal of Robert Boyle's residuary estate', *NRRS* 10 (1952): 15–27

MADDISON, R. E. W., 'Studies in the life of Robert Boyle, 4: Robert Boyle and some of his foreign visitors', *NRRS* 11 (1954): 38–53

MADDISON, R. E. W., 'Studies in the life of Robert Boyle, 5: Boyle's operator: Ambrose Godfrey Hanckwitz, F.R.S.', *NRRS* 11 (1955): 159–88

MADDISON, R. E. W., 'Notes on some members of the Hanckwitz family in England', *Ann. Sci.* 11 (1955): 64–73

MADDISON, R. E. W., 'A summary of former accounts of the life and work of Robert Boyle', *Ann. Sci.* 13 (1957): 90–108

MADDISON, R. E. W., 'A tentative index of the correspondence of the honourable Robert Boyle', *NRRS* 13 (1958): 128–201

MADDISON, R. E. W., 'Robert Boyle and the Irish bible', *Bulletin of the John Rylands Library* 41 (1958): 81–101

MADDISON, R. E. W., 'The portraiture of the honourable Robert Boyle', *Ann Sci.* 15 (1959): 141–214

MADDISON, R. E. W., 'The plagiary of Francis Boyle', *Ann. Sci.* 17 (1961): 111–20

MADDISON, R. E. W., 'The earliest published writing of Robert Boyle', *Ann. Sci.* 17 (1961): 165–73

MADDISON, R. E. W., 'The first edition of Robert Boyle's *Medicinal Experiments*', *Ann. Sci.* 18 (1962): 43–7

MADDISON, R. E. W., 'Studies in the life of Robert Boyle, 6: the Stalbridge period, 1645–55, and the invisible college', *NRRS* 18 (1963): 104–24

MADDISON, R. E. W., 'Boyle's hell', *Ann. Sci.* 20 (1964): 101–10

MADDISON, R. E. W., 'Studies in the life of Robert Boyle, 7: the grand tour', *NRRS* 20 (1965): 51–77

MADDISON, R. E. W., 'Galileo and Boyle: a contrast', in *Saggi su Galileo Galilei*, ed. C. Maccagni, 2 vols (Florence: G. Barbèra, 1967), ii, 348–61

MADDISON, R. E. W., *The Life of the Honourable Robert Boyle, F.R.S.* (London: Taylor & Francis, 1969)

MADDISON, R. E. W., 'Notes and discussions: the life of Robert Boyle: addenda; the portraiture of Robert Boyle: addenda', *Ann. Sci.* 45 (1988): 193–8

MANDELBAUM, MAURICE, *Philosophy, Science and Sense Perception* (Baltimore: Johns Hopkins Press, 1964), ch. 2

MARKLEY, ROBERT, 'Objectivity as ideology: Boyle, Newton and the languages of science', *Genre* 16 (1983): 355–72

MARKLEY, ROBERT, 'Robert Boyle on language: *Some Considerations Touching the Style of the Holy Scriptures*', *Studies in Eighteenth-century Culture* 14 (1985): 159–71

METZGER, HELENE, *Chemistry*, trans. and annotated by C. V. Michael (West Cornwall, Ct.: Locust Hill Press, 1991), pp. 21–30

MILTON, JOHN, 'The influence of the nominalist movement in the thought of Bacon, Boyle and Locke', University of London Ph.D. thesis, 1982

MIRO, J., 'Spinoza and Boyle: examination of the polemics on homogeneity', *Anales de la Seccion de Ciencias del Colegio Universitario de Gerona (Universidad Autonomia de Barcelona)* 2 (1977): 39–67 (in Catalan)

MORE, L. T., 'Boyle as alchemist', *JHI* 2 (1941): 61–76

MORE, L. T., *The Life and Works of the Honourable Robert Boyle* (New York: Oxford University Press, 1944)

MOULTON, F. R., 'Robert Boyle 1627–91', *Scientific Monthly* 53 (1941): 380–2

MULTHAUF, R. P., 'The beginning of mineralogical chemistry', *Isis* 49 (1958): 50–3

MULTHAUF, R. P., 'Some non-existent chemists of the seventeenth century: remarks on the use of the dialogue in scientific writing', in A. G. Debus and R. P. Multhauf, *Alchemy and Chemistry in the Seventeenth Century* (Los Angeles: William Andrews Clark Memorial Library, 1966), pp. 31–50

NEVILLE, R. G., 'Query no. 155: was Boyle the first to use spot test analysis?', *Isis* 49 (1958): 438–9; with replies in *ibid.* 55 (1964): 91; 58 (1965): 210–11

NEVILLE, R. G., '"The Sceptical Chymist", 1661: a tercentenary tribute', *Journal of Chemical Education* 38 (1961): 106–9

NEVILLE, R. G., 'The discovery of Boyle's law, 1661–2', *Journal of Chemical Education* 39 (1962): 356–9

NEWMAN, WILLIAM, 'Newton's *Clavis* as Starkey's *Key*', *Isis* 78 (1987): 564–74

O'BRIEN, J. J., 'Samuel Hartlib's influence on Robert Boyle's scientific development', *Ann. Sci.* 21 (1965): 1–14, 257–76

OSBORNE, E. A., 'Cancellandum R2 in Boyle's "Sceptical Chymist", 1661', *The Book Collector* 11 (1962): 215

OSLER, MARGARET J., 'John Locke and some philosophical problems in the science of Boyle and Newton', Indiana University Ph.D. thesis, 1968

OSLER, MARGARET J., 'The intellectual sources of Boyle's philosophy of nature: Gassendi's voluntarism and Boyle's physico-theological project', in *Philosophy, Science and Religion in England 1640–1700*, eds Richard Kroll, Richard Ashhcraft and Perez Zagorin (Cambridge University Press, 1992), pp. 178–98

OSTER, MALCOLM, '"The Beame of Divinity": animal suffering in the early thought of Robert Boyle', *BJHS* 22 (1989): 151–79

OSTER, MALCOLM, 'Nature, ethics and divinity: the early thought of Robert Boyle', University of Oxford D.Phil. thesis, 1990

OSTER, MALCOLM, 'The scholar and the craftsman revisited: Robert Boyle as aristocrat and artisan', *Ann. Sci.* 49 (1992): 255–76

OSTER, MALCOLM, 'Biography, culture and science: the formative years of Robert Boyle', *Hist. Sci.* 31 (1993): 177–226

OSTER, MALCOLM, 'Millenarianism and the new science: the case of Robert Boyle', in *The Advancement of Learning in the Seventeenth Century: the World of Samuel Hartlib*, eds M. Greengrass, M. P. Leslie and T. Raylor (Cambridge University Press, forthcoming)

O'TOOLE, R. J., 'Qualities and powers in the corpuscular philosophy of Robert Boyle', *Journal of the History of Philosophy* 12 (1974): 295–315

PACCHI, ARRIGO, *Cartesio in Inghilterra: da More a Boyle* (Rome/Bari: Editori Laterza, 1973)

PALMER, D., 'Boyle's corpuscular hypothesis and Locke's primary–secondary quality distinction', *Philosophical Studies* 29 (1976): 181–9

PAOLINELLI, MARCO, *Fisico-teologia e Principio di Ragion Sufficiente: Boyle, Maupertius, Wolff, Kant* (Milan: Vita e Pensiero, 1971); originally published in *Rivista di Filosofia Neoscolastica* 62 (1970), 63 (1971)

PARADIS, JAMES, 'Montaigne, Boyle and the essay of experience', in *One Culture: Essays in Science and Literature*, ed. George Levine (Madison: University of Wisconsin Press, 1987), pp. 59–91

PARTINGTON, J. R., *A History of Chemistry*, 4 vols (London: Macmillan, 1961–70), ii, 486–549

PATTERSON, L. D., 'Thermometers of the Royal Society, 1663–1768', *American Journal of Physics* 19 (1951): 523–35

PEREZ-RAMOS, ANTONIO, *Francis Bacon's Idea of Science and the Maker's Knowledge Tradition* (Oxford: Clarendon Press, 1988), pp. 166–79

PHILLIPS, PAUL, 'Robert Boyle – a shoulder for Newton', *Chemistry in Britain*, Oct. 1992, 906–8

PIGHETTI, CLELIA, 'Aspetti della metodologia storica americana in una recente interpretazione dell'opera di Boyle', in *Atti del Convegno sui Problemi Metodologici di Storia della Scienza, Torino 29–31 Marzo 1967* (Florence, 1967), pp. 116–23

PIGHETTI, CLELIA, 'Boyle, il virtuoso cristiano', *Cultura e Scuola* 25 (1968): 231–5

PIGHETTI, CLELIA, 'Boyle e il corpuscularismo inglese del seicento', *Cultura e Scuola* 10 (1971): 213–19

PIGHETTI, CLELIA, *Robert Boyle e la Scienza Virtuosa*, Quaderni di Storia della Scienza e della Medicina, 13 (Ferrara, 1974)

PIGHETTI, CLELIA, *Opere di Robert Boyle* (Turin: Utet, 1977)

PIGHETTI, CLELIA, *Boyle: La Vita, il Pensiero, le Opere* (Milan: Accademia, 1978)

PIGHETTI, CLELIA, 'L'opera pneumatica di Robert Boyle', *Cultura e Scuola* 20 (1981): 244–51

PIGHETTI, CLELIA, *L'Influsso Scientifico di Robert Boyle nel Tardo '600 Italiano* (Milan: Franco Angeli, 1988)

PILKINGTON, ROGER, *Robert Boyle: Father of Chemistry* (London: John Murray, 1959)

POYNTER, F. N. L., 'Rysbrack's bust of Robert Boyle', *Journal of the History of Medicine* 24 (1969): 475–8

PRINCIPE, LAWRENCE, 'The gold process: directions in the study of Robert Boyle's alchemy', in *Alchemy Revisited*, ed. Z. R. W. M. von Martels (Leiden: E. J. Brill, 1990), pp. 200–5

PRINCIPE, LAWRENCE, 'Robert Boyle's alchemical secrecy: codes, ciphers, and concealments', *Ambix* 39 (1992), 63–74

RANCKE-MADSEN, E., 'A painting of the Hon. Robert Boyle in Danish possession', *Ann. Sci.* 19 (1963): 147–8

REAL, H.-J., 'A book from Robert Boyle's library', *The Book Collector* 19 (1970): 527–9

REILLY, D., 'Robert Boyle and his background', *Journal of Chemical Education* 28 (1951): 178–83

RENALDO, J. J., 'Bacon's empiricism, Boyle's science, and the Jesuit response in Italy', *JHI* 37 (1976): 689–95

RETI, L., 'Van Helmont, Boyle and the alkahest', in L. Reti and W. C. Gibson, *Some Aspects of Seventeenth-century Medicine and Science* (Los Angeles: William Andrews Clark Memorial Library, 1969), pp. 1–19

ROGERS, G. A. J., 'Boyle, Locke and reason', *JHI* 27 (1966): 205–16

ROGERS, G. A. J., 'Descartes and the method of English science', *Ann. Sci.* 29 (1972): 237–55

ROWBOTTOM, M. E., 'The earliest published writing of Robert Boyle', *Ann. Sci.* 6 (1950): 376–89

ROWBOTTOM. M. E., 'The chemical studies of Robert Boyle and his place in the history of chemistry', University of London Ph.D. thesis, 1955

ROWBOTTOM, M. E., 'Some Huguenot friends and acquaintances of Robert Boyle (1627–91)', *Proceedings of the Huguenot Society of London* 20 (1959–60): 177–94

SAMBURSKY, S., 'The influence of Galileo on Boyle's philosophy of science', in *Actes du Symposium International des Sciences Physiques et Mathématiques dans la Première Moitié du XVIIe Siècle, 16–18 June 1958* (Florence/Paris, 1960), pp. 142–6

SARGENT, ROSE-MARY, 'Robert Boyle's Baconian inheritance: a response to Laudan's Cartesian thesis', *Studies in History and Philosophy of Science* 17 (1986): 469–86

SARGENT, ROSE-MARY, 'Explaining the success of science', *PSA 1988*, 2 vols (Proceedings of the 1988 Biennial Meeting of the Philosophy of Science Association: East Lansing, Michigan, 1988), i, 55–63

SARGENT, ROSE-MARY, 'Scientific experiment and legal expertise: the way of experience in seventeenth-century England', *Studies in History and Philosophy of Science* 20 (1989): 19–45

SARTON, GEORGE, 'Boyle and Bayle: the sceptical chemist and the sceptical historian', *Chymia* 3 (1950): 155–89

SCHAFFER, SIMON, 'Godly men and mechanical philosophers: souls and spirits in Restoration natural philosophy', *Science in Context* 1 (1987): 55–85

SCHOFIELD, M., 'Robert Boyle, scientist', *Chemistry and Industry*, 59 (1940): 615–19

SHANAHAN, TIMOTHY, 'God and nature in the thought of Robert Boyle', *Journal of the History of Philosophy* 26 (1988): 547–69

SHAPIN, STEVEN, 'Pump and circumstance: Robert Boyle's literary technology', *Social Studies of Science* 14 (1984): 481–520

SHAPIN, STEVEN, 'The house of experiment in seventeenth-century England', *Isis* 79 (1988): 373–404

SHAPIN, STEVEN, 'Robert Boyle and mathematics: reality, representation and experimental practice', *Science in Context* 2 (1988): 23–58

SHAPIN, STEVEN, 'The invisible technician', *American Scientist* 77 (1989): 554–63

SHAPIN, STEVEN, 'Who was Robert Hooke?', in *Robert Hooke: New Studies*, eds Michael Hunter and Simon Schaffer (Woodbridge: Boydell Press, 1989), pp. 253–85

SHAPIN, STEVEN, 'Personal development and intellectual biography: the case of Robert Boyle', *BJHS* 26 (1993): 335–45

SHAPIN, STEVEN, and SCHAFFER, SIMON, *Leviathan and the Air-Pump: Hobbes, Boyle and the Experimental Life* (Princeton University Press, 1985)

SHUGG, W., 'Humanitarian attitudes in early animal experiments', *Ann. Sci.* 24 (1968), 227–38

SINA, M., 'Robert Boyle ed il problema dell' "above reason"', *Rivista di Filosofia Neoscolastica* 65 (1973): 746–70

SMEATON, W. A., 'Theories of the chemical elements from Robert Boyle to Marie Curie', *Monografie z Dziejow Nauki i Techniki* 51 (1970): 69–83

SOOTIN, HARRY, *Robert Boyle, Founder of Modern Chemistry* (London: Chatton & Windus, 1963)

STANTON, M. E., 'J.F.F. and Robert Boyle', *Journal of the History of Medicine* 17 (1962): 189–90

STENECK, NICHOLAS, 'Greatrakes the stroker: the interpretations of historians', *Isis* 73 (1982): 161–77

STEWART, M. A., 'The authenticity of Robert Boyle's anonymous writings on reason', *Bodleian Library Quarterly* 10 (1978–82): 280–9

STEWART, M. A., ed., *Selected Philosophical Papers of Robert Boyle* (Manchester University Press, 1979)

STEWART, M. A., 'Locke's professional contacts with Robert Boyle', *Locke Newsletter* 12 (1981): 19–44

STIEB, E. W., 'Robert Boyle's *Medicina Hydrostatica* and the detection of adulteration', *Actes du dixième Congrès International d'Histoire des Sciences, Ithaca 1962* (Paris, 1964), pp. 841–5

STRUBE, I., 'The role of ancient atomism in the revolution of chemical research in the second half of the seventeenth century', *Organon* 4 (1967): 127–32

SZABADVARY, F., 'Early eminent English contributions to the development of analytical chemistry from Boyle to Ure', *Periodica Polytechnica, Chemical Engineering* 19 (1975): 339–45

TANAKA, M., 'Founders of chemistry: Robert Boyle', *Kagaku* 13 (1958): 7–9

TAZAWA, Y., 'Robert Boyle (1627–91). Skepticism in chemistry', *Kagaku No Ryoiki* 20 (1966): 357–64, 462–70, 525–33, 599–606 (in Japanese)

TEAGUE, B. C., 'The origins of Robert Boyle's philosophy', University of Cambridge Ph.D. thesis, 1971

THORNDIKE, LYNN, *A History of Magic and Experimental Science*, 8 vols (New York: Columbia University Press, 1923–59), viii, 170–201

TURNER, H. D., 'Robert Hooke and Boyle's air pump', *Nature* 184 (1959): 395–7

VELDE, A. J. J. VAN DE, 'Robert Boyle, 250 jaren ne zijn dood op 30 December 1691, herdacht', *Jaarboek der K. Vlaamse Academie voor Wetenschappen, Letteren en Schone Kunsten van Belgie*, 1941: 166–211

WALDEN, P., 'Robert Boyle (1627–91) und die physischen Wissenschaften', *Naturwissenschaften* 37 (1950): 169–72

WALLIS, P. J., 'English books in Dutch libraries: Newton and Boyle', *The Library* 4th series 26 (1971): 60–2

WALLIS, P. J., 'English books in Moscow–Leningrad libraries: Newton, Boyle and others', *The Library* 4th series 27 (1972): 51–3

WALTON, M. T., 'Boyle and Newton on the transmutation of water and air from the root of Helmont's tree', *Ambix* 27 (1980): 11–18

WEBB, C. C. J., 'Some Bodleian memories', *Bodleian Library Record* 4 (1953): 277–9

WEBSTER, CHARLES, 'Richard Towneley and Boyle's law', *Nature* 197 (1963): 226–8

WEBSTER, CHARLES, 'The discovery of Boyle's law and the concept of the elasticity of air in the seventeenth century', *Archive for History of Exact Sciences* 2 (1965): 441–502

WEBSTER, CHARLES, 'Water as the ultimate principle of nature: the background to Boyle's *Sceptical Chymist*', *Ambix* 13 (1965): 96–107

WEBSTER, CHARLES, 'New Light on the invisible college: the social relations of English science in the mid seventeenth century', *Transactions of the Royal Historical Society* 5th series 24 (1974): 19–42

WEBSTER, CHARLES, *The Great Instauration: Science, Medicine and Reform 1626–60* (London: Duckworth, 1975)

WEST, MURIEL, 'Notes on the importance of alchemy to modern science in the writings of Francis Bacon and Robert Boyle', *Ambix* 9 (1961): 102–14

WEST, N., 'Robert Boyle (1627–91) and the vacuum pump', *Vacuum* 43 (1992): 283–6

WESTFALL, R. S., 'Unpublished Boyle papers relating to scientific method', *Ann. Sci* 12 (1956): 63–73, 103–17

WESTFALL, R. S., *Science and Religion in Seventeenth-Century England* (New Haven: Yale University

Press, 1958; reprinted with new preface and supplemental bibliography, Ann Arbor: University of Michigan Press, 1973)

WILLIAMS, E., 'Some experiments on the expansive force of freezing water', *Ann. Sci.* 10 (1954): 166–71

WOJCIK, JAN W., 'Robert Boyle and the limits of reason: a study in the relationship between science and religion in seventeenth-century England', University of Kentucky Ph.D. thesis, 1992

WOOD, P. B., 'Behemoth v. the Sceptical Chymist', *Hist. Sci.* 26 (1988): 103–9

YOSHIMOTO, H., 'Science and religion in the thought of Robert Boyle', *Kagakusi Kenkyu* 23 (1984), 193–200; 24 (1985), 10–17 (in Japanese)

Index